Unusual DNA Structures

R.D. Wells   S.C. Harvey
Editors

Unusual DNA Structures

Proceedings of the First Gulf Shores Symposium,
held at Gulf Shores State Park Resort, April 6–8 1987,
sponsored by the Department of Biochemistry, Schools
of Medicine and Dentistry, University of Alabama at
Birmingham, Birmingham, Alabama

With 104 Illustrations

Springer-Verlag
New York Berlin Heidelberg
London Paris Tokyo

R.D. Wells and S.C. Harvey
Department of Biochemistry
Schools of Medicine and Dentistry
University of Alabama at Birmingham
Birmingham, Alabama 35294
USA

LCCN 87-31387

Text prepared by the editors in camera-ready form.
Printed and bound by Arcata Graphics/Halliday, West Hanover, Massachusetts.
Printed in the United States of America.

9 8 7 6 5 4 3 2 1

ISBN 0-387-96631-5 Springer-Verlag New York Berlin Heidelberg
ISBN 3-540-96631-5 Springer-Verlag Berlin Heidelberg New York

Table of Contents

Preface

This book contains the papers which were presented at the First Gulf
Shores Symposium on Unusual DNA Structures. The meeting was held April
6-8, 1987, in Gulf Shores, Alabama.

A veritable explosion has taken place in recent years regarding our
understanding of unusual DNA structures. This symposium was dedicated to
enhancing our understanding of the biology and chemistry of these
important structural features.

The symposium was supported by funds provided by the Department of
Biochemistry, University of Alabama at Birmingham, Schools of Medicine and
Dentistry.

We wish to express our appreciation to Ms. Patti Guyton for her expert
organizational skills and assistance in organizing the meeting and
preparation of this book.

Robert D. Wells and
Stephen C. Harvey
Department of Biochemistry
Schools of Medicine and Dentistry
University of Alabama at Birmingham
Birmingham, Alabama   35294

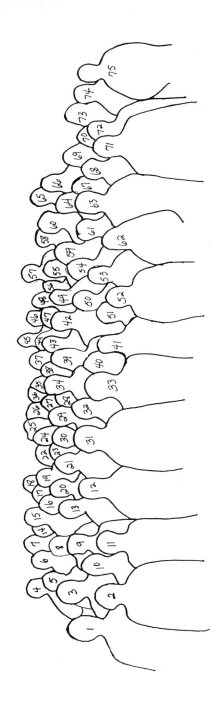

1. Pete Dickie
2. Wang-Ting Hsieh
3. Mark Glover
4. David Pulleyblank
5. Robert Wells
6. Mark Richardson
7. A. Argirokastritis
8. Marc Leng
9. Guoxing Zheng
10. Eliane Taillandier
11. Charles Cantor
12. Jim Hartley
13. Richard Sinden
14. John Sullivan
15. Eric Henderson
16. Brian Loeffler
17. Jacob Lebowitz
18. Gilberto Fronza
19. Sanjeev Jain

20. Dave Wilson
21. David Pettijohn
22. Micaela Caserta
23. Adam Jaworski
24. Han Htun
25. Franz Wohlrab
26. Gregg Bellomy
27. Wolfgang Zacharias
28. Tom Tullius
29. DeWitt Sumners
30. Christopher Marzec
31. Uwe Muller
32. Andrzej Stasiak
33. David Lilley
34. David Collier
35. Michael Kilpatrick
36. Scott Umlauf
37. Mick McLean
38. Horace Gray

39. Paul Hagerman
40. Jacquelynn Larson
41. Johanna Griffin
42. Jeremy Lee
43. Robert Hopkins
44. Patricia Biesiot
45. Stephen Harvey
46. Raymond Gesteland
47. Bill Suggs
48. Shiao Wang
49. Jeffery Hanvey
50. Nicole Theriault
51. Anna DiGabriele
52. Sorour Amirhaeri
53. Mary Reaban
54. Shashi Rao
55. Richard Dickerson
56. Luigi Marzilli
57. Thomas Record

58. Robert Shapiro
59. Angel Garcia
60. Richard Lavery
61. Norma Wills
62. Debra Milton
63. Chang-Shung Tung
64. John Blaho
65. Dinshaw Patel
66. Bill Christens-Barry
67. Dorothy Skinner
68. Robert Tan
69. Sarah Elgin
70. Mary Barkley
71. Edward Trifonov
72. M. Prabhakaran
73. Kathleen Morden
74. T.J. Thomas
75. David Stollar

Cantor, Charles, Department of Human Genetics & Development, Columbia University, 701 W. 168th Street, New York, NY 10032.

Dickerson, Richard E., Molecular Biology Institute, University of California, Los Angeles, CA 90024.

Elgin, Sarah C.R., Washington University, St. Louis, MO 63130.

Hagerman, Paul, Department of Biochemistry & Biophysics, University of Colorado Medical Center, 4200 East 9th Avenue, Denver, CO 80262.

Harvey, Stephen C., Department of Biochemistry, University of Alabama at Birmingham, Birmingham, AL 35294.

Kollman, Peter, Department of Pharmaceutical Chemistry, University of California, San Francisco, CA 94143.

Lavery, Richard, Institut de Biologie Physico-Chimique, Fondation Edmond de Rothschild, 13, Rue Pierre et Marie Curie, 75005 Paris, France.

Levy, Ronald M., Department of Chemistry, Wright and Rieman Laboratories, P.O. Box 939, Piscataway, NJ 08854.

Lilley, David M., Department of Biochemistry, The University of Dundee, DD1 4HN, United Kingdom.

Patel, Dinshaw, Columbia University, Department of Biochemistry, New York, NY.

Leng, Marc, Centre Biophys. Molec., 1A, Avenue Rech. Scientif., 45071 Orleans, Cedex 2, France.

Pettijohn, David, Department of Biochemistry and Biophysical Genetics, University of Colorado Health Science Center, 4200 E. 9th Avenue, Denver, CO 80262.

Pulleyblank, David, Department of Biochemistry, University of Toronto, Toronto, Canada, M5S 1A8.

Record, Thomas, Department of Chemistry, University of Wisconsin, Madison, WI 53706.

Stollar, B. David, Department of Biochemistry and Pharmacology, Tufts University School of Medicine, 136 Harrison Avenue, Boston, MA 02111.

Trifonov, Edward, Department of Polymer Research, The Weizmann Institute of Science, Rehovot 76 100, Israel P.O.B. 26, 76100.

Wells, Robert D., Department of Biochemistry, University of Alabama at Birmingham, Birmingham, AL 35294.

Sorour Amirhaeri, Department of Biochemistry, University of Alabama at Birmingham, Birmingham, Alabama 35294.

Alexandros Argirokastritis, Department of Biophysics, Cell and Molecular Biology, King's College London, University of London, 26-29 Drury Lane, London WC2B 5RL, England.

Mary D. Barkley, Department of Chemistry, Louisiana State University, Baton Rouge, LA 70803-1804.

Patricia Biesiot, Biology Division, Oak Ridge National Laboratory, Post Office Box Y, Oak Ridge, TN 37831.

John Blaho, Department of Biochemistry, University of Alabama at Birmingham, Birmingham, AL 35294.

Micaela Caserta, Centro Acidi Nucleici, Consiglio Nazionale delle Richerche, Universita di Roma "La Sapienza", Roma, Italy.

Bill Christens-Barry, Department of Physics, University of Alabama at Birmingham, Birmingham, AL 35294.

David Collier, Department of Biochemistry, University of Alabama at Birmingham, Birmingham, AL 35294.

Anna DiGabriele, Yale University, Department of Molecular Biophysics & Biochemistry, KBT 242, P.O. Box 6666, 260 Whitney Avenue, New Haven, CT 06511.

Gilberto Fronza, Johns Hopkins University, School of Hygiene and Public Health, Department of Biochemistry, 615 N. Wolfe Street, Baltimore, MD 21205.

Angel Garcia, T10 MS K710, Los Alamos National Lab, Los Alamos, NM 87545.

Raymond Gesteland, Howard Hughes Medical Institute, Room 743 Wintrobe Building, University of Utah, Salt Lake City, UT 84132.

Mark Glover, Department of Biochemistry, University of Toronto, Toronto, Ontario, Canada M5S 1A8.

Horace Gray, Jr., Department of Biochemical and Biophysical Sciences, University of Houston, Houston, TX 77004.

Johanna Griffin, Department of Microbiology, University of Alabama at Birmingham, Birmingham, AL 35294.

Jack Griffith, 127 Lineberger Cancer Research Center, University of North Carolina, Chapel Hill, NC 27514.

Jeff Hanvey, Department of Biochemistry, University of Alabama at Birmingham, Birmingham, AL 35294.

Jim Hartley, Life Technologies, Inc., 8717 Grovemont Circle, Gaithersburg, MD 20877.

Eric R. Henderson, University of California, Berkeley, CA 94720.

Robert C. Hopkins, University of Houston, Clear Lake, 2700 Bay Area Boulevard, Houston, TX 77058

Wang-Ting Hsieh, Department of Biochemistry, University of Alabama at Birmingham, Birmingham, AL 35294.

Han Htun, Department of Physiological Chemistry, University of Wisconsin - Madison, 1300 University Avenue, Madison, WI 53706.

Sanjeev Jain, Department of Biochemistry, University of Wisconsin - Madison, 420 Henry Mall, Madison, WI 53703.

Adam Jaworski, Department of Biochemistry, University of Alabama at Birmingham, Birmingham, AL 35294.

M.W. Kilpatrick, Department of Clinical Genetics, University of Birmingham, Birmingham Maternity Hospital, Birmingham B15 2TG, England.

Jacquelynn Larson, Department of Biochemistry, University of Alabama at Birmingham, Birmingham, AL 35294.

Jacob Lebowitz, Department of Microbiology, University of Alabama at Birmingham, Birmingham, AL 35294.

J.S. Lee, Department of Biochemistry, University of Saskatchewan, A3 Health Sciences Building, Saskatoon, Canada S7N 0W0.

Marc Leng, Centre Biophys. Molec., 1A, Avenue Rech. Scientif. 5071 Orleans Cedex 2, France.

Brian Loeffler, Pharmacia P-L Biochemicals, 2202 N. Bartlett Avenue, Milwaukee, WI 53202.

Christopher Marzec, Public Health Research Institute, Department of Developmental and Structural Biology, Room 1033, 455 1st Avenue, New York, NY 10016.

Luigi G. Marzilli, Department of Chemistry, Emory University, Atlanta, GA 30322.

Michael McLean, Department of Biochemistry, University of Alabama at Birmingham, Birmingham, AL 35294.

Debra Milton, Howard Hughes Medical Institute, Room 743 Wintrobe Building, University of Utah, Salt Lake City, UT 84132.

Kathleen Morden, Department of Biochemistry, Louisiana State University, Baton Rouge, LA 70803-1806.

A. Richard Morgan, Department of Biochemistry, University of Alberta, Edmonton, Alberta, T6G 2H7, Canada.

Uwe Muller, Microbiology and Immunology, East Carolina University, Brody Building, Room 5E106A, Moye Boulevard, Greenville, NC 27858.

M. Prabhakaran, Department of Biochemistry, University of Alabama at Birmingham, Birmingham, AL 35294.

Shashidhar N. Rao, Department of Pharmaceutical Chemistry, University of California, San Francisco, CA 94143.

Mary E. Reaban, Department of Microbiology, University of Alabama at Birmingham, Birmingham, AL 35294.

Mark Richardson, Department of Molecular Biology, University of California, Berkeley, CA 94720.

Ida Ruberti, Centro Di Studio Per Gli Acidi Nucleici, c/o Dipartimento di Genetica e Biologia Molecolare, Universita di Roma "La Sapienza", 00100 Roma, Italy.

Robert Shapiro, Department of Chemistry, New York University, 4 Washington Place, New York, NY 10003.

Richard Sinden, Department of Biochemistry & Molecular Biology, University of Cincinnati, 231 Bethesda Avenue, Cincinnati, OH 45267-0522.

Dorothy M. Skinner, Biology Division, Oak Ridge National Laboratory, Post Office Box Y, Oak Ridge, TN 37831.

Andrzej Stasiak, Institute for Cell Biology, Swiss Federal Institute of Technology, CH-8093, Zurich, Switzerland.

Bill Suggs, Department of Chemistry, Brown University, Box H, Providence, RI 02912.

John Sullivan, Department of Microbiology, University of Alabama at Birmingham, Birmingham, AL 35294.

De Witt Sumners, Department of Mathematics, Florida State University, Tallahassee, FL 32306.

Eliane Taillandier, Universite Paris Word, UFR de Medecine, 76 Rue Marcel Cachin, F93012, Bobigny, Cedex, France.

Robert Tan, Department of Biochemistry, University of Alabama at Birmingham, Birmingham, AL 35294.

Nicole Theriault, The Upjohn Company, Kalamazoo, MI 49001.

T.J. Thomas, Department of Medicine, University of Minnesota, Minneapolis, MN 55455.

Tom Tullius, Department of Chemistry, Johns Hopkins University, Baltimore, MD 21218.

Chang-Shung Tung, T-10, MS K710, Los Alamos National Lab, Los Alamos, NM 87545.

Scott Umlauf, Department of Biochemistry, University of Wisconsin – Madison, 420 Henry Mall, Madison, WI 53706-1569.

Shiao Y. Wang, Biology Division, Oak Ridge National Laboratory, Post office Box Y, Oak Ridge, TN 37831.

Norma Wills, Howard Hughes Medical Institute, Room 743 Wintrobe Building, University of Utah, Salt Lake City, UT 84132.

W. David Wilson, Department of Chemistry, Georgia State University, University Plaza, Atlanta, GA 30303.

Franz Wohlrab, Department of Biochemistry, University of Alabama at Birmingham, Birmingham, AL 35294.

Wolfgang Zacharias, Department of Biochemistry, University of Alabama at Birmingham, Birmingham, AL 35294.

Guoxing Zheng, Department of Biochemistry and Molecular Biology, University of Cincinnati, 231 Bethesda Avenue, Cincinnati, OH 45267-0522.

UNUSUAL DNA STRUCTURES AND THE PROBES USED FOR THEIR DETECTION

Robert D. Wells, Sorour Amirhaeri, John A. Blaho, David A. Collier,
Jeffery C. Hanvey, Wang-Ting Hsieh, Adam Jaworski, Janusz Klysik,
Jacquelynn E. Larson, Michael J. McLean, Franz Wohlrab and Wolfgang
Zacharias

Department of Biochemistry
Schools of Medicine and Dentistry
University of Alabama at Birmingham
Birmingham, Alabama 35294

Introduction

A veritable explosion in our knowledge of unusual DNA structures has
occurred in recent years. Our enhanced knowledge about specific DNA
sequences via DNA sequencing and cloning has been a contributory factor.
Furthermore, the establishment of diverse types of methods for the
detection and analysis of unusual DNA structures has also been important.
For example, in the early days of left-handed Z-DNA (1979-1982), the only
techniques available were x-ray diffraction on short oligonucleotides,
circular dichroism, phosphorus NMR, and laser Raman spectroscopy. Today,
at least 15 different types of physical, enzymatic, immunological,
chemical, and spectroscopic methods are available. Studies have been
conducted on recombinant plasmids, short synthetic oligonucleotides, as
well as purified restriction fragments.

Fig. 1 schematically shows several types of unusual DNA structures.
Left-handed Z-DNA, polypurine·polypyrimidine (pur·pyr) structures, and
anisomorphic DNA are described below in greater detail. Cruciforms occur
at inverted repeat sequences in duplex DNA and are stabilized by modest
amounts of negative supercoiling (1-12). Substantial information has been
obtained on the mechanism and kinetics of formation, structure of the
branch points, effects of DNA sequence, temperature, stem and loop
lengths, ionic conditions, superhelical density and other important
related questions. These studies are significant because of the
similarity of the branch points to Holliday structures in recombination
(13). Bent DNA (Fig. 1) was discovered at the structural interface
between a DNA A conformation and a B structure (14-16). Bent or curved
DNA has been detected in restriction fragments (17-32) from kinetoplasts
as well as certain regulatory regions from both prokaryotic and eukaryotic

organisms. Some of these sequences appear to contain runs of A's ($\sim$ 4–6 in length) which exhibit a periodicity of approximately one turn of helix (10–11 bp) which gives rise to a sytematically bending helix.

Slipped structures (Fig. 1) have been postulated (33–35) to occur at direct repeat sequences, in general upstream from important regulatory sites such as the promoters for mouse $\alpha 2(1)$ collagen genes. Whereas it is possible to draw plausible structures consistent with the data obtained by nuclease cleavage, these slipped structures are comparatively poorly characterized.

Other interesting properties of DNA include long range thermodynamic and structural effects such as telestability (reviewed in 36–41). No attempt is made in this review to be exhaustive; instead, a critical analysis of new developments in this emerging field is presented.

Left-Handed Z-DNA

Substantial progress has occurred in the past several years in our understanding of several facets of left-handed Z-DNA. These include: the conditions which stabilize the structure, the effect of methylation of cytosine residues, the types of sequences that adopt Z helices, the properties of B-Z junctions, the development of new assays for Z structures, features of the immunological properties of different left-handed structures, characteristics of proteins which bind to Z helices, thermodynamics and kinetics of transitions, and interactions with small ligands such as carcinogens and mutagens, etc. (42–54 and references cited therein).

Determination of the types of sequences which easily adopt left-handed Z conformations has been the subject of substantial recent investigation. Early hypotheses suggested the necessity of strictly alternating purine-pyrimidine types of sequences. However, recent demonstrations have revealed that this type of sequence is neither necessary nor sufficient. When a BamHI site (GGATCC) is present within a tract of 58 bp of strictly alternating CG, this site adopts a non-B structure which may be fully left-handed as stabilized by negative supercoiling (55). Furthermore, the IVS2 sequences from human fetal globin genes (56) contain GTTTG and GACTG sequences which were proven to adopt left-handed (presumably Z) structures by supercoil relaxation studies as well as two other determinations. Hence, an alternating purine-pyrimidine sequence is not necessary for a Z helix. Also, the alternating $(A-T)_n$ sequence (approximately 30–70 bp in length), which is an alternating purine-pyrimidine region, does not adopt a Z structure (9,12,57), but forms a cruciform under the influence of negative supercoiling.

We and others have conducted a series of investigations on cloned synthetic oligonucleotides which differ from each other in the number and orientation of TA or GC base pairs in the middle of a run of alternating CG (58-61). The energetics of the B to Z transitions for these helices (as well as other possible secondary structures) were studied as a function of supercoil density. In general, a systematic investigation (58) revealed that consecutive A·T pairs can adopt Z helices irrespective of strand orientation (TTTT or TATA). Other workers (59) found similar results but concluded that the AT rich segment between the flanking Z-helices was in a non-B but not Z-structure. Obviously, a goal of this effort is to establish rules for identifying potential Z helices in natural genomes. However, we have recently discovered (McLean, Lee, and Wells, unpublished work) that the symmetry of bases on the complementary strands, in addition to the types of sequences involved, has a major influence on the types of structures which are formed. For example, a $(TG)_{12}$ sequence is not equivalent to a $(TG)_6(CA)_6$ sequence which may not be equivalent to a $(TG)_6(AC)_6$ sequence. Hence, attempts (60) to establish simple rules which do not take account of these as well as possibly other factors will fail. Indeed, most or all sequences may have the capacity to adopt left-handed helices under the appropriate conditions which may, in some cases, be quite extreme.

Pur·Pyr Structures

Recent studies have shown that a variety of naturally occurring nuclease hypersensitive sites, when cloned into plasmids, can retain their unusual structure when the plasmids are negatively supercoiled (33-35,62-73). These regions of nuclease hypersensitivity are often found within the regulatory regions of active genes and these unusual DNA structures have been implicated as having an important role in the regulation of transcription of these genes. Although no consensus sequence can be established for these sites, these segments of DNA often contain a pur·pyr strand bias. Prior determinations with DNA polymers with purines in one strand and pyrimidines in the other strand revealed unusual spectroscopic and physical properties (reviewed in 40,74,75). For example, the polymer $(G-A)_n \cdot (T-C)_n$, which is a common naturally occurring sequence, exists in two metastable states as determined by density gradient centrifugation (75). In addition, when this sequence is contained within a supercoiled plasmid, it is cleaved by S1 nuclease (68,72,76,77). Furthermore, recent studies (78,79) revealed S1 nuclease sensitivity in the major late promoter of adenovirus; we (78) interpreted this behavior as due to the presence of a flanking pur•pyr sequence.

The unusual structures adopted by the pur•pyr sequences are likely to be a family of non-B, right-handed conformations (Fig. 1). Different structures may exist for different sequences. DNA structural probes which were employed included S1, P1, and BAL31 nucleases, DNase I, gene 3 product of phage T7, phage T4 endonuclease VII, bromoacetaldehyde, diethylpyrocarbonate, osmium tetroxide, hydroxylamine, dimethylsulfate, sodium bisulfite, supercoil induced relaxation determinations, the amount of supercoiling required for different lengths of sequences, crosslinking studies, helical repeat determinations, and antibody binding measurements. In general, lower pHs (~ 4.5 - 6.0) had a stabilizing influence. These aggregate studies revealed the presence of some characteristics of non-paired bases.

At least three models have been proposed to explain the unusual structures of pur•pyr sequences, including the two strands having different backbone conformations (77), formation of an intramolecular triplex structure (72,80) as shown to form with pur•pyr polymers (74), and a duplex in which A-T Watson-Crick base pairs alternate with Hoogsteen (syn)G-C pairs (for an AG•TC sequence) (76,81). The second and third models require protonation of cytosines for the formation of the (unusual) structure, whereas the first proposal does not require protonation. A systematic evaluation of the effects of pH and chain length of inserts in recombinant plasmids will be required with a family of homologous sequences in order to evaluate the factors required for stabilization.

Anisomorphic DNA

Recent studies (82) on recombinant plasmids containing the site of segment inversion in herpes simplex virus type 1 revealed the presence of a previously unrecognized DNA conformation. The 12 bp direct repeat sequences (DR2) at the joint region (a sequence) of strain F revealed a new conformation which is stabilized by negative supercoiling. Formation of the novel conformation depended on the number of repeats; the 19mer (228 bp) and the 14mer (168 bp) readily undergo the structural transition whereas a pentamer, trimer, and dimer repeat show somewhat different properties. S1 and P1 nuclease studies demonstrate that the new

4

conformation has a major structural aberration at the center and conformational periodicities which are not identical on the complementary strands. Also, the effect of salt and pH, the site of reactions with bromoacetaldehyde and chloroacetaldehyde, the type of sequence (direct repeat) involved, and the nature and extent of supercoil induced relaxations demonstrate that this structure differs from previously recognized conformations including Z-DNA, cruciforms, bent DNA, and slipped structures (Fig. 1).

This novel conformation was termed anisomorphic DNA to designate the unequal structures in different directions. It has different structures on the complementary strands which elicit structural aberrations at the physical center of the tandem sequences. Since the pur•pyr sequence may be inherently inflexible, this supercoil induced structural change and the physical stress on these inserts in recombinant plasmids may deform (crack) the DR2 sequence at their centers.

Fig. 1 shows two representations of the DR2 sequence under the stress of negative supercoiling; the structure at the right may account for the unusually large extent of supercoil relaxation which was observed (82). The other representation at the base pair level of anisomorphic DNA shows a buckling of the pyrimidine rich strand of the DR2 sequences caused by this deformation and possibly a length dependent structural enhancement due to an unequal rise per residue or a different angle of rotation or certain types of slipped pairing between the complementary strands. At present, evidence is not available to distinguish between these possibilities.

Biological Functions

In general, the biochemical community has little doubt that unusual DNA conformations exist in vivo and serve important biological functions (40,74, 75,83). However, it has been particularly difficult to obtain convincing evidence on these questions. Perhaps a reason why the in vivo existence and properties of unusual structures has been recalcitrant to yielding definitive conclusions is that structural questions must be answered using genetic or other noninvasive analyses. In general, structural questions with macromolecules such as proteins or nucleic acids are solved by physical analyses such as x-ray diffraction, spectroscopy, hydrodynamic measurements, etc. Obviously, these determinations are of little value when investigating tiny segments of DNA chromosomes in vivo. Genetic analyses may be required in order to not perturb cellular superstructures and functions. Thus, a superficial analysis may suggest that it is not possible to address macromolecular structural questions _in vivo_ since these types of analyses are incompatible. Indeed, I am unaware of any cases where the three dimensional configuration of a complex macromolecule (such as a protein or a tRNA) has been elucidated in a living cell. Hence, alternate methodology must be designed and implemented.

Progress has been made, however, in certain areas, for example with respect to the possible role of left-handed Z-DNA and anisomorphic DNA in genetic recombination (52,82-88). Also, it is generally acknowledged (83) that cruciforms may not serve as primary signals for genetic regulation but, instead, may have a secondary role in modulation of activities of different genomes.

Methods Used to Detect Unusual Structures

Many of the procedures which are generally used to analyze for left-handed Z-DNA, cruciforms, bent DNA, unusual structures adopted by pur•pyr sequences, and anisomorphic DNA are similar. However, the results which are obtained with each probe are quite different for each conformation. This section will outline the general techniques which are routinely used for the characterization of unusual DNA structures. The probes to be described are applicable for studies with recombinant plasmids and, to some extent, with restriction fragments. X-ray crystallographic analyses and antibody determinations are not described herein, but are the subjects of other reviews (42-45,50) and are treated in detail by other authors in this book, especially Drs. Dickerson and Stollar.

a. Nucleases

S1 nuclease is widely used as a probe for unusual DNA structures since it rapidly and sensitively detects a variety of different conformations. This enzyme from A. oryzae was originally isolated as a single strand specific nuclease (89), but also cleaves double stranded DNA at specific sites, such as the junctions between right and left-handed sequences (55,56, 90-92) and the center of inverted repeats (potential cruciform structures) (2-12). In addition, the enzyme seems to recognize some direct repeats (33-35), polypyrimidine•polypurine stretches (62-68), and anisomorphic DNA (82). Many of these hypersensitive sites seem to lie

6

proximal to eukaryotic active genes (33–35,62–71), a property shared with
DNAse I.

Determinations have also been conducted with nuclease P1 (93), mung
bean nuclease, BAL 31 nuclease (94), DNAse I, and the gene 3 product of
phage T7. All of these enzymes except mung bean nuclease can be employed
at neutral pH in contrast to S1 nuclease. In general, these enzymes have
given results which are rather similar to those found with S1 nuclease.
However, it is important to conduct studies with more than one enzyme
since it is possible to gain information regarding the state of
protonation of the DNA. Some unusual DNA structures which embody
Hoogsteen base pairing schemes require the hemi-protonation of C residues
to form a G·2C complex. In general, DNA supercoiling has been used as a
structural perturbant for these nuclease sensitivity studies as well as
other analyses. Investigators do not necessarily propose that
supercoiling in vivo is the stabilizing influence for the unusual
structures. However, supercoiling is utilized to mediate DNA localized
conformational transitions which may be similar to those induced in vivo
by proteins, ionic gradients, etc.

b. Other Enzymatic Probes

The target sites for the HhaI methylase, as well as the HhaI and BssH
II restriction nucleases, consist of alternating (dG–dC) sequences. None
of these enzymes are active on left-handed DNA (95,96). Furthermore, it
is likely that other unusual DNA structures (cruciforms, slipped
structures) may inhibit the methylation or cleavage of specific sites.
Plasmids with different supercoil densities containing interesting inserts
have been studied with these and other enzymes to determine the presence
of non–DNA B structures. Furthermore, investigations have been conducted
(55) on plasmids containing sequences which adopt left-handed Z helices
with a restriction site (BamHI) at the interface between the left-handed Z
and right-handed B helices. When the insert is left-handed, BamHI does
not recognize or cleave the GGATCC site (55). Similar studies have been
conducted with other types of inserts and other restriction sites (Larson,
Hsieh, Jaworski, Blaho, Zacharias, and Wells, unpublished observations).
These studies demonstrate that the EcoRI methylase and restriction enzyme
will not recognize or act on the GAATTC site when it is at the interface
between the vector DNA and the insert which adopts an unusual
conformation.

The phage T4 endonuclease VII specifically cleaves Holliday-like junctions in DNA (8,97). Cruciforms are formally equivalent, at the branch points, to Holliday recombination intermediates. The topic of enzymatic probes for left-handed Z-DNA and other unusual structures has been recently reviewed (98).

c. Chemical Probes.

Bromoacetaldehyde (BAA) has been used to identify single-stranded regions of 3 or more nucleotides in duplex DNA. BAA reacts preferentially with the exocyclic N and a ring nitrogen function of cytosine, forming a 4-membered ring, but also forms a similar adduct with adenine. For the reaction to occur, the DNA cannot be base paired in a Watson-Crick duplex. We have previously reported (99) that BAA does not react with the junctions occurring at the ends of stretches of left-handed Z-DNA, but reacts readily with cruciforms. Further investigations under different conditions with BAA (M. McLean, F. Wohlrab, J.E. Larson, and R.D. Wells, submitted) reveal that BAA will react at B-Z junctions under some conditions. Investigations are currently under way to evaluate the capacity of this probe to react with pur·pyr regions and anisomorphic DNA (J. Hanvey, F. Wohlrab, M. McLean, and R.D. Wells, unpublished work).

Diethylpyrocarbonate (DEP) has been shown to react preferentially with alternating (dC-dG)$_n$ or (dA-dC)$_n$ sequences in the left-handed state (46,100). Purines at the N-7 position (101,102) are carbethoxylated in a conformationally dependent manner. It has been proposed that the syn conformation of the purine nucleotides within the left-handed helix are the reason for the observed hyperreactivity. Also, this chemical reacts preferentially at the loop regions of the supercoil-induced cruciforms (103,104) which are single stranded in nature. Proteins, such as anti-Z DNA antibodies which are attached to the left-handed sequence, have been footprinted using this chemical (105). The conformational features of the purines which give rise to DEP hyperreactivity are not known at present. This chemical seems to be a very powerful probe since the DNA chain can be cleaved at the modified bases by treatment with hot piperidine and the modification sites can be mapped at the nucleotide level. The use of this probe with other types of unusual DNA conformations remains to be determined.

OsO$_4$ in the presence of pyridine has been shown to be a site specific probe which is hyperreactive with supercoil induced cruciforms and B-Z junctions (46,106-108). OsO$_4$ adds to the 5,6 double bond of pyrimidine residues (109,110) forming osmate ester derivatives. T residues are hyperreactive toward this chemical in a conformationally dependent manner (46,100,111,112). Our preliminary results indicate that C, A and G residues can also be detected in certain cases (113,114). Furthermore, our results indicate that OsO_4 is hyperreactive with T residues in the loop regions of cruciforms. The conformational requirements for hyperreactivity with OsO_4 are not known at present. However, as in the case of DEP, it is possible to cleave the DNA chain at the modified bases, thus facilitating the fine mapping of the modified sites by DNA sequence analyses. We are enthusiastic about the use of this probe since our preliminary studies on the conformation of pur·pyr stretches in supercoiled plasmids revealed that the polymer block insert is very specifically modified in the center as well as at one end of the sequence, but not at the other end. This result, in combination with DEP, BAA, and DMS, may give new insights into the structural peculiarity adopted by these types of sequences.

Chemical approaches to study the conformations of unusual sequences are of special interest since these sequences are frequently present in the vicinity of functional genes (33-35,62-71).

Other probes used to study supercoil induced conformational changes in duplex structures include hydroxylamine (46), dimethylsulfate (DMS) (46,77,81), and sodium bisulfite (113). Hydroxylamine was shown to modify C residues at the B-Z junctions in supercoiled molecules. The disadvantage of using this chemical as a conformational probe is that a very high concentration of the reagent must be employed (above 1M). Thus, the high concentration of the chemical may influence the properties of the structure under investigation. However, the modified C residues can be fine mapped by DNA sequencing after treatment with piperidine. Dimethylsulfate, which is widely used in the Maxam and Gilbert DNA sequencing method, has also been used to probe DNA structures (46,77,81). This chemical specifically methylates the N-7 position of guanine. There are, however, examples of conformationally dependent methylation of A residues (77) and C residues (J. Klysik, J. Hanvey, and R.D. Wells, unpublished observations). Sodium bisulfite (113) has also been used in studies of DNA conformation. A major disadvantage of sodium bisulfite as a probe is that the DNA chain cannot be chemically cleaved directly at the modified bases. Thus, either S1 nuclease or some other steps must be employed to map the modification sites.

It may be appropriate to test the use of thallium—catalyzed
iodination as a probe for unusual structures. This chemical method has
been carefully explored in the mid-1970's as a probe for small regions of
disrupted secondary structure in duplex DNAs (114). However, it has not
been studied in recent years on the more subtle unusual structures which
are the subject of this review. Because the iodination proceeds 100-200
times faster on random coil DNA than for native DNA, we believe that it is
likely that this may be a useful reagent for investigation of cruciforms,
B—Z junctions, pur•pyr sequences, and other unusual conformations. Since
5—iodocytidylic acid is the only stable product which is formed and since
radioactive iodine may be employed, rather sophisticated mapping
experiments can be conducted.

Chemical probes represent an extremely powerful approach to the
investigation of unusual DNA structures since mapping of reactive sites
can be conducted at the base pair level. However, much work remains to be
conducted in order to understand the DNA structural requirements for
reactivity.

    d.   Supercoil Relaxation.

The basis of this assay is the fact that certain structural
transitions (cruciform extrusion, right- to left-handed flipping, etc.)
will result in a net reduction of negative superhelical density
(4,8,55,56,90,91,95,97,115,117-121). The first step in the assay is the
construction of a population of topoisomers differing in linking number by
1 (118). These populations will then be subjected to 2-dimensional
agarose gel electrophoresis (56,90,117,119,120), which is a 2-dimensional
display of the relaxation previously characterized (115) in a single
dimension. The extent of relaxation and the free energy needed to drive a
structural transition will be derived from the gels and the results will
be correlated to the sequence data. Accordingly, the presence of
left-handed versus cruciform structures can be rigorously tested.

These analyses give extremely important results since it is possible
to rigorously evaluate the types of unusual conformations which are
formed. The rationale of the analysis was previously described (91,122).
To summarize, if one knows the length of inserts or, more appropriately,
if a series of inserts is available in a family of recombinant plasmids,
it is possible to calculate quite rigorously the rise per residue in a
left-handed Z helix or the type of cruciform which is formed in an
appropriate insert. Furthermore, these determinations have given

important results on the thermodynamics of these transitions (6,49,122). The energetic values for causing the transition of one base pair from a B structure to a Z helix and/or for a B–Z junction have been determined (91,122,123) from this method.

e. Conditions Which Stabilize Unusual Structures.

Conditions that stabilize unusual conformations have been evaluated. This is particularly important since an understanding of these conditions provides important insights regarding the types of conformations which are formed. Negative supercoiling stabilizes several types of unusual structures (left-handed Z-DNA, cruciforms, pur·pyr structures, and anisomorphic DNA). Furthermore, the unusual S1 nuclease sensitive properties of the major late promoter of adenovirus (78,79) require negative supercoiling.

Substantial previous studies have been conducted on the influence of supercoiling, ionic conditions, methylation of the 5 position of cytosine, and the role of ligands and carcinogens in stabilizing left-handed Z-DNA and cruciforms. Negative supercoiling (0.04-.055) is sufficient to stabilize the majority of types of left-handed Z-DNA structures (58-61 and references therein) and cruciforms (1-12).

The role of alternate ionic conditions in stabilizing unusual structures will be evaluated employing a variety of ionic strengths, types of salts including divalent metal ions, and dehydrating agents as previously described for left-handed Z-DNA (42-45,116,122,124-129). In some cases, a given salt (such as sodium acetate for Z-DNA) can have very unusual effects (126).

The role of methylation at the 5 position of cytosine (122,130,131 and references therein) has been evaluated. It may be important in the future to evaluate the role of methylation at other positions (such as the 7 position of G (42)). Also, almost no work has been performed to evaluate the role of carcinogens, mutagens, and/or other types of ligands on the conformation and properties of a number of types of unusual DNA structures (42-45,129). Furthermore, it is possible that multi-strand structures (132,133) may play an important role in stabilizing some types of alternate DNA structures such as described above (72). The general methodology to be employed for these types of determinations has been established. However, substantial additional work will be required to perform and evaluate these questions.

f. Anomalous Gel Mobility: Bent DNA.

As mentioned above, certain types of DNA sequences are thought to
exist in a compact structure which may consist of a systematically bending
helix. These sequences appear to contain runs of A's which exhibit a
periodicity of approximately one turn of helix and occur in kinetoplast
DNAs as well as other regions from both prokaryotic and eukaryotic
organisms (17-32). Gel electrophoretic experiments on a series of
acrylamide (5%-16%) and agarose (0.6%-2.0%) gels at room temperature and
at 5° have been diagnostic for the presence of bent DNA. Very few
determinations have been employed in general for evaluating the properties
and functions of bent DNA. In addition to the gel mobility determinations
just described, these include the following: large circular dichroism
changes at unusually low ethanol concentrations (25,32); rates of ring
closure of restriction fragments of varying lengths (18); and electron
microscopy on certain fragments (J. Griffith, P. Englund, unpublished
data). All of these determinations have been employed for a variety of
fragments. However, in general, assays do not exist for determining the
presence of bent DNA in recombinant plasmids. Prior studies have not
revealed an influence of negative supercoiling on determinations which can
be applied for bent DNA (25,32). However, a recent investigation (121)
reveals that chemical and/or certain enzymatic probes may be useful for
these types of determinations.

g. Crosslinking of DNA.

In certain relatively rare cases where multistrand complexes are
suspected, it may be appropriate to attempt crosslinking studies with
psoralen derivatives. Exposure to psoralens followed by ultraviolet
irradiation results in crosslinking of pyrimidines across the strands of a
duplex DNA (134). The irradiated samples may then be analyzed by
restriction enzyme digestion and subsequent gel electrophoresis. This
type of experiment is designed to freeze potential slipped structures for
other determinations (electron microscopy). Also, if multistranded DNA
molecules occur, crosslinking should yield band patterns different than
expected for canonical B-DNA. Similarly, cruciform structures in
supercoiled molecules will be fixed by trimethylpsoralen crosslinking
(135). If unusual structures are detected by this method, electron
microscopy may be used to visualize and quantitate the cross-linked
elements in a population of DNA molecules.

Also, diepoxybutane (DEB) specifically crosslinks left-handed Z-DNA
(125,136). Multistranded structures might have an appropriate distance
and orientation of the N7 sites on G's to permit crosslinking, although
this type of structure has not been tested yet. We believe that
crosslinking studies have the potential to give important new information
in certain restricted cases.

h. Electron Microscopy.

Electron microscopy has been widely used in the past for studies of
protein-DNA complexes as well as certain DNA structural studies (137-157).
However, this technique has been virtually unexploited with respect to
analyses of unusual DNA conformations at the base pair level such as small
cruciforms, Z-DNA, multistranded complexes, and bent DNA. Investigations
are just beginning in the area of bent DNA (J. Griffith, personal
communication). Methodology has been developed for analyzing unusual DNA
features in fragments as small as approximately 100 base pairs (J.
Griffith, unpublished data). New procedures for rapid freezing of samples
for electron microscopy which allows the study of DNA fragments as small
as 50 base pairs are important for these determinations. Furthermore, new
procedures which enable the ultrahigh vacuum (cryopumped) freeze etch and
freeze slamming procedures have been of value for these innovations. The
study of unusual DNA conformations such as those described above by
electron microscopy is essentially virgin territory. Part of the reason
for this lack of information has been the inability to obtain an
appropriate state of resolution to enable unequivocal conclusions. We are
hopeful that these technical developments will substantially improve our
ability to utilize this technique.

i. Helical Repeat.

Two types of helical repeat experiments have been utilized with
different types of unusual DNA structures. First, the procedure of Rhodes
and Klug (158) provides information on the helical repeat of certain types
of DNA polymers and fragments. In these analyses, the DNA molecules are
isolated and labelled at their 5'-ends. After fixation of the fragment on
hydroxyapatite, the resulting suspension is treated with DNAse I and the
products analyzed on denaturing gels. The accessibility of the DNA helix
to the endonuclease is determined in this case by the helical repeat.
Molecules have been analyzed to determine if they contain a significant
deviation from the established B-DNA helical repeat of 10.4 bp per turn.

13

A second method for determining helical repeats with recombinant plasmids uses the band shift method (159,160). The helical repeat (h) (i.e., the number of bp per complete helix turn) is an important structural parameter of the DNA helix from a physicochemical point of view. Obviously, this factor may determine the specificity of certain DNA-protein interactions. In the future, it is likely that determinations will be performed on both methylated and unmethylated states of both right-handed and left-handed helical structures of short inserts cloned into recombinant plasmid vectors. Also, it is likely that these types of measurements will be performed on other unusual DNA structures. The band shift method described by Wang (161-163) has been used to determine the h-values for a series of homopolymer and alternating copolymer inserts in plasmids by one-dimensional agarose gel electrophoresis. In the situation where the structural transition is accompanied by supercoil relaxation at increasing negative superhelical density (B-Z transition or cruciform formation), the interpretation of the band patterns is complicated by overlapping regions of topoisomers before and after the structural transition. Also, possible intermediate states may further complicate the interpretation. Thus, it will be necessary to embellish the band shift method by employing two-dimensional gel electrophoretic analyses (164). This may enable the analysis of positive topoisomers as well as negative topoisomers before and after the B to Z transition on the same gel.

In general, it is hoped that these procedures will give important results regarding the types of helices which are adopted by unusual DNA structures.

Prospects for the Future

The future of investigations on unusual DNA structures is very bright indeed since the field is in its infancy. All of the structures shown in Fig. 1 were discovered since approximately 1980 and two of the new conformations were first described quite recently. It is likely that biological functions for DNA microheterogeneity will be identified in the near future. Since DNA is the site of action of replication complexes, regulatory proteins, mutagenic and carcinogenic insults, recombination processes, etc., and because these unusual conformations are likely to play (at least in some cases) an intimate role in these interactions, our investigations have just begun.

Acknowledgments

This work was supported by grants from the National Institutes of Health (GM 30822) and the National Science Foundation (86 07785). A portion of this chapter was published previously as a minireview (165).

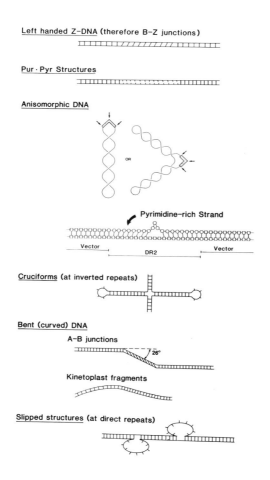

Fig. 1.   Cartoon of some unusual secondary structures of DNA.

REFERENCES

1. Gellert, M., Mizuuchi, K., O'Dea, M.H., Ohmori, H., and Tomizawa, J., Cold Spring Harbor Symp. Quant. Biol., 43, 35-40 (1979).
2. Lilley, D.M.J., Proc. Natl. Acad. Sci., USA, 77, 6468-6472 (1980).
3. Panayotatos, N., and Wells, R.D., Nature (London), 289, 466-470 (1981).
4. Singleton, C.K., and Wells, R.D., J. Biol. Chem., 257, 6292-6295 (1982).
5. Lilley, D.M.J., Nature (London), 305, 276-277 (1983).
6. Singleton, C.K., J. Biol. Chem., 258, 7661-7668 (1983).
7. Lilley, D.M.J., and Hallam, L.R., J. Mol. Biol., 180, 179-200 (1984).
8. Lilley, D.M.J., and Kemper, B., Cell, 36, 413-422 (1984).
9. Greaves, D.R., and Patient, R.K., EMBO J., 4, 2617-2626 (1985).
10. Gough, G.W., Sullivan, K.M., and Lilley, D.M.J., EMBO J., 5, 191-196 (1986).
11. Muller, U.R., and Wilson, C.L., J. Biol. Chem., 262, 3730-3738 (1987).
12. Greaves, D.R., Patient, R.K., and Lilley, D.M.J., J. Mol. Biol., 185, 461-478 (1985).
13. Holliday, R., Genet. Res., 5, 282-304 (1964).
14. Selsing, E., Wells, R.D., Early, T.A., and Kearns, D.R., Nature, 275, 249-250 (1979).
15. Selsing, E. and Wells, R.D., J. Biol. Chem., 254, 5410-5416 (1979).
16. Selsing, E., Wells, R.D., Alden, C.J., and Arnott, S., J. Biol. Chem., 254, 5417-5422 (1979).
17. Challberg, S.S., and Englund, P.T., J. Molec. Biol., 138, 447-442 (1980).
18. Trifonov, E.N., and Sussman, J.L., Proc. Natl. Acad. Sci., USA, 77, 3816-3820 (1980).
19. Ross, W., Shulman, J., and Landy, A., J. Molec. Biol., 156, 505-529 (1982).
20. Simpson, L., Proc. Natl. Acad. Sci., USA, 76, 1585-1588 (1979).
21. Kidane, G.Z., Hughes, D., and Simpson, L., Gene, 27, 265-277 (1984).
22. Wu, H.-M., and Crothers, D.M., Nature, 308, 509-513 (1984).
23. Hagerman, P.J., Proc. Natl. Acad. Sci., USA, 81, 4632-4636 (1984).
24. Hagerman, P.J., Biochemistry, 24, 7033-7037 (1985).
25. Marini, J.C., Levene, S.D., Crothers, D.M., and Englund, P.T., Proc. Natl. Acad. Sci., USA, 79, 7664-7668 (1982).
26. Levene, S.D., and Crothers, D.M., J. Biomolec. Struct. Dyn., 1, 429-435 (1983).
27. Stellwagen, N.C., Biochemistry, 22, 6186-6193 (1983).
28. Bossi, L., and Smith, D.M., Cell, 39, 643-652 (1984).
29. Zahn, K., and Blattner, F.R., Nature, 317, 451-453 (1985).
30. Ulanovsky, L., Bodner, M., Trifonov, E.N., and Choder, M., Proc. Natl. Acad. Sci., USA, 83, 862-866 (1986).
31. Hagerman, P.J., Nature, 321, 449-450 (1986).
32. Kitchin, P.A., Klein, V.A., Ryan, K.A., Gann, K.L., Raush, C.A., Kang, D.S., Wells, R.D., and Englund, P.T., J. Biol. Chem., 261, 11302-11309 (1986).
33. Hentschel, C.C., Nature, 295, 714-716, (1982).
34. Mace, M.A.F., Pelham, H.R.B., and Travers, A.A., Nature, 304, 555-557 (1983).
35. McKeon, C., Schmidt, A., and deCrombrugghe, B., J. Biol. Chem., 259, 6636-6640 (1984).

36. Burd, J.F., Wartell, R.M., Dodgson, J.B., and Wells, R.D., J. Biol. Chem., 250, 5109-5113 (1975).
37. Burd, J.F., Larson, J.E., and Wells, R.D., J. Biol. Chem., 250, 6002-6007 (1975).
38. Wartell, R.M., and Burd, J.F., Biopolymers, 15, 1461-1480 (1976).
39. Early, T.A., Kearns, D.R., Burd, J.F., Larson, J.E., and Wells, R.D., Biochemistry, 16, 541-551 (1977).
40. Wells, R.D., Goodman, T.C., Hillen, W., Horn, G.T., Klein, R.D., Larson, J.E., Muller, U.R., Neuendorf, S.K., Panayotatos, N., and Stirdivant, S.M., Progress in Nucleic Acid Research and Molecular Biology, 24, 167-267 (1980).
41. Sullivan, K.M., and Lilley, D.M.J., Cell, 47, 817-827 (1986).
42. Rich, A., Nordheim, A., and Wang, A. H.-J, Ann. Rev. Biochem., 53, 791-846, (1984).
43. Wells, R.D., Erlanger, B.F., Gray, H.B., Jr., Hanau, L.H., Jovin, T.M., Kilpatrick, M.W., Klysik, J., Larson, J.E., Martin, J.C., Miglietta, J.J., Singleton, C.K., Stirdivant, S.M., Veneziale, C.M., Wartell, R.M., Wei, C.F., Zacharias, W., and Zarling, D. (1983). Left-handed Z-DNA helices, cruciforms, and supercoiling. In Gene Expression, UCLA Symposia on Molecular and Cellular Biology, New Series, Vol. 8, 3-18, eds. Dean Hamer and Martin Rosenberg, Alan R. Liss, Inc., New York, New York.
44. Zimmerman, S.B., Ann. Rev. Biochem., 51, 395-427 (1982).
45. Cold Spring Harbor Symp. Quant. Biol., 47 (1983).
46. Johnston, B.H., and Rich, A., Cell, 42, 713-724 (1985).
47. Wang, A.H.-J., Gessner, R.V., van der Marel, G.A., van Boom, J.H., and Rich, A., Proc. Natl. Acad. Sci., USA, 82, 3611-3615 (1985).
48. Peck, L.J., Wang, J.C., Nordheim, A., and Rich, A., J. Mol. Biol., 190, 125-127 (1986).
49. O'Connor, T.R., Kang, D.S., and Wells, R.D., J. Biol. Chem., 261, 13302-13308 (1986).
50. Dickerson, R.D., Drew, H.R., Conner, B.N., Wing, R.M., Fratini, A.V., and Kopka, M.L., Science, 216, 475-485 (1982).
51. Kmiec, E.B., and Holloman, W.K., Cell, 44, 545-554 (1986).
52. Blaho, J.A. and Wells, R.D., J. Biol. Chem., 262, 6082-6088 (1987).
53. Azorin, F., and Rich, A., Cell, 41, 365-374 (1985).
54. Stollar, D., Crit. Rev. Biochem., 20, 1-36 (1986).
55. Singleton, C.K., Klysik, J., and Wells, R.D., Proc. Natl. Acad. Sci., 80, 2447-2451 (1983).
56. Kilpatrick, M.W., Klysik, J., Singleton, C.K., Zarling, D., Jovin, T.M., Hanau, L.H., Erlanger, B.F., and Wells, R.D., J. Biol. Chem., 259, 7268-7274 (1984).
57. Haniford, D.B., and Pulleyblank, D.E., Nucleic Acids Res., 13, 4343-4362 (1985).
58. McLean, M.J., Blaho, J.A., Kilpatrick, M.W., and Wells, R.D., Proc. Natl. Acad. Sci., 83, 5884-5888 (1986).
59. Ellison, M.J., Feigon, J., Kelleher, R.J. III, Wang, A. H.-J., Habener, J.F., and Rich, A., Biochemistry, 25, 3648-3655 (1986).
60. Ho, P.S., Ellison, M.J., Quigley, G.J., and Rich, A., The EMBO Journal, 5 (10), 2737-2744 (1986).
61. Ellison, M.J., Kelleher, R.J. III, Wang, A. H.-J., Habener, J.F., and Rich, A., Proc. Natl. Acad. Sci., 82, 8320-8324 (1985).
62. Selleck, S.B., Elgin, S.C.R., and Cartwright, I.L., J. Mol. Biol., 178, 17-33 (1984).
63. Larsen, A., and Weintraub, H., Cell, 29, 609-622 (1982).
64. Elgin, S.C.R., Nature, 309, 213-214 (1984).
65. Dybvig, K., Clark, D.D., Aliperti, G., and Schlesinger, M.J., Nucleic Acids Res., 11, 8495-8505 (1983).
66. Nickol, J.M. and Felsenfeld, G., Cell, 35, 467-477 (1983).

67. Schon, E., Evans, T., Welsh, J., and Efstratiadis, A., Cell, 35, 838-848 (1983).
68. Htun, H., Lund, E., and Dahlberg, J.E., Proc. Natl. Acad. Sci., USA, 81, 7288-7292 (1984).
69. Reynolds, W.F. and Gottesfeld, J.M., Proc. Natl. Acad. Sci., USA, 82, 4018-4022 (1985).
70. Shen, C-K.J., Nuc. Acids Res., 11, 7899-7910 (1983).
71. Glikin, G.C., Garuils, G., Rena-Descalzi, L., and Worcel, A., Nature, 303, 770-774 (1983).
72. Lyamichev, V.I., Mirkin, S.M., and Frank-Kamenetskii, M.D., J. Biomol. Struct. & Dyn., 3, 327-338 (1985).
73. Kohwi-Shigematsu, T., and Kohwi, Y., Cell, 43, 199-206 (1985).
74. Wells, R.D., Blakesley, R.W., Burd, J.F., Chan, H.W., Dodgson, J.B., Hardies, S.C., Horn, G.T., Jensen, K.F., Larson, J., Nes, I.F., Selsing, E., and Wartell, R.M., Critical Reviews in Biochemistry, 4, 305-340 (1977).
75. Wells, R.D., Larson, J.E., Grant, R.C., Shortle, B.E., and Cantor, C.R., J. Mol. Biol., 54, 465-497 (1970).
76. Pulleyblank, D.A., Haniford, D.B., and Morgan, A.R., Cell, 42, 271-280 (1985).
77. Evans, T., and Efstratiadis, A., J. Biol. Chem., 261, 14771-14780 (1986).
78. Kilpatrick, M.W., Torri, A., Kang, D.S., Engler, J.A., and Wells, R.D., J. Biol. Chem., 261, 11350-11354 (1986).
79. Yu, Y.-T., and Manley, J.L., Cell, 45, 743-751 (1986).
80. Lyamichev, V.I., Mirkin, S.M., and Frank-Kamenetskii, M.D., J. Biomol. Struct. & Dyn., 3, 667-669 (1986).
81. Pulleyblank, D.E., and Haniford, D.B., Biomolecular Stereodynamics IV. Proceedings of the Fourth Conversation in the Discipline Biomolecular Stereodynamics, State University of New York, Albany, NY, June 04-09, 1985, Eds., R.H. Sarma and M.H. Sarma, ISBN 0-940030-18-7, Adenine Press, 1986.
82. Wohlrab, F., McLean, M.J., and Wells, R.D., J. Biol. Chem., 262, 6407-6416 (1987).
83. Wells, R.D., and Harvey, S.C., editors, (1987) "Unusual DNA Structures"; Springer-Verlag Publishing Company, New York, NY.
84. Klysik, J., Stirdivant, S.M., and Wells, R.D., J. Biol. Chem., 257, 10152-10158 (1982).
85. Kilpatrick, M.W., Klysik, J., Singleton, C.K., Zarling, D., Jovin, T.M., Hanau, L.H., Erlanger, B.F., and Wells, R.D., J. Biol. Chem., 259, 7268-7274 (1984).
86. Kmiec, E.B., and Holloman, W.K., Cell, 36, 593-598 (1984).
87. Kmiec, E.B., Angelides, K.J., and Holloman, W.K., Cell, 40, 139-145 (1985).
88. Murphy, K.E., and Stringer, J.R., Nucleic Acids Res., 14, 7325-7340 (1986).
89. Vogt, W.M., Methods Enzymol., 65, 248-255 (1980).
90. Singleton, C.K., Klpatrick, M.W., and Wells, R.D., J. Biol. Chem., 259, 1963-1967 (1984).
91. Singleton, C.K., Klysik, J., Stirdivant, S.M., and Wells, R.D., Nature, 299, 312-316 (1982).
92. Hayes, T.E., and Dixon, J.E., J. Biol. Chem., 260, 8145-8156 (1985).
93. Camilloni, G., Della Seta, F., Negri, R., Ficca, A.G., and DiMauro, E., The EMBO Journal, 5 (4), 763-771 (1986).
94. Kilpatrick, M.W., Wei, C.-F., Gray, H.B., Jr., and Wells, R.D., Nucleic Acids Research, 11, 3811-3822 (1983).
95. Zacharias, W. Larson, J.E., Kilpatrick, M.W., and Wells, R.D., Nucleic Acids Res,. 12, 7677-7692 (1984).

96. Vardimon, L. and Rich, A., Proc. Natl. Acad. Sci., USA, 81, 3268-3272 (1984).

97. Kemper, B. and Garabett, M., Eur. J. Biochem., 115, 123-131 (1981).

98. Wohlrab, F., and Wells, R.D., Gene Amplification and Analysis, Volume V, in the press (1986) (J.G. Chirikjian, ed., Elsevier Science Publishing Co., Inc.).

99. Kang, D.S. and Wells, R.D., J. Biol. Chem., 260, 7783-7790 (1985).

100. Herr, W., Proc. Natl. Acad. Sci., USA, 82, 8009-8013 (1985).

101. Leonard, N.J., McDonald, F.F., Henderson, E.L., and Reichman, M.E., Biochemistry, 10, 3335-3342 (1971).

102. Vincze, A., Henderson, R.E.L., McDonald, F.F., and Leonard, N.J., J. Am. Chem. Soc., 95, 2677-2682 (1973).

103. Furlog, J.C., and Lilley, D.M.J., Nucl. Acids Res., 14, 3995-4007 (1986).

104. Scholten, P.M., and Nordheim, A., Nucl. Acids Res., 14, 3981-3993 (1986).

105. Runkel, L., and Nordheim, A., J. Mol. Biol., 189, 487-501 (1986).

106. Lilley, D.M.J., and Palecek, E., EMBO J., 3, 1187-1192 (1984).

107. Naylor, L.H., Lilley, D.M.J., and van de Sande, J.H., EMBO J., 5, 2407-2413 (1986).

108. Glikin, G.C., Vojtiskova, M., Rena-Descalzi, L., and Palecek, E., Nucl. Acids Res., 12, 1725-1735 (1984).

109. Neidle, S., and Stuart, D.I., Biochim. Biophys. Acta., 418, 226-231 (1976).

110. Beer, M., Stern, S., Carmalt, D., and Mohlhenrich, K.H., Biochemistry, 5, 2283-2288 (1966).

111. Glazka, G., Palecek, E., Wells, R.D., and Klysik, J., J. Biol. Chem., 261, 7093-7098 (1986).

112. Nejedly, K., Kwinkowski, M., Galazka, G., Klysik, J., and Palecek, E., J. Biomolec. Struct. Dyn., 3, 467-478 (1985).

113. Gough, G.W., Sullivan, K.M., and Lilley, D.M.J., EMBO J., 5, 191-196 (1986).

114. Jensen, K.F., Nes, I.F., and Wells, R.D., Nucleic Acids Research, 3, 3143-3155 (1976).

115. Klysik, J., Stirdivant, S.M., Larson, J.E., Hart, P.A., and Wells, R.D., Nature, 290, 672-677 (1981).

116. Zacharias, W., Larson, J.E., Klysik, J., Stirdivant, S.M., and Wells, R.D., J. Biol. Chem., 257, 2775-2782 (1982).

117. Haniford, D.B., and Pulleyblank, D.E., Nature, 302, 632-634 (1983).

118. Singleton, C.K., and Wells, R.D., Anal. Biochemistry, 122, 253-257 (1982).

119. Wang, A.H.-J., Gessner, R.V., van der Marel, G.A., van Boom, J.H., and Rich, A., Proc. Natl. Acad. Sci., USA, 82, 3611-3615 (1985).

120. Wang, J.C., Peck, L.J., and Becherer, K., Cold Spring Harbor Symp. Quant. Biol., 47, 85-92 (1983).

121. Krause, H.M., Kilpatrick, M.W., Collier, D.A., Wells, R.D., and Higgins, N.P., submitted (1987).

122. Klysik, J., Stirdivant, S.M., Singleton, C.K., Zacharias, W., and Wells, R.D., J. Mol. Biol., 168, 51-71 (1983).

123. Peck, L.J., and Wang, J.C., Proc. Natl. Acad. Sci., USA, 80, 6206-6210 (1983).

124. Stirdivant, S.M., Klysik, J., and Wells, R.D., J. Biol. Chem., 257, 10159-10165 (1982).

125. Kang, D.S., Harvey, S.C., and Wells, R.D., Nucleic Acids Res., 13, 5645-5656 (1985).

126. Zacharias, W., Martin, J.C., and Wells, R.D., Biochemistry, 22, 2398-2405 (1983).

127. Wartell, R.M., Klysik, J., Hillen, W., and Wells, R.D., Proc. Natl. Acad. Sci., USA, 79, 2549-2553 (1982).

128. Klysik, J., Stirdivant, S.M., and Wells, R.D., J. Biol. Chem., 257, 10152-10158 (1982).
129. Wells, R.D., Miglietta, J.J., Klysik, J., Larson, J.E., Stirdivant, S.M., and Zacharias, W., J. Biol. Chem., 257, 10166-10171 (1982).
130. Behe, M., and Felsenfeld, G., Proc. Natl. Acad. Sci., USA, 78, 1619-1623 (1981).
131. Behe, M., Zimmerman, S., and Felsenfeld, G., Nature, 293, 233-235 (1981).
132. Morgan, A.R., and Wells, R.D., J. Mol. Biol., 37, 63-80 (1968).
133. Chamberlin, M.J., and Patterson, D.L., J. Mol. Biol., 12, 410-436 (1965).
134. Song, P.S. and Tapley, K.J., Photochem. Photobiol., 29, 1177-1197 (1979).
135. Sinden, R.R., Broyles, S.S., and Pettijohn, D.E., Proc. Natl. Acad. Sci., USA, 80, 1797-1801 (1983).
136. Castleman, H., Hanau, L.H., and Erlanger, B.F., Nucleic Acids Res., 11, 8421-8425 (1983).
137. Sperrazza, J. M., Register III, J.C., and Griffith, J., Gene, 31, 17-22 (1984).
138. Griffith, J.D., and Nash, H.A., Proc. Natl. Acad. Sci., USA, 82, 3124-3128 (1985).
139. Chrysogelos, S., Register III, J.C., and Griffith, J., J. Biol. Chem., 258, 12624-12631 (1983).
140. Register III, J.C. and Griffith, J., J. Biol. Chem., 260, 12308-12312 (1985).
141. Register III, J.C., and Griffith, J., Proc. Natl. Acad. Sci., USA, 83, 624-628 (1986).
142. Griffith, J.D., Christiansen, G., Ann. Rev. Biophys. Bioeng., 7, 19-35 (1978).
143. Griffith, J., Hochschild, A., and Ptashne, M., Nature, 322, 750-752 (1986). 144. Griffith, J.D., Science, 187, 1202-1203 (1975).
145. Griffith, J.D., Science, 201, 525-527 (1978).
146. Griffith, J.D., and Formosa, T., J. Biol. Chem., 260, 4484-4491 (1985).
147. Griffith, J., Bleyman, M., Rauch, C.A., Kitchin, P.A., and Englund, P.T., Cell, 46, 717-724 (1986).
148. Stasiak, A., Stasiak, A.Z., and Koller, T., Cold Spring Harbor Symp. Quant. Biol., 49, 561-570 (1984).
149. Stasiak, A., DiCapua, E., and Koller, T., J. Mol. Biol., 151, 557-564 (1981).
150. Dunn, K., Chrysogelos, S., and Griffith, J., Cell, 28, 757-756 (1982).
151. DiCapua, E., Engel, A., Stasiak, A., and Koller, T., J. Mol. Biol., 157, 82-103 (1982).
152. Koller, T., DiCapua, E., and Stasiak, A., In Mechanisms of DNA Replication and Recombination, (N. Cozzarelli, ed.), Alan K. Liss, New York, 723-729 (1983).
153. Register III, J.C., and Griffith, J., Mol. Gen. Genet., 199, 415-420 (1985).
154. Egelman, E.H., and Stasiak, A., J. Mol. Biol., 191, 677-697 (1986).
155. Griffith, J., and Shores, C.G., Biochemistry, 24, 158-162 (1985).
156. Griffith, J.D., Harris, L.D., and Register III, J., Cold Spring Harbor Symp. Quant. Biol., 49, 553-559 (1984).
157. Chrysogelos, S., and Griffith, J., Proc. Natl. Acad. Sci., USA, 9, 5803-5807 (1982).
158. Rhodes, D., and Klug, A., Nature, 286 573-578 (1980).
159. Wang, J.C., Proc. Natl. Acad. Sci., USA, 76, 200-203 (1979).
160. Peck, L.J., and Wang, J.C., Nature, 292, 375-378 (1981).
161. Wang, J.C., Proc. Natl. Acad. Sci., USA, 76, 200-203 (1979).

162. Peck, L.J., and Wang, J.C., _Nature_, _292_, 375-378 (1981).
163. Strauss, F., Gaillard, C., and Prunell, A., _Eur. J. Biochem._, _118_, 215-222 (1981).
164. Zacharias, W., O'Connor, T.R., and Larson, J.E., submitted for publication (1987).
165. Wells, R.D., a Minireview for _Journal of Biological Chemistry_, in the press (1987).

The Specificity of "Single Strand Specific Endonucleases": Probes of
phosphodiester conformation in double stranded nucleic acids. Left-Handed
Polypurine/polypyrimidine structures. Long range transmission of
conformational information in DNA.

David E. Pulleyblank, Mark Glover, Chuck Farah and David B. Haniford
Dept. of Biochemistry, University of Toronto, Toronto, Ontario,
CANADA M5S-1A8.

ABSTRACT

The family of zinc dependent endonucleases (EC 30.1.30.x, herein
referred to as SS nucleases, reviewed in 1) exemplified by S1 and mung
bean nucleases, have been widely regarded as being specific for single
stranded nucleic acids. These enzymes have recently been shown to
recognize a variety of non-B structures in double stranded DNA [2-14].
The basis for the selectivity of these enzymes is discussed with
reference to their cleavage of cruciform loops in poly d(AT)n.d(AT)n and
of protonated polypurine/polypyrimidine structures. Evidence is presented
for a novel left handed form of d(TC)n.d(GA)n and for unexpectedly long
range interactions between regions of different structure in plasmid DNA
molecules. The existence of these interactions indicate that even the
"normal" B-form of double stranded DNA has properties which are not
predicted by classical theories of nucleic acid structure.

INTRODUCTION

Perhaps the most remarkable characteristic of the class of zinc
dependent SS endonucleases is the efficiency with which they discriminate
against the B and Z conformations of DNA and the A conformations of
double stranded RNA and RNA/DNA hybrids. Under optimal conditions S1 and
mung bean nucleases cleave single stranded nucleic acids about 4 orders
of magnitude more rapidly than double stranded substrates. Various lines
of evidence now indicate that these selectivities do not depend on a
unique property of single stranded nucleic acids but are instead a
consequence of the ability of these enzymes to recognize discrete
conformations of phosphodiester bonds that are rare in double stranded

nucleic acids. Palindromic, polypurine/polypyrimidine and alternating purine/pyrimidine sequences in closed circular DNA undergo conformational transitions to underwound conformations when the DNA is sufficiently supercoiled [2-4, 13-22]. In the cruciform and protonated structures formed by the first two categories of sequence phosphodiester bonds within the sequence become sensitive to the action of the SS nucleases. In the case of alternating purine/pyrimidine sequences which adopt a Z conformation the most dramatic enhancements of cleavage rate are observed in junctional sequences between B and Z forms [13-15].

In this article we discuss the SS nuclease cleavage patterns in the a series of different non-B double stranded structures and examine the implications of these patterns for models of the cleavage mechanisms of these nucleases.

A B-Cruciform transition in poly d(AT)n.d(AT)n

Poly d(AT)n.d(AT)n tracts respond to unwinding torsion by being extruded as cruciforms instead of by adopting the Z-conformation [19-22]. The extrusion reaction differs from those of other palindromic sequences in that no measurable kinetic barrier has to be overcome at 23°. This property is probably related to the very small contribution by stacking forces to stabilization of the B-form of poly d(AT)n.d(AT)n [45]. Chemical and enzymatic probes reveal further unusual characteristics of the d(AT)n.d(AT)n cruciform arms in a plasmid, pAT34[19]*, which contains a 34 bp tract of this polymer. Bromoacetaldehyde, diethylpyrocarbonate and permanganate ion each reveal a pattern of enhanced reactivity on the 3' side of the center of symmetry of the extruded d(AT)n.d(AT)n cruciform as illustrated in figure 1a. The reaction of adenine with bromoacetal-dehyde is particularly significant since it requires simultaneous exposure of the N-1 and the N-6 NH_2 groups. It is therefore a useful probe for breakage of pairing between adenine and its complementary thymine base. In pAT34 it appears that only one base pair is completely broken and that breakage of a second base pair makes only a 25% contribution to the average structure. Permanganate and diethylpyro-carbonate cannot be considered to be truly "single strand specific" reagents since they react with the C5=C6 bond of thymine and with the N-7 atoms of adenine respectively†. Neither of these reagents react at a significant rate with B-DNA but do react with a number of deformed DNA structures in which the stacking of the bases is disturbed. The patterns

24

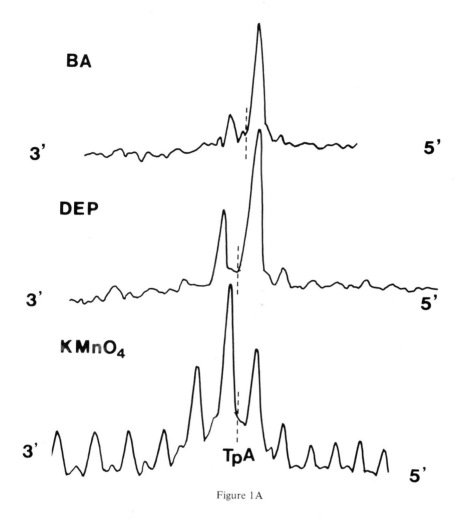

Figure 1A

Figure 1. <u>Chemical</u> <u>and</u> <u>Enzymatic</u> <u>reactivity</u> <u>of</u> <u>residues</u> <u>near</u> <u>the</u> <u>end</u> <u>of</u> <u>the</u> <u>cruciform</u> <u>loop</u> <u>in</u> <u>pAT34.</u>

a). Densitometer traces showing enhanced chemical reactivity of residues near the centre of symmetry of pAT34. Supercoiled plasmid DNA was treated in 50 mM NaAc pH 5.0, 0.1 mM $ZnCl_2$ with, BA: Bromoacetaldehyde 1% final concentration, 2 Hrs 23°, DEP: Diethylpyrocarbonate, 0.15% final concentration 7 minutes at 23°, $KMnO_4$: Potassium permanganate, final concentration 50 uM, 10 minutes at 23°. Reactions were terminated by (BA and DEP) extraction with butanol, $KMnO_4$ addition of 0.1M NaI and extraction with butanol. Samples were 3' end labelled at the unique Xba 1 site in the polylinker and cleaved with 1M piperidine, 30 minutes, 90°. Electrophoresis as in figure 1. The centre of symmetry is marked "TpA".

of cleavage by these reagents within the d(AT)n tract of pAT34
substantiate the view obtained with bromoacetaldehyde and SS nucleases
(see below) since they indicate that usual stacking of bases is disturbed
over a slightly larger region of the cruciform arm than is actually
unpaired.

Cleavage of pAT34 by SS nucleases.

The patterns of SS nuclease sensitivity of the extruded cruciform
arm of pAT34 are shown in figure 1b. Although each of these nucleases
cleave the plasmid in the vicinity of the broken base pair in the
cruciform loop they do so in strikingly different ways. Only mung bean
nuclease yields a pattern of cleavage that approximates those obtained
with the chemical reagents. It is possible to order the nucleases in
order of their recognition specificity starting with Mung bean nuclease
which cleaves selectively in the region of maximal distortion through S1,
P1 and Aspergillus niger nuclease and ending with Neurospora crassa
endonuclease (not shown) which cleaves selectively at the end which is
most B-like. In figure 1b the patterns of cleavage in each case follow
the pattern of dinucleotide repeat characteristic of the d(AT)n.d(AT)n
sequence. This is due to an alternation of the phosphodiester bond
conformation rather than to a direct effect of the bases on the action of
the enzyme. 5'TpA3' instead of 5'ApT3' linkages within the loop are most
susceptible in a plasmid where the maximum phosphodiester deformation
between the unpaired bases occurs at an 5'TpA3' dinucleotide instead of
an 5'ApT3' dinucleotide (Glover unpublished).

Polypurine/polypyrimidine structures.

Tracts of d(TC)n.d(GA)n and other polypurine/polypyrimidine
sequences in supercoiled plasmids undergo unwinding transitions
associated with selective protonation and the appearance of SS nuclease
sensitivity at pH's that approach 7.0 [23,24]. In [23] data were
presented which support a model for the protonated structures in which
dG:dCH$^+$ Hoogsteen pairs are interleaved between dA:dT Watson-Crick pairs.
Circular dichroism spectra of protonated linear polymers have been
interpreted in terms of a similar model by Courtois et al. [25]. Other
workers have presented evidence suggesting that polypurine/polypyrimidine
tracts can adopt SS nuclease sensitive conformations without accompanying
protonation [26,27]. Here we show that a second novel conformation which
is not protonated can arise in association with a protonated form. This

26

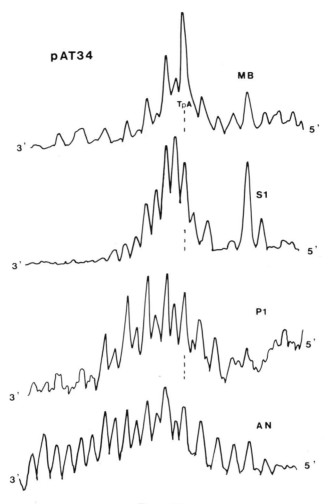

Figure 1B

b). Densitometer traces showing 3' end-labelled fragments generated by SS nuclease cleavage of pAT34. Supercoiled samples of pAT34 were nicked by addition of MB: mung bean nuclease, Sl: Sl nuclease, Pl: Pl nuclease, AN: Aspergillus niger nuclease (Farah and Pulleyblank, unpublished), to a solution containing supercoiled pAT34 in 0.05 M NaAc pH 4.4, 0.1 mM $ZnCl_2$. Reactions were terminated when ~60% of the molecules had recieved a single stranded break. Samples were end labelled and electrophoresed as in 2a.

second form is probably left-handed. Unlike the previously described protonated form this novel form is sensitive to SS nucleases on both strands.

In figure 2 the low pH induced structural transition in a plasmid containing 34 bp of d(TC)n.d(GA)n (pGA34) is compared with that previously reported in pTC45, a plasmid containing a 45 bp tract of the same polymer. In each case the transitions are driven by a combination of low pH and unwinding torsion, as indicated by lack of mobility transition at pH's greater than 7.0 (not shown, see refs. 17,23,24). In the example gel the resolution of the pTC45 topoisomers is somewhat superior to those shown in the previous reports and a new property of these transitions is revealed. Topoisomers which have just sufficient negative supercoiling to drive the transition in the pH 5 electrophoresis buffer each resolve into two discrete electrophoretic variants. The difference in mobility between the members of each variant pair corresponds to approximately one superhelical turn. The mobility variants are metastable in the electrophoresis buffer since they can be observed after 18 hrs. of electrophoresis. Results not shown indicate that the variants represent the products of alternate structural transitions in which the plasmid inserts accept differing numbers of protons and exhibit differing degrees of helical unwinding.

The unwinding observed in pTC45 and pGA34 is not proportional to the length of the polymer insert. This contrasts with examples of torsionally driven structural transitions in other simple DNA sequences (e.g. 16-20). In pTC45 a total of 3+0.5 negative superhelical turns are unwound during the transition. Assuming that in its initial state the d(TC)n.d(GA)n insert is a right handed 10.5 bp/turn helix [28], the average helical winding of the plasmid insert following the transition is 1 right handed turn/20-30 bp. A total of 5+0.5 turns are unwound in pGA34. Since the d(TC)n.d(GA)n insert in pGA34 is only 34 bp long, the average helical winding in the less mobile series of electrophoretic variants after the transition is 1 left handed turn/22 residues. It is probable that most of the unwinding is concentrated in the region of the insert (domain 2) corresponding to the 5' end of the purine strand. The other region (domain 1) is probably in an underwound right handed state similar to that observed in pTC45. In this case the left handed portion of the plasmid insert would have a helical pitch of approximately 1 left handed turn/10-15 residues.

Chemical and Enzymatic mapping experiments

The experiments described below were performed on samples of DNA with the relatively high negative superhelix densities (~-0.06) characteristic of native plasmids. Under these conditions of supercoiling only the less supercoiled family of electrophoretic variants are present in significant concentration. The results obtained therefore emphasize the most underwound transition forms of the inserts.

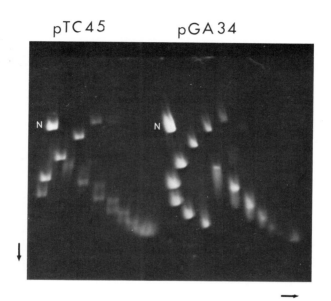

pTC45   pGA34

Figure 2. Two dimensional electrophoresis of topoisomers of GA34, and pTC45.

pTC45 is described in (23). pGA34 was constructed by inserting d(TC)n.d(GA)n into the SalI site of vector p914 with XhoI linkers. pTC45 and pGA34 are identical except for the length, site of insertion and orientation of the insertion. Families of plasmid topoisomers were prepared as described in (23). First dimension: 1.5% agarose in A: 0.05 M Tris.acetate, 5 mM sodium acetate, pH 7.0, B: 0.05 M Tris.acetate, 5 mM sodium acetate pH 6.0, C: 0.05 M Tris.acetate, 5 mM sodium acetate pH 5.0 18 hrs. 23°C. Second dimension 0.05M Tris.acetate 5mM sodium acetate pH 8.0, 10 ug/mL chloroquine phosphate. Plasmids pTC45 and pGA34 were co-electrophoresed on the same gel to emphasize the reproducibility of the differences between the transitions in these plasmids. Mobility variants a and b are marked for topoisomer -8 and -9 for plasmids pTC45 and pGA34.

Note: 1) the differences in degree of topological unwinding associated with each transition 2) differences in superhelical torsion required to initiate the transitions and 3) mobility splitting of topoisomers which have undergone the transitions.

Exposure of Purine N-7 atoms.

At native superhelix density at pH 5.0 dG residues of a zone corresponding to the 3' half of the polypurine strand (domain 1) of pGA34
become protected against alkylation by dimethylsulphate as shown in
figure 3. This property of domain 1 is consistent with the presence of
protonated Hoogsteen dCH$^+$:dG pairs within this region [23]. At pH 5.0
rearranged base pairs of this type are not likely to be present in domain
2 since dG residues here remain exposed to N-7 alkylation. Diethylpyrocarbonate also reacts at purine N-7 atoms† but shows less intrinsic
preference for guanine than dimethylsulphate and requires a less hindered
environment than that found in B-DNA [22,29-31]. The N-7 atoms of dA
residues in domain 2 of pGA34 become exposed to diethylpyrocarbonate as
the pH is lowered below neutrality (figure 3) while vector sequences and
residues in domain 1 showed no alteration in sensitivity. Although the N-
7 atoms of the dG residues of this domain are sensitive to dimethylsulphate they do not show enhanced reactivity to diethylpyrocarbonate.
These sites are therefore exposed to about the same degree as in B-DNA.

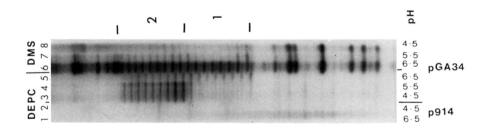

Figure 3. Chemical reactivity of purine residues in the d(TC)n.d(GA)n
tract of pGA34.

Reactivity of the polylinker segment of p914 and of the d(GA)n tract of
pGA34 to diethylpyrocarbonate and dimethylsulphate as a function of pH.
Domains 1 and 2 are indicated.

p914: Channels 1: 0.1 M NaAc adjusted to pH 6.5, 2: 0.1 M NaAc adjusted
to pH indicated with HAc. DEPC 5 ug supercoiled plasmid in 500 uL of
buffer was shaken with 5 uL diethylpyrocarbonate 15 min at 23°C.
Reactions were terminated by extraction 2 x with water saturated butanol.
Note the lack of reactivity of p914 vector sequences at either pH.

pGA34 Channels 3-8: 0.1 M NaAc adjusted to pH indicated with HAc. DEPC
reaction conditions as for p914, DMS reactions: 5 ug supercoiled plasmid
in 500 uL of the buffer was shaken for 3 minutes with 1 uL Dimethylsulphate, Reactions were terminated by making the solution 0.1 M in b-
mercaptoethanol.

Single Strand Specific endonucleases.

In pTC45 only CpT linkages of the polypyrimidine strand are sensitive to SS nucleases[23]. In contrast: both strands of pGA34 are hypersensitive to SS endonucleases. As in the case of the d(AT)n cruciform arm, the patterns of cleavage depend on the nature of the nuclease and can be ordered in the same progressive series with respect to their cleavage characteristics as illustrated in figure 4. For S1 nuclease the two domains of the polypyrimidine strand of pGA34 are comparable in sensitivity but differ in character. TpC linkages are sensitive in both domains 1 and 2 but CpT linkages are sensitive only in

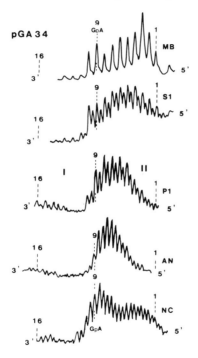

Figure 4. Enzymatic reactivity of the polypurine strand of pGA34

5 ug samples of pGA34 dissolved in 0.5 mL of the buffer indicated were cleaved with single strand specific endonucleases from MB: mung bean, S1: Aspergillus oryzae (BRL), Pl: Penicillium citrinum (Pharmacia), AN: Aspergillus niger, NC: Neurospora crassa in 0.05 M NaAc pH 5.0, 0.05 M NaCl, 0.1 mM ZnSO$_4$. Reactions were terminated when approximately 60% had recieved a single stranded break. The plasmids were end-labelled and the single stranded fragments analysed by electrophoresis as described in the legend to figure 1. Densitometer traces illustrate the progressive change in character of the cleavage reaction from MB to NC. The central GpA linkage is marked. Essentially all cleavage occurred within domain 2 defined by chemical cleavage (Figure 3)

domain 2. All of the nucleases selectively cleave domain 2 of the
polypurine strand but in dramatically different ways. Mung bean nuclease
specifically cleaves the ApG linkages with a peak of activity near the 5'
end of the domain while Aspergillus niger and Neurospora crassa nucleases
cleave both GpA and ApG phosphodiester bonds in the 3' half of the
polypurine domain 2 while exhibiting less activity towards the 5' half of
this domain. S1 and P1 nucleases have an intermediate character of
cleavage specificity.

Long Range Junctional Effects near d(TC)n.d(GA)n inserts in plasmids.

Previous reports have indicated that regions extending up to 30 bp
away from the ends of a region of Z-conformation can exhibit enhanced
sensitivity to S1 nuclease [13-15]. Although the intrinsic hypersensitivity of d(TC)n.d(GA)n inserts makes observation difficult, a highly
ordered pattern of cleavage can be discerned in the junctions upon
overexposure of autoradiograms of mapping experiments such as that shown
in figure 4. A densitometer trace showing S1 cleavages in the region of
the polypyrimidine strand immediately 3' to the d(TC)n strand of pTC45 is
shown in figure 5. In plasmids containing the polymer d(TC)n.d(GA)n the
waves of enhanced S1 nuclease sensitivity show a characteristic
dinucleotide motif of higher and lower sensitivity superimposed on an ~40
bp period. On occasion it has been possible to follow these waves through
as many as three cycles of the basic ~40 bp period. The vector without
insert does not exhibit any comparable pattern of S1 cleavage in the
vicinity of the insertion site.

DISCUSSION

Patterns of SS nuclease cleavage within the examples of sensitive
double stranded structures described here make it unlikely that SS
nucleases discriminate against A, B and Z forms on the basis of steric
occlusion of the active site by the second strand. The discrimination
must instead be based on the conformations of the phosphodiester backbone
that appear in these structures. According to this view the greater
mobility of single stranded nucleic acids permits them to enter the
conformations recognized by these enzymes and therefore to be cleaved
efficiently. In double stranded nucleic acids absolute sensitivities of
susceptible bonds vary over at least two orders of magnitude. In the
present work the most sensitive sites, which are described as being

32

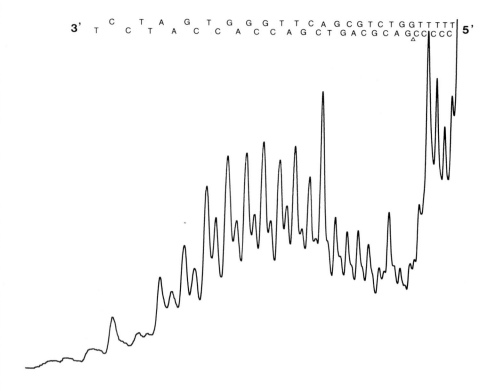

Figure 5. Densitometer trace showing junctional sensitivity near a 45 bp tract of d(TC)n.d(GA)n.

The densitometer trace shows a pattern of S1 nuclease cleavage in a 40 bp region adjoining the 45 bp d(TC)n.d(GA)n tract of pTC45 in the region near the 3' end of the polymer insert. Reaction conditions and labelling method as described in figure 4. Note that the cleavage by S1 nuclease in this region of the plasmid does not exceed 5% of that observed within the protonated polypurine/polypyrimidine tract. The ordered pattern of cleavage bears no relationship to the sequence of bases in this region of the plasmid and is therefore not the result from a sequence dependent structural transition.

"hypersensitive", probably conform closely to the active sites of the enzymes. The less susceptible sensitive bonds may owe their partial sensitivity to relatively minor distortions in the phosphodiester bond torsion angles away from those found in the B-conformation. These distortions could either directly permit attack by SS nucleases, or might result in transient adoption of the sensitive conformation as the DNA experiences thermal motion.

SS Nuclease Cleavage of the d(AT)n cruciform arm of pAT34.

Haasnoot et al. [32] have discussed the relationship between the form of a helix in a cruciform arm and the number of pairs that must be broken in order to attain structural continuity at the 180° bend. In the case of a B-form helix the shortest distance between phosphodiester bonds on opposite strands is between those that are staggered 3 residues in a 3' direction with respect to each other. This lead to the proposal that the optimum model for a cruciform loop in a B-form helix is one in which 4 bases are unpaired, three of which are stacked on the 5' side of the cruciform arm, while the remaining one connects the two sides of the helix. Experimental observations of chemical reactivity of bases in cruciform arms are generally consistent with this model for a loop in B-DNA [4,19,31,33]. In the case of pAT34 the loop in which only a single base pair appears to be broken is unexpectedly short. In order to close a loop with only a single broken base pair it is necessary for the phospho-diester backbones to be brought closer together than in B-DNA. The deformation required for closure of this loop is likely to be cause of the SS nuclease sensitivity of the 3' side of the bend [*]. The energy of formation of only 1 dA:dT base pair at 23° (i.e. ~5 kcals) must be sufficient to drive the adjustment of the 10 bp segment in pAT34 to the SS nuclease sensitive state. As discussed below, local reductions of the helix diameter are likely to underlie the SS nuclease sensitivity observed in protonated polypurine/polypyrimidine structures as well as that observed in the poly d(AT)n cruciform arm.

SS Nuclease sensitive Polypurine/polypyrimidine structures.

Protonated polypurine/polypyrimidine tracts are among the most complex double stranded polynucleotide structures ever observed. Although the available data on the transition states are not sufficiently unambiguous to yield a proof of structure (for different models of these structures see those proposed in [24,26]), we assume here that the principal features of the interleaved Hoogsteen/Watson-Crick model for the protonated structure presented in [23] are correct.

The systematic fluctuations of enzymatic and chemical reactivity described above for pGA34 and for other plasmids containing polypurine-polypyrimidine tracts in [5-11,23,24,27] show that the bases and backbone make separate contributions to progressive changes in conformation throughout the SS nuclease sensitive tracts. In non-symmetrical models such as the interleaved model [23] the preferred conformations of the separate strands are likely to be non-commensurate. The periodicity (squirm) indicated by the variations in nuclease sensitivity and chemical reactivity throughout these inserts may therefore arise from mismatch in optimal helical periodicities of the non-equivalent strands. Junctional deformations are also required at the ends of the regions of non-B structure where they are matched to the adjoining B-DNA. In the extreme case exemplified by pGA34 the combined strain arising from these deformations may be sufficient to result in the stable co-existence of two different non-Watson-Crick states in the same plasmid insert. Other experimental observations that can be explained on the basis of non-commensurate conformations of the strands of the transition structures are:

1) That the degree of unwinding associated with protonation of polypurine/polypyrimidine DNA cannot be predicted by a simple relationship between insert length an total unwinding such as previously found in B-Z and B-cruciform transitions.

2) That the degenerate orthogonal states shown in figure 3 may result from a trade off between the energy of superhelical coiling and the internal energy due to strain within the protonated tract.

The nature of SS nuclease sensitive bonds.

The principal driving force for the structural transitions observed in polypurine/polypyrimidine tracts is likely to be improvements in the stacked arrangement of the bases (in a single crystal structure of d(GGGGCCCC) the dCCCC segments were found to be essentially unstacked [34]). We therefore assume that in protonated structures the planes of Hoogsteen base pairs are parallel to those of neighboring Watson–Crick pairs and that the planes of the deoxyribose moieties are (as in other forms of DNA) approximately perpendicular to the bases. The Cl'-Cl' distance in a Hoogsteen base pair (8.46 Å) is ~2Å shorter than the corresponding distance in a Watson-Crick pair [35]. Adjustments are therefore required in the phosphodiester backbone in order to maintain connectivity in interleaved structures. Since, for each narrowing of the helix there must be a subsequent widening, two distinct adjustments are required at the location of each Hoogsteen pair. Although several different types of adjustment could contribute to this reduction in Cl'-Cl' distance, two classes of adjustment avoid major changes in the orientation of deoxyribose moieties and bases of the Hoogsteen pair relative to the axis of the helix as illustrated in figure 6. In the first the distance between the Cl' atom and the 5' phosphorous atoms can be increased by approximately 1 Å through a coupled (crankshaft) change of the C4'-C5' and O5'-P torsion angles from the g,g and g- ranges characteristic of B-DNA to either the g,t and t or t,g and g+ ranges respectively. Of these the C4'-C5 t,g, O5'-P g+ adjustment is less likely since it introduces a steric clash between the 5' methylene group and the 2' methylene group of the adjacent residue. Adjustments of the second class result in an inward displacement by approximately 0.8 Å of the deoxyribose of a residue in a Hoogsteen pair relative to the deoxyribose of the Watson–Crick pair on the 3' side through a t to g+ conversion of the O3'-C3'-C4'-C5' torsion angle of the Hoogsteen paired residue (i.e the deoxyribose changes from a C2'-endo to a C3'-endo conformation).

As illustrated in figures 6b and 6d two adjustments of the same class located at the same base pair but on opposite strands accommodate a local reduction in Cl'-Cl' distance by offsetting the strands in a parallel fashion. Alternatively both inward and outward adjustments can be achieved on the same strand if they are of opposite classes as illustrated in figure 6c. Since neither type of adjustment results in a full 2Å displacement by itself the two classes of adjustment may also be mixed in a variety of ways in response to variations in local strains within the protonated structure (the C3' endo conformation of deoxyribose is not

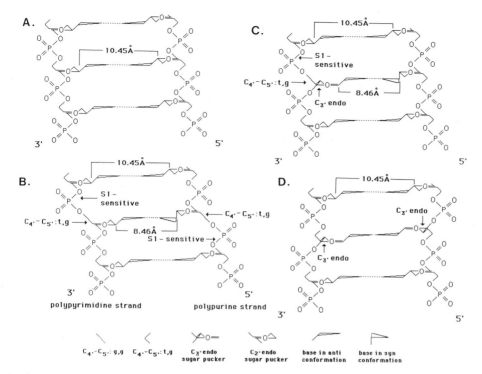

Figure 6. Schematic illustration of the types of adjustment that are expected in the phosphodiester backbone in response to narrowing of the helix to accommodate a Hoogsteen base pair.

6a. Schematic of B-DNA with all C4'-C5' bonds in the g,g range and all deoxyribose residues in C2' endo conformation.

6b. Type I conformational adjustment to accommodate the 2Å reduction in helix diameter at the location of a Hoogsteen base pair. The C4'-C5' bonds of each of the bases are shown in a g,t conformation which results in the movement of the C1' atoms towards each other. Since equivalent adjustments are present on opposite strands the local narrowing of the helix results in an offset in the helix axis as the normal helix diameter is resumed. The change in torsion angle about the C4'-C5' bond is proposed to result in these bonds becoming sensitive to the action of SS nucleases (see text).

6c. Type II adjustment to accommodate a Hoogsteen pair in a DNA helix. At the location of the pyrimidine the C4'-C5' bond is in the g,t range and the deoxyribose is shown in a C3' endo conformation. Since both inward and outward adjustments occur on the polypyrimidine strand, only this strand contains bonds that are sensitive to the action of SS nucleases (cf pTC45).

6d. Type III adjustment to accommodate helix narrowing. In this case the compensating adjustments occur on opposite strands of the DNA helix through change in the pucker of the the deoxyribose at the narrow base pair to C3' endo. This type of adjustment is not expected to occur at base pairs in which one of the bases is in a (syn) conformation.

fully compatible with a syn conformation of a purine bases involved in a Hoogsteen pair [36]). Varying the mix of the different types of adjustment in response to local changes in strain permits the interleaved model to account for the progressive changes in phosphodiester conformation that are observed in the non-B polypurine/polypyrimidine structures.

It is important to note that each of these adjustments can be accommodated within a Watson-Crick base paired double helix when appropriately coupled on opposite strands as shown in figure 7. The nuclease sensitivity of the pAT34 plasmid alternates between adjacent residues in a manner which suggests that the shortening of the inter-strand phosphodiester distance required for closure of the cruciform loop is achieved in a zig-zag manner. In crystals of d(pATAT) the deoxyaden-osine residues have a C3'-endo pucker while the thymidine residue have a C2'-endo pucker [45]. On the basis of this observation a model has been proposed for d(AT)n.d(AT)n in which the sugar pucker alternates between the two forms [46] (figure 7a).

The cleavage specificities of SS endonucleases.

Changes in torsion angle about the C4'-C5' bond alter the access pathway for trans-attack on the phosphorous atom required for elimination of the 3'OH of the next residue. Since neither the g,t not the t,g isomers about this bond are present in A or B forms of nucleic acids, detection of either the t,g or g,t isomers by SS nucleases would provide a sufficient basis for their ability to discriminate against the A and B forms of double stranded nucleic acids. In the case of the Z conformation the C4'-C5 bonds to the 5' side of (syn) purine residues fall within the g,t range. In this case the inversion of the orientation of the 5' adjacent sugar places its O3' oxygen atom in a different location from that proposed for protonated polypurine/polypyrimidine structures. Adjustments of the second type which involve changes in sugar pucker are not by themselves sufficient to explain the high degree of discrimination by SS nucleases against double stranded nucleic acids, since the C3' endo conformation is present in the A forms of double stranded RNA and of RNA/DNA hybrids, neither of which are cleaved by theseeenzymes. Although we regard the C4'-C5' torsion angle as being the most likely candidate for the structural adjustment that permits the SS nucleases to discriminate against A and B forms of the double srranded nucleic acids

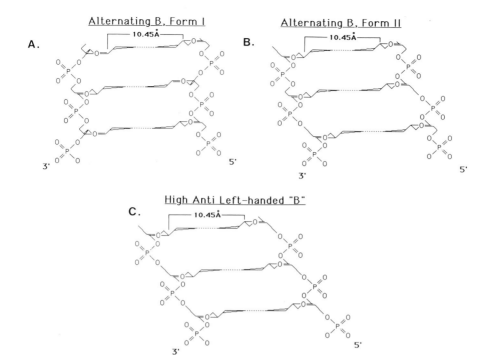

A. Alternating B, Form I

B. Alternating B, Form II

C. High Anti Left-handed "B"

10.45Å

3'   5'

Figure 7. Distorted B-forms containing C4'-C5' g,t bonds and/or C3 endo deoxyribose residues.

7a. Alternating B type I. Alternate residues have C2' endo and C3' endo conformations. Models of this type have been proposed for the B-form of d(AT)n.d(AT)n and d(TG)n.d(CA)n [46]. Note that the normal diameter of the Watson-Crick base pair is maintained by offsetting the inward adjustment of the C1' atoms by one residue on opposite strands.

7b. Alternating B type II. In this case inward and outward adjustments are achieved by alternation between C4'-C5' g,g and C4'-C5' g,t conformations. Adjustments of this type are proposed here to be responsible for the long range junctional sensitivity shown in figure 5.

7c. The high anti-Left-Handed-B conformation. Schematic showing that in order to reverse the handedness of the helix while maintaining the usual stacking of Watson-Crick pairs found in B-DNA it is necessary to stretch the phosphodiester bonds. This can be achieved by changing the C4'-C5' torsion angles from g,g to g,t.

the pucker of the sugar ring aay serve a secondary role in determining
the different cleavage specificities of the different SS nucleases since
each enzyme may have its own optimum value for each of the two critical
sets of torsion angles.

The left handed polypurineppolypyrimidine structure in pGA34.

To date the only well characterized left handed conformation of base
paired DNA is the Z conformation in which the bases are Watson-Crick
paired but alternate between syn- and anti- conformations on each strand
[37]. A number of other geometrically feasible models for left handed DNA
structures have been proposed [38-41]. Two factors make a simple Z form
an unlikely alternative for the left handed domain of pGA34.

1) Recent estimates indicate that the energy costs of a B-Z transition in
d(G)n.d(C)n, although not totally prohibitive, are at least 1.2 kcal/bp
[42]. The energy required to drive the transition to the most underwound
state in pGA34 is 4.5 kcal greater than that required to drive the
transition in pTC45 under identical conditions. Assuming the additional
energy is required to drive 17 bp to a left handed state and to form the
internal junction between states, the additional energy cost of the left
handed form of the polypurine/polypyrimidine is 0.27 kcal/bp. This
quantity is 5 fold lower than that required to drive a theoretical B-Z
transition in a polypurine/polypyrimidine tract. The relatively small
energy cost of the transition also rules out models for S1 sensitive
structures in pTC45 and pGA34 in which the S1 sensitive residues are
unpaired [5].

2) In pGA34 all phosphodiester bonds of domain 2 are sensitive to SS
nucleases and therefore exist in conformations which are similar to each
other and which differ from those found in the Z-conformation. In another
plasmid pGGA45, which exhibits similar domain behaviour to pGA34 but
which contains the repeating trinucleotide d(TCC)n.d(GGA)n, the only
motifs of chemical and enzymatic reactivity that can be discerned are
trinucleotide repeats. If a Z form were present in one of the domains of
the latter plasmid a hexanucleotide motif should be apparent.

Uesugi et al. [41] reported that two dG.dC pairs in a self complementary oligonucleotide conform to a novel type of left handed helix when
surrounded by conformationally restricted cyclonucleoside residues.

40

Reversal of the normal right handed helical character is a consequence a "High-anti" conformation about the glycosidic bond together with the stretching of the phosphodiester backbone by conversion of the C4'-C5' bonds to a g,t conformation. Although the "high anti" conformational range is a relatively high energy state for pyrimidine nucleosides this type of model illustrated schematically in figure 7c could account for the nuclease sensitivity of all phosphodiester bonds in the left-handed polypurine/polypyrimidine domain in pGA34.

Junctional Sensitivity to SS nucleases.

The rates of cleavage at individual residues in the junctional regions surrounding protonated polypurine/polypyrimidine structures are clearly independent of the sequence. Since we have not been able to detect any marked changes in the chemical reactivities of bases in these junctional domains we believe that the SS nuclease susceptibility of these junctions is the result of relatively minor deformations of the structure away from the B-conformation. Two possibilities which are not mutually exclusive are that the deformation could be static (c.f. the long range periodic structures found in martensitic alloys) or alternatively might be associated with a dynamic standing wave in which the nodes of a thermally activated oscillation are defined by the boundaries of the non-B polypurine/polypyrimidine structure and by the elastic properties of the plasmid as a whole. In either case the relatively low absolute sensitivities of the junctional phosphodiester bonds to S1 nuclease, together with the lack of detectable change in chemical reactivities of the bases in this region, indicate that the deformation has a small amplitude.

As discussed above, two separate conformational adjustments are required when a Hoogsteen base pair is interleaved between Watson-Crick base pairs. Each of the proposed adjustments are ones that could be accommodated within Watson-Crick base paired DNA if appropriately coupled on opposite strands. The observations of long range junctional sensitivity in plasmids containing inserts of d(TC)n.d(GA)n support the suggestion that candidates for the torsion angle adjustments leading to SS nuclease sensitivity are ones that could occur in a Watson-Crick paired helix without major reorganization of the structure. An alternating B conformation which has the normal helix diameter but in which every other C4'-C5' bond is in a g,t conformation is illustrated in

41

figure 7b. The systematic variation in the relative cleavage rates of each dinucleotide pair shown in figure 6 suggests that the observed ~40 bp cycle in cleavage rate is due to a beat between two superimposed conformational adjustments which compensate for each other imperfectly.

Footnotes

* In other examples of d(AT)n cruciforms reactivity to S1 nuclease and diethylpyrocarbonate have been reported to be skewed to the 5' side of the center of symmetry [20-22]. The difference between these results and those presented here may be related to the 7 base pairs of GC rich sequence at the base of the cruciform arms of pAT34. This region of GC rich sequence is likely adopt a rigid B-form and may impose this DNA helix form on the adjacent d(AT)n.d(AT)n in the cruciform arm.

† Purine N-7 atoms in Z-DNA are more exposed than those of B-DNA to reaction with diethylpyrocarbonate [29,30]. Mendel and Dervan in a study of purine bases flanking echinomycin binding sites in DNA [43] suggested that purine bases in Hoogsteen base pairs are also reactive to this reagent. This interpretation requires the reagent to attack sites other than N-7. Unpublished results from our laboratory indicate that purine residues flanking echinomycin binding sites of the 5'PyCGPu3' type do not form Hoogsteen base pairs. Other binding sites (5'PuCGPy3') which are more likely to contain Hoogsteen pairs [47] do not exhibit unusual reactivity to diethylpyrocarbonate.

ACKNOWLEDGEMENTS

We wish to thank the Medical Research Council of Canada for an operating grant under which this work was performed. MG is recipient of an MRC studentship, CF is recipient of a University of Toronto open Fellowship. DH is recipient of an MRC fellowship.

REFERENCES

1. Shishido, K. and Ando, K. (1982) in "Nucleases", pp 155-185, Eds. Linn and Roberts Cold Spr. Harb. Lab.
2. Courey A. J. and Wang, J. C., Cell, 33, 817-829 (1983)
3. Lilley, D. Proc. Nat. Acad. Sci USA, 77, 6468 (1980)
4. Panayotatos and Wells, Nature 289, 466 (1981)
5. Hentschel, C. C. Nature 295, 714-716 (1982).
6. Weintraub, H. Cell 32,1191-1203 (1983).
7. Htun, H., Lund, E. and Dahlberg, J. E. Proc. Nat. Acad. Sci. U. S. A. 81, 7288-7292 (1984).
8. Nickol, J. M. and Felsenfeld, G. Cell 35, 467-47728 Cell 35, 467-477 (1983).
9. Kohwi-Shigematsu, T, Gelinas, R. and Weintraub, H. Proc. Nat. Acad. Sci. USA. 80, 4389-4393 (1983).
10. Schon, E., Evans, T., Welsch, J. and Efstradiatis, A. Cell 35, 837-848 (1983).
11. Ruiz-Carillo, A. Nucleic Acids Res. 12, 6473-6492 (1984).
12. Kowhi-Shigematsu T. and Kowhi, Y., Cell 43, 199-206 (1985).
13. Klysik, J., Stirdivant, S. M., Larson, J. E., Hart, P. A. and Wells R. D. Nature 290, 627-677 (1982).
14. Singleton, C. K., Klysik, J., Stirdivant, S. M. and Wells, R.D. Nature 299,312-316 (1982).
15. Kang, D. S. and Wells, R. D. J. Biol. Chem. 260, 7783-7790 (1985).
16. Peck, L. J., Nordheim, A., Rich, A. and Wang, J. C. Proc. Nat. Acad. Sci. U. S. A. 79, 4560-4564 (1982).
17. Haniford, D.B., and Pulleyblank, D. E. Nature 302,632-634 (1983).
18. Haniford, D.B., and Pulleyblank, D.E. J. Biomol. Str. Dyn.1,593-609 (1983) .
19. Haniford, D. B. and Pulleyblank, D. E. Nucleic Acids Res. 13, 4343-

4363 (1985).

20. Greaves D. R., Patient, R. K. and Lilley, M. J. J. Mol. Biol. 185, 461-478 (1985)

21. Panyutin, I., Lyamichev, V. and Mirkin, S. J. Biomol. Struct. and Dynam. 2, 1221-1233 (1985).

22. Furlong, J. D. and Lilley, D. M. J. Nucleic Acids Res. 14, 3995-4007 (1986)

23. Pulleyblank, D. E., Haniford, D. B. and Morgan, A. R. Cell 42,271-280 (1985)

24. Lyamichev, V. I., Mirkin, S. M. and Frank-Kamanetskii, M. D. J. Biomol. Struc. and Dynam. 3, 327-338 (1985)

25. Courtois, Y., Framgeot, P. and Guschlbauer, W. Eur. J. Biochem. 6, 493-501 (1968).

26. Cantor, C. R. and Efstradiatis, A. Nucleic Acids Res. 12, 8059-8072 (1984).

27. Evans, T. and Efstradiatis, A. J. Biol. Chem. 261, 14771-14780 (1986)

28. Wang, J. C. Proc. Nat. Acad. Sci. U. S. A. 76, 200-203 (1979)

29. Herr, W. Proc. Nat. Acad. Sci. 82, 8009-8013 (1985)

30. Johnson, B. D. and Rich, A. Cell 42, 713-742 (1985)

31. Scholten, P. M. and Nordheim, A. Nucleic Acids Res. 14, 3981-3993 (1986)

32. Haasnoot, C. A. G., Hilbers, C. W., van der Marel, G. A., van Boom, J. H., Singh, U. C., Pattabiraman, N. and Kollman, P. A. in "Biomolecular Stereodynamics" Ed. Sarma, R. H. & Sarma, M. H., Adenine Press pp 101-115 (1986)

33. Lilley, D. M. J., Nucleic Acids Res. 11, 3097-3112 (1983)

34. McCall, M., Brown, T. and Kennard, O. J. Mol. Biol. 183, 385-396 (1985)

35. Hoogsteen, K., Acta Cryst. 16, 907-916 (1963)

36. Saran, A., Perahia, D. and Pullman, B., Theor. Chim. Acta. 30,31-44 (1973)

37. Wang, A. H. J., Quigley, G. J., Kolpack, F. J., Crawford, J. L., van Boom, J. H., van der Marel, G. and Rich, A. Nature 282,680-686 (1979).

38. Mitsui, Y., Langridge, R., Shortle, B. E., Cantor, C. R., Grant, R. C., Kodama, M. and Wells R. D. Nature 228, 1166 (1970)

39. Sasisekharan, V., and Pattabiraman, N. Nature 275, 159-162 (1978).

40. Drew, H. R., and Dickerson, R. E., EMBO J. 1, 663-667 (1982).

41. Uesugi, K., Lee, B. L., Ikehara, M., Kobayashi, Y. and Kyogoku, Y. J. Biomol. Struct. and Dynam. 3, 339-347 (1985).

42. Ellison, M. J., Kellegher, R. J. III, Wang, A. H.-J., Habener, J. F. and Rich, A. Proc. Nat. Acad. Sci. U. S. A. 82, 8320-8324 (1985).

43. Mendel, D. and Dervan, P. B. Proc. Nat. Acad. Sci. USA 84, 910-914 (1987)

44. Hosur, R. V., Govil, G., Hosur, M. V. and Viswamitra, M. A. J. Mol. Struct. 72, 261-276 (1981).

45. Viswamitra, M. A., Shakked, Z., Jones, P. G., Sheldrick, G. M., Salisbury, S. A. and Kennard, O. Biopolymers 21, 513-553 (1982)

46. Klug, A., Jack, A., Viswamitra, M. A., Kennard, O., Shakked, Z. and Steitz, T. A. J., Mol. Biol. 131, 669-680 (1979)

47. Wang, A. H.-J., Ughetto, G., Quigley, G. J., Hakoshima, T., van der Marel, G. , van Boom, J. H. and Rich, A. Science 225, 1115-1121

Chromatin Structure and DNA Structure
at the *hsp 26* Locus of *Drosophila*

S.C.R. Elgin, I.L. Cartwright*, D.S. Gilmour
E. Siegfried, G.H. Thomas

Washington University
Department of Biology
St. Louis, MO 63130

*current address: Department of Biochemistry and Molecular Biology
University of Cincinnati College of Medicine
Cincinnati, Ohio 45267

The heat shock genes of Drosophila provide a very convenient system for the study of gene activation. These genes are activated in response to elevated temperature, metabolic inhibitors, and other stresses, apparently using a mechanism common to all cell types. The genes encoding the small heat shock proteins, *hsp 22*, *hsp 23*, *hsp 26* and *hsp 28* are also activated in specific Drosophila tissues at specific times according to a program of developmental regulation. (For a review of work on the organization and expression of Drosophila heat shock genes see Southgate et al., 1985.) We have chosen to focus our analysis on *hsp 26*, because of certain interesting features of the DNA in the 5' regulatory region.

CHROMATIN STRUCTURE OF *hsp 26*

The organization of the heat shock genes at locus 67B in Drosophila is diagrammed in Figure 1. We have carried out an analysis of the protein-DNA interactions in this region of the genome using DNase I and methidiumpropyl-EDTA·Fe(II) [abbreviated MPE·Fe(II)]. These cleavage reagents, being relatively sequence neutral, tend to cut the DNA preferentially in regions that are not protected by protein-DNA interactions, generating a pattern that allows one to map protein-free and protein-associated regions. Following brief digestion of the chromatin (in isolated nuclei from normal and heat-shocked Drosophila embryos), the DNA is purified and the fragments analyzed using the indirect end-labeling technique. In this approach, the purified DNA is cut to completion with a restriction enzyme having sites bracketing the region of interest. The DNA is then size-separated by agarose gel electrophoresis, the fragments transferred to nitrocellulose by Southern

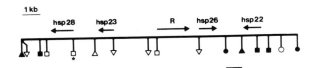

Figure 1. Partial restriction map showing the region of 67B1 that encodes the four small heat shock proteins of *D. melanogaster*. The genes are designated by arrows above the map; the direction of transcription is shown by the arrowheads. The DNA fragment used as a probe in the analysis of *hsp 26* shown in Figure 2 is indicated by a thick line under the map. Symbols: ▲ , *Bam*HI; ▽ , *Eco*RI; ■ , *Hind*III; ☐ , *Sal*I; △ , *Xba*I; ● , *Sst*II; ○ , *Bgl*I. Not all sites for each enzyme are shown.

blot, and the filter hybridized with a small labelled subfragment which abuts one of the restriction endpoints. The autoradiogram then reveals a set of fragments having a common endpoint (labelled by the hybridized probe), whose length indicates the position of cleavage sites along the chromatin fiber. The results of such an experiment on *hsp 26* are shown in Figure 2.

Analysis with DNase I shows the presence of two preferentially accessible regions of DNA, or "DNase I hypersensitive sites" 5' to the gene. Each site appears to be a doublet. Following gene activation by heat shock, a new "footprint" is observed in the center of these regions, indicating a new protein-DNA association which prevents DNA cleavage. Similar results are obtained with MPE·Fe(II). DNase I cannot detect the nucleosome array, apparently being too bulky to gain access to the linker regions. MPE·Fe(II), however, detects a nucleosome array across the inactive gene, indicating that the nucleosomes in this region are precisely positioned. Following gene activation, dramatic changes occur. The chromatin fiber appears to be "unfolded", allowing access (cleavage) by DNase I. This is a very specific effect; note that no such change has occurred across the adjacent R gene, which is not transcribed at high levels following heat shock. Similarly the pattern of nucleosomes detected by MPE·Fe(II) across the gene becomes smeared, i.e., a greater frequency of cutting in the previously protected regions is observed. The results suggest a perturbation of the histone-DNA interactions along the chromatin fiber associated with transcription (Cartwright and Elgin, 1986).

As mapped in Figure 2, the DNase I hypersensitive sites appear to encompass the heat shock consensus sequences (HSCS's) defined by Pelham

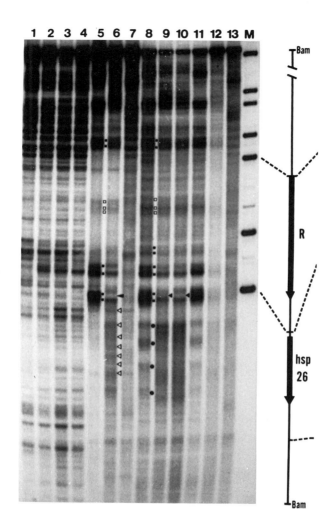

Figure 2. Chromatin fine structure in the vicinity of *hsp 26*. Nuclei
from non-heat-shocked (NHS) and heat-shocked (HS) Drosophila embryos (at
5×10^8 to 10^9 nuclei/ml) and DNA were digested at 25°C with micrococcal
nuclease (lanes 1-4), DNase I (lanes 5-7) or MPE·Fe(II) (lanes 8-13).
The DNA was purified and digested to completion with *Bam*HI. DNA samples
(9 μg) were fractionated on a 1.2% agarose gel; the gel was blotted to
nitrocellulose and the filter hybridized to the probe (diagrammed in
Figure 1). Micrococcal nuclease digestion of protein-free genomic DNA
(lane 1), NHS nuclei (lane 2), and HS nuclei (lanes 3-4); DNase I
digestion of NHS nuclei (lane 5), HS nuclei (lane 6), and genomic DNA
(lane 7); and MPE·Fe(II) digestion of NHS nuclei (lane 8), HS nuclei
(lanes 9-10), 0.35M KCl-extracted NHS nuclei (lane 11), 0.5M KCl-
extracted NHS nuclei (lane 12), and genomic DNA (lane 13). M denotes
molecular weight markers. Black squares denote hypersensitive sites,
while black circles mark linker regions in the nucleosomal array. Black
arrowheads denote the heat shock induced footprint in the proximal
hypersensitive site. Adapted from Cartwright and Elgin, 1986.

(1982) which have been mapped upstream of *hsp 26*. The protein binding in these regions following heat shock can reasonably be interpreted as the binding of the heat shock transcription factor reported by Parker and Topol (1984), in analogy with the results reported by Wu (1984) for protein binding at the HSCS upstream of *hsp 70*. This interpretation has recently been confirmed by sequence-level mapping [using the Church and Gilbert (1984) gel blotting technique] of the proximal site, which shows binding of a protein to position -41 to -76 after, but not before, heat shock (G.H. Thomas and S.C.R. Elgin, unpublished information). The HSCS in this region is at postion -47 to -70.

These results show that formation of the DH sites precedes gene activation, and suggest that this may facilitate the process by placing the HSCS in a very accessible conformation within the chromatin. Formation of such a chromatin structure may reflect the binding of other NHC proteins, the properties of the DNA sequence itself, or both. It is interesting to note that in the case of *hsp 70* there is evidence from UV-photocrosslinking studies that a molecule of RNA polymerase II is associated with the 5' region prior to gene activation (Gilmour and Lis, 1986). Binding of protein to the TATA box and immediate downstream region could have an effect on the structure over the HSCS region. Note also the conspicuous "footprint" over the region of -100 to -250. Much further work is needed to determine the means of forming specific DNase I hypersensitive sites.

DNA STRUCTURE AROUND *hsp 26*

One of the interesting features of *hsp 26* is a ~60 bp homopurine-homopyrimidine segment found adjacent (upstream) to the proximal HSCS, as diagrammed in Figure 3. Analysis of the segment 88B13 (the major *Bam*HI fragment shown in Figure 1), cloned into pBR322, shows that this site is very sensitive to cleavage by S1 nuclease when the DNA is supercoiled, indicating a localized perturbation of DNA structure (Selleck et al., 1985). Sequence-level analysis shows that the S1 sensitive site lies toward the 3' end of the homopurine-homopyrimidine tract. Initial cleavage is on the GA strand, in the region -84 to -96, followed by cleavage on the CT strand, predominantly at -84 and -86 (Figure 4). Interestingly, this site is pH dependent, suggesting that base protonation can play a role in stabilizing the S1-sensitive configuration (Siegfried et al., 1986).

A deletion which removes the homopurine-homopyrimidine block from -82 to -132 has recently been constructed; this deletion results in a

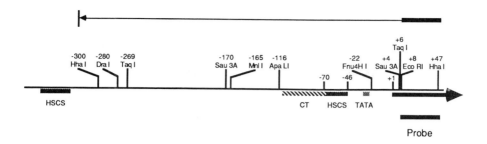

Figure 3. A detailed restriction map of the region 5' of *hsp 26*.
Sequences indicated include the heat shock consensus sequence (HSCS),
the homopurine-homopyrimidine stretch (CT), and the TATA box. Numbering
is from the start site of transcription (+1); transcription proceeds to
the right. The fragment transcribed to generate the RNA probes used for
Figure 4 is indicated by the bar below the map.

plasmid that no longer has an S1-sensitive site in this region.
Preliminary experiments using this *hsp 26* gene (marked by an insertion
of adenovirus DNA) in a transient expression assay indicate a loss of
expression under both normal and heat shock conditions (E. Siegfried and
S.C.R. Elgin, unpublished observations). The work of others has shown
that both major upstream HSCS's are needed for efficient induction of
hsp 26 (Cohen and Meselson, 1985; Pauli et al., 1986; Lis et al., 1987).
The HSCS sites themselves are not altered in the above manipulations.
We suggest that deletion of the homopurine-homopyrimidine site in some
way interferes with the interaction of the HSCS sites, and that this
failure leads to a loss of *hsp 26* expression.

ASSOCIATION OF TOPOISOMERASE I WITH *hsp 26*

We have analyzed the distribution of topoisomerase I (topo I) in the
polytene chromosomes of Drosophila using an immunofluorescence staining
assay. While there is little or no topo I associated with the inactive
heat shock genes, significant amounts are found in association with the
induced, active genes (Fleischmann et al., 1984). Gilmour et al. (1986)
have subsequently shown by UV-photocrosslinking of intact diploid cells
that this interaction is concentrated in the region of transcription.
Recently we have been able to map the sites of interaction at a
resolution of ± 20 bp by using camptothecin. This drug apparently
increases the stability of the DNA-topo I covalent intermediate,

49

S1 site 5' of <u>hsp 26</u>

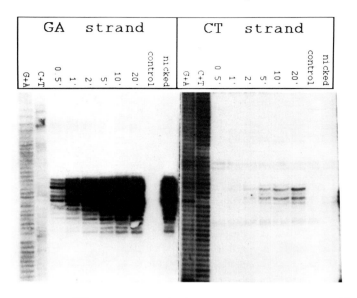

-110   -65

S1   TGCAAGAGAGAGAAGAGAAGAGAGAGAGATGAGAAAGGAAAAAGACAGTG

Figure 4. High resolution mapping of S1 cleavage sites 5' to *hsp 26*.
Samples of supercoiled 88B13 DNA were incubated with S1 nuclease for the
times indicated in buffer (100 mM NaCl, 50 mM Na acetate pH 5.0, 1 mM
ZnSO$_4$, 0.5% glycerol). The control sample was incubated in buffer for
20' without S1. The sample designated "nicked" was open circle DNA
electrophoretically purified from an S1-treated preparation of 88B13.
DNA samples were purified, digested to completion with *Hha*I, size
separated on a sequencing gel, blotted, and hybridized with the
appropriate labeled strand from the probe fragment shown in Figure 3.
Samples labelled GA and CT are sequencing ladders used as markers. See
Siegfried et al., 1986, for additional details.

allowing one to isolate the DNA with a nick to mark the site of topo I
interaction *in vivo* (Hsiang et al., 1985). The position of the nicks
can be mapped on alkaline (denaturing) gels using the indirect-end label
technique with strand-specific probes.

    The results shown in Figure 5 illustrate that the active heat shock
genes *hsp 26* and *hsp 22* are prominently associated with topo I across
the regions of transcription. This occurs on both the transcribed and
nontranscribed strands of the DNA double helix. Neither the spacers nor
the nontranscribed gene (*R*, labeled *gene 1* in this figure) shows such an
association. The interaction with topo I is seen after, but not before,

50

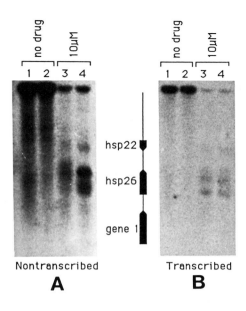

Figure 5. Association of topoisomerase I with the active *hsp 26* gene.
Heat-shocked cells were treated with either 10 μM camptothecin (lanes 3
and 4) or an equivalent volume of carrier (dimethyl sulfoxide) (lanes 1
and 2) for 3 min at 36° and were then collected by a 5 min. spin in an
microcentrifuge. Cells were lysed in 1% SDS and the DNA was extracted
with chloroform and ethanol precipitated. The DNA was cut to completion
with *Sal*I in the presence of 0.5 mM PMSF; one portion was treated with
proteinase K and one was not. Both portions were ethanol precipitated,
and then taken up and size separated on alkaline gels. Camptothecin-
induced cuts were detected by hybridizing an appropriate strand-specific
probe from a fragment abutting the *Sal*I site at the 5' end of the R gene
(gene 1) (see Figure 1). Lanes 1 and 3 are samples that were not
treated with protease; lanes 2 and 4 are samples that were treated with
protease after restriction digestion and just before electrophoresis.
The probe used in panel A detects cleavage fragments (on the
nontranscribed strand) whose 3' ends were generated with camptothecin;
the probe used in panel B detects those (on the transcribed strand)
whose 5' ends were generated with camptothecin. Figure reproduced from
Gilmour and Elgin, 1987, with permission.

activation by heat shock. Similar results have been obtained for the
other small heat shock genes. In looking at the transcripts of *hsp 28*
from heat-shocked cells on a Northern blot, one observes reduced levels
of the full-length transcript from cells treated with the drug, but no
significant amount of smaller RNA (Gilmour and Elgin, 1987). This
suggests a dynamic interaction of topo I with the chromatin fiber in
which the topo I may cause the RNA polymerase II to pause, but does not
cause significant dissociation of the polymerase or truncation of the
transcript.

The results suggest a very dynamic model of the active gene. Whether or not the topo I plays a role in unfolding the chromatin structure, or whether it passively relieves local torsional strain as it arises during transcription, remains to be seen. Note that all of the available studies point to an interaction of the topo I with the region of transcription after, but not before, gene activation. Thus the above results are quite compatible with the utilization of torsional tension in establishing the initiation complex, a possibility suggested by many lines of experimentation. However, once transcription has been established, topo I is abundantly present, and the relaxation of *unconstrained* torsional tension can be anticipated. The coincidence of the region of topo I interaction (see Figure 5) with the region of nucleosome smearing (see Figure 2) suggests the possibility that topo I might facilitate the "switching out" or unfolding of nucleosomes if required for high levels of transcription. Whether or not topo II can substitute for topo I in the swivel functions associated with transcription in higher eukaryotes [as has been shown for yeast (Goto and Wang, 1985; Uemura and Yanagida, 1984)] remains to be seen.

CONCLUSIONS

The basic chromatin structure of the *hsp 26* gene is well defined, two DNase I hypersensitive sites being found over the two HSCS regulatory sites, and a regular array of nucleosomes being observed across the gene. On heat shock activation, one sees an association of protein (presumably the heat shock transcription factor) with the HSCS sites. The active gene adopts a more open structure, with a "smearing" of the nucleosome array; this change maps across the region of transcription, as does an association with topoisomerase I. A complex set of protein-DNA interactions around the 5' end of the gene appears to be involved in establishing the pre-transcription complex; this structure may utilize the conformational flexibility of the homopurine-homopyrimidine tract. Thus *hsp 26* appears to be an excellent gene for the continuing analysis of DNA and chromatin structure in relation to gene expression.

REFERENCES

Cartwright, I.L. and S.C.R. Elgin (1986). Nucleosomal instability and induction of new upstream protein-DNA associations accompany activation of four small heat shock protein genes in *Drosophila melanogaster*. Molec. Cell. Biol. 6: 779-791.

Church, G.M., and W. Gilbert (1984). Genomic sequencing. Proc. Natl. Acad. Sci. USA 81: 1991-1995.

Cohen, R.S., and M. Meselson (1985). Separate regulatory elements for the heat-inducible and ovarian expression of the *Drosophila hsp 26* gene. Cell 43: 737-746.

Gilmour, D.S. and J.T. Lis (1986). RNA polymerase II interacts with the promoter region of the noninduced *hsp 70* gene in *Drosophila melanogaster* cells. Mol. Cell. Biol. 6: 3984-3989.

Gilmour, D.S., and S.C.R. Elgin (1987). Localization of specific topoisomerase I interactions within the transcribed region of active heat shock genes by using the inhibitor camptothecin. Molec. Cell. Biol. 7: 141-148.

Gilmour, D.S., G. Pflugfelder, J.C. Wang, and J.T. Lis (1986). Topoisomerase I interacts with transcribed regions in *Drosophila* cells. Cell 44: 401-407.

Goto, T. and J.C. Wang (1985). Cloning of yeast *TOP1*, the gene encoding topoisomerase I, and the construction of mutants defective in both DNA topoisomerase I and DNA topoisomerase II. Proc. Natl. Acad. Sci. USA 82: 7178-7182.

Hsiang, Y.-H., R. Hertzberg, S. Hecht, and L.R. Lui (1985). Camptothecin induces protein-linked DNA breaks via mammalian DNA topoisomerase I. J. Biol. Chem. 260: 14873-14878.

Parker, C.S. and J. Topol (1984). A *Drosophila* RNA polymerase II transcription factor specific for the heat shock gene binds to the regulatory site of an *hsp 70* gene. Cell 37: 273-283.

Pauli, D., A. Spierer, and A. Tissieres (1986). Several hundred base pairs upstream of *Drosophila hsp 23* and *26* genes are required for their heat induction in transformed flies. EMBO J. 5: 755-761.

Pelham, H.R.B. (1982). A regulatory upstream promoter element in the *Drosophila hsp 70* heat shock gene. Cell 30: 517-528.

Selleck, S.B., S.C.R. Elgin, and I.L. Cartwright (1984). Supercoil-dependent features of DNA structure at *Drosophila* locus 67B1. J. Mol. Biol. 178: 17-33.

Siegfried, E., G.H. Thomas, U.M. Bond, and S.C.R. Elgin (1986). Characterization of a supercoil-dependent S1 sensitive site 5' to the *Drosophila melanogaster hsp 26* gene. Nuc. Acids. Res. 14: 9425-9444.

Simon, J.A. and J.T. Lis (1987). A germline transformation analysis reveals flexibility in the organization of heat shock consensus elements. Nuc. Acids Res. 15: in press.

Southgate, R., M.-E. Mirault, A. Ayme, and A. Tissieres (1985). Organization, sequences, and induction of heat shock genes, in "Changes in Eukaryotic Gene Expression in Response to Environmental Stress," ed. by B.G. Atkinson and D.B. Walden (Academic Press, N.Y.), pp. 3-30.

Uemura, T. and M. Yanagida (1984). Isolation of type I and II DNA topoisomerase mutants from fission yeast: single and double mutants show different phenotypes in cell growth and chromatin organization. EMBO J. 31: 1737-1744.

Wu, C. (1984) Activating protein factor binds *in vitro* to upstream control sequences in heat shock gene chromatin. Nature (London) 311: 81-84.

Cruciform Extrusion in Supercoiled DNA - Mechanisms and Contextual Influence.

David MJ Lilley, Karen M Sullivan, Alastair IH Murchie and Judy C Furlong.

Department of Biochemistry
The University
Dundee DD1 4HN
United Kingdom

Cruciform structures in supercoiled DNA.

Two factors contribute to structural diversity in DNA molecules. These are base sequence and DNA supercoiling. Certain DNA sequence arrangements have an inherrent propensity to adopt a geometry which is recognisably different from B-DNA, which in the present context means an inverted repeat having a two-fold symmetry axis. However, many perturbed DNA structures are unstable relative to B-DNA in the absence of an energy source, which may be provided by negative supercoiling. Cruciform structures (paired stem-loop structures) were first recognised experimentally about eight years ago [Gellert *et al*, 1979; Lilley, 1980; Panayotatos & Wells, 1981], twenty five years after their first theoretical description [Platt, 1955; Gierer, 1966]. The crucial role of supercoiling was overlooked in earlier studies. Cruciform formation has subsequently been demonstrated for many inverted repeat sequences in a variety of supercoiled plasmids and phage.

The experimental study of cruciforms - a fusion of chemistry and molecular genetics.

Structural and physical chemical investigation of supercoil-stabilised features is difficult by conventional methods, because in general the feature of interest, eg the cruciform, is but a tiny part of the entire circular plasmid molecule. For example, in the plasmid pColIR315

Probe	Target	Result	Reference
S1 nuclease	loop	cleavage	Lilley,1980; Panayotatos & Wells, 1981
Micrococcal nuclease	loop	cleavage	Dingwall *et al*, 1982; Lilley, 1982
Bal31 nuclease	loop	cleavage	Lilley & Hallam, 1984
P1 nuclease	loop	cleavage	Haniford & Pulleyblank, 1985
Mung bean nuclease	loop	cleavage	Sheflin &Kowalski, 1984
Bromoacetaldehyde	loop (A,C)	etheno adduct	Lilley, 1983
Osmium tetroxide	loop (T,C)	*cis* diester	Lilley & Paleček, 1984
Glyoxal	loop (G)	etheno adduct	Gough, 1986; Lilley, 1986
Bisulphite	loop (C)	deamination to dU	Gough *et al*, 1986
Diethyl pyrocarbonate	loop (A)	carbethoxylation	Furlong & Lilley, 1986; Scholten & Nordheim, 1986
T4 endonuclease VII	junction	cleavage	Mizuuchi *et al*, 1982a; Lilley & Kemper, 1984
T7 endonuclease I	junction	cleavage	de Massey *et al*, 1984
Yeast resolvase	junction	cleavage	Symington & Kolodner, 1985; West & Korner, 1985

Table 1. Enzyme and chemical probes which have been employed in the study of cruciform structures.

[Lilley, 1981] the ColE1 inverted repeat (31 bp) is contained in a molecule of 3.75 kb, and thus the region of interest represents less than 1%. This would be quite invisible to NMR for example. We have therefore been obliged to develop new methods with which to attack these problems.

Our principal inovation was the procedure of nuclease cleavage on supercoiled DNA molecules [Lilley, 1980; Panayotatos & Wells, 1981]. In such probing methods we rely on some feature of the novel structure being recognisably different to an enzyme or chemical probe. In the case of the cruciform there are two such features, the single-stranded loops and the four-way junction. Most of our probes select the former as their target, being either single-strand specific enzymes such as S1 nuclease, or single-strand selective chemicals such as bromoacetaldehyde. Probes of the junction are presently fewer in number, but are structurally more discriminating. A tabulation of the probes presently available for cruciform structures is presented in Table 1.

The methods of molecular genetics have also been imported into these studies. Plasmids may be disected almost at will using restriction enzymes, and cloned to yield chimæric molecules.

The sequence dependence of any aspect of novel conformation may most easily be approached using a combined oligonucleotide synthesis and cloning method [Lilley & Markham, 1983], now made very easy with the advent of automated DNA synthesisers. Using these methods we can ask some very specific questions about the geometry and physical chemistry of cruciform structures.

The cruciform structure.

The cruciform is an inherrently unstable entity, having a free energy of formation of between 13 and 18 kcal mole^{-1}, see Table 2. This may be compensated for by the relaxation of supercoiling resulting from the local negative change in twist (ΔTw) on formation of the structure.

Plasmid	Free energy (kcal mol^{-1})	Reference
pUC7	17	Mizuuchi et al, 1983b
pAC102	18	Courey & Wang, 1983
pColIR515 {ColE1}	18.4	Lilley & Hallam, 1984
pXG540 {(AT)$_{34}$}	13.7	Greaves et al, 1985
pXGAT23 {(AT)$_{23}$}	13.2	McClellan et al, 1986
pLNc40 {(CATG)$_{10}$}	16.2	Naylor et al, 1986

Table 2. Some free energies of cruciform formation measured experimentally.

The structure of the cruciform is defined by the two aforementioned features, viz the loops and the four-way junction. As the latter is formally equivalent to the Holliday [1966] junction, its study is important to a full understanding of homologous genetic recombination. Chemical probing experiments [Gough et al, 1986] have indicated that the optimal loop size lies between four and six nucleotides, and that under normal circumstances the junction is probably fully base paired. NMR studies of an isolated junction [Wemmer et al, 1985] support the conclusion of full base pairing. Detailed analysis of diethyl pyrocarbonate modification patterns in the ColE1 cruciform loop have suggested to us [Furlong & Lilley, 1986] that the loop possesses a well defined structure, probably involving base stacking. Studies of isolated DNA hairpins by Hilbers and co-workers [Haasnoot et al, 1986] have indicated that bases may stack as if to continue the helical structure on the 5' side of the loop. Investigation of a pseudo-cruciform construct [Gough & Lilley, 1985] revealed a strongly position-sensitive gel migration retardation akin to that seen for kinetoplast DNA [Wu & Crothers, 1984], and led us to propose that the cruciform junction introduces a pronounced bend into the molecule. Structural studies of cruciform geometry are continuing in this laboratory.

The cruciform extrusion reaction.

Thermodynamic stability does not guarentee kinetic accessibility. The first kinetic studies of the extrusion of a cruciform in supercoiled DNA [Mizuuchi et al, 1982b] revealed a substantial kinetic barrier. This came as something of a surprise, although perhaps it should not have in view of the substantial reorganisation of DNA structure which must be involved. It was reported that the extrusion reaction could be very slow, even at elevated temperatures [Gellert et al, 1983; Courey & Wang, 1983]. However, it was found in different systems that extrusion could occur rather more easily [Panyutin et al, 1984; Sinden & Pettijohn, 1984; Lilley 1985]. We therefore decided to embark on a systematic comparison of cruciform extrusion by two different plasmids.

Two classes of kinetic behaviour.

We studied [Lilley,1985] the kinetics of cruciform extrusion in two plasmids, pColIR315 and pIRbke8. Restriction maps and the sequences of their inverted repeats are presented in Figure 1. pColIR315 contains a 440 bp insert of ColE1 (thicker line), which includes the ColE1 inverted repeat and about 100 bp and 300 bp of left and right flanking DNA respectively. pIRbke8 was constructed [Lilley & Markham, 1983] by cloning synthetic oligonucleotides into the BamH1 site of pAT153 to generate a perfect 32 bp inverted repeat (termed bke), flanked by vector sequences.

Figure 1 summarises the kinetic data obtained for the two plasmids. The properties are strikingly different, in two main respects:

NaCl dependence - The ColE1 cruciform of pColIR315 extrudes maximally in the absence of salt, and the extrusion is supressed as NaCl is added to the buffer. At 50 mM NaCl the rate is reduced below 20%. By contrast, the bke cruciform of pIRbke8 totally fails to extrude in the absence of added salt, and exhibits a maximal extrusion rate at 50 - 60 mM NaCl.

Temperature dependence - The temperature dependence of the extrusion of pColIR315 is very marked. The rate constant for the extrusion reaction increased by a factor greater than 2000 in an 8 degree temperature interval. The reaction is characterised by a huge Arrhenius activation energy (Ea) in excess of 180 kcal mol^{-1}. The temperature dependence for the extrusion of pIRbke8 is rather lower, corresponding to an Ea of around 40 kcal mol^{-1}. One point should perhaps be emphasised. Despite the fact that the activation barrier for the ColE1 cruciform in pColIR315 is over four times higher than that

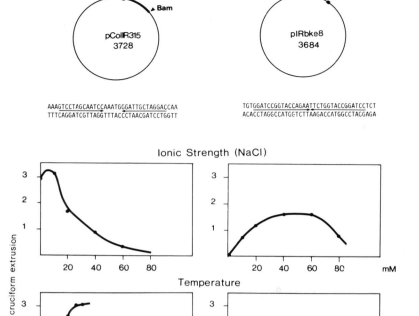

Figure 1. Two plasmids having inverted repeats of contrasting kinetic character. Plasmid maps and inverted repeat base sequences for pColIR315 (ColE1 inverted repeat) and pIRbke8 (bke inverted repeat). Below each are shown the dependence of extrusion kinetics upon ionic strength and temperature.

of pIRbke8, the extrusion actually can occur at rather lower temperatures. For example, the rate of extrusion of pColIR315 in 0 mM NaCl at 28°C is rapid, whilst at the same temperature the rate of extrusion of pIRbke8 in 50 mM NaCl is very slow. This indicates that the extrusion of pColIR315 must have a large entropy of activation associated.

So different are these kinetic properties that we have made them members of two classes of cruciform. We term these C-type (for ColE1) and S-type (for Salt-dependent). The only natural member of the C-type class is the ColE1, subclones like pColIR315 [Lilley, 1981] and the deletant pAO3 [Panyutin *et al*, 1984]. The great majority of sequences behave as S-type cruciforms, including pUC7 [Gellert *et al*, 1983], pAC102 [Courey & Wang, 1983], pOCE12 [Sinden & Pettijohn, 1984], with lower temperature dependences, a requirement for salt and in general extrusion occuring at higher temperatures. The kinetic properties of C- and S-type cruciforms are summarised in Table 3.

	C-type	S-type
Example	ColE1	pIRbke8
Occurence	rare	common
NaCl optimum	0 mM	50 mM
Ea	180 kcal mol^{-1}	40 kcal mol^{-1}
$\Delta H^{\neq}$	180 kcal mol^{-1}	40 kcal mol^{-1}
$\Delta S^{\neq}$ (at 37°C)	400 cal deg^{-1}mol^{-1}	60 cal deg^{-1}mol^{-1}

Table 3. Kinetic characteristics of C-type and S-type cruciforms.

Alternative mechanisms of cruciform extrusion.

The strongly contrasting kinetic properties of C- and S-type cruciforms obliges us to consider the possibility that they reflect alternative mechanistic pathways for the extrusion process. We had previously considered the possibilities from a purely theoretical point of view [Lilley & Markham, 1983], and proposed two possible pathways. These are presented schematically in Figure 2. In the upper path, a large region of DNA is unpaired in the transition state, which then forms the fully extruded cruciform. In the lower path, the extent of initial disruption is smaller. A relatively short proto-cruciform is formed as an intermediate, followed by branch migration [sequential transfer of base pairing to the growing cruciform stem] to the completely extruded cruciform. The transition state is harder to identify in this pathway, but is likely to be intermediate between the initial unpairing and the proto-cruciform. We have proposed (Lilley, 1985) that the C-type cruciforms extrude *via* the top pathway, while the S-type sequences proceed *via* the lower mechanism.

The major differences between these two postulated mechanisms may be summarised:
1. There is a much larger disruption of base-pairing in the upper pathway.

2. The transition state in the lower mechanism resembles the forming four-way junction, as opposed to a simple melted 'bubble'.

The upper pathway would be expected give rise to large values of enthalpy and entropy of activation, and to be achieved most readily at low ionic strength. Since the extent of disruption will be smaller in the lower pathway, this will reduce the enthalpy and entropy of activation, but since phosphate-phosphate repulsion is likely to be significant in a more structured transition state then the energy of the activated complex will be reduced by cations. These sets of properties are exactly those observed for the C- and S-type cruciforms.

We have recently examined the S-type mechanism in greater detail [Sullivan & Lilley 1987]. We took pIRxke/vec, a typical S-type molecule which is rather similar to pIRbke8, and studied the extrusion as a function of the nature of the cation present. While the extrusion of S-type cruciforms is totally dependent upon the presence of exogenously added salts, ions vary greatly in the efficiency with which they promote the extrusion. Cations may be divided into four classes:

1. Maximal extrusion rate at 50-60 mM, falling at higher concentrations. These are the monovalent ions, chiefly of Group Ia and $(CH_3)_4N^+$.
2. Maximal rates by 100-200 μM, plateauing up to 5 mM. These are the divalent ions of Group IIa, together with Mn^{++} and (poorly) Co^{++}.
3. Peak of rate at 15-50 μM. $[Co(NH_3)_6]$ (III) and polyamines.
4. Ineffective ions. Most transition metals are completely ineffective in promoting extrusion.

Despite the very different efficiency with which the different ions facilitate cruciform extrusion, it is likely to be the same reaction mechanism which is used. We have measured Arrhenius activation energies for the reaction in the presence of Na^+ and Mg^{++}, which are the same within experimental error. We may make a number of observations concerning the significance of these data. Our central hypothesis is that the observed kinetic results of using various cations results from differential stabilisation of the transition state. The nature of this leads to some clues on the structure of the activated complex.

1. Groups 1 to 3 show that the optimal ion concentration falls with charge. This must reflect ionic binding.
2. Polyamines are very effective, basic amino acids are totally ineffective. The spacing and distibution of positive and negative charges on these ions must affect transition state binding.
3. Many transition metals are totally ineffective in promoting extrusion. Most of the ions in this group are soft, binding preferentially to nitrogen or exocyclic keto substituents ie they will be better at binding bases than phosphate groups. Clearly phosphate binding is required in order to reduce the free energy of the transition state.

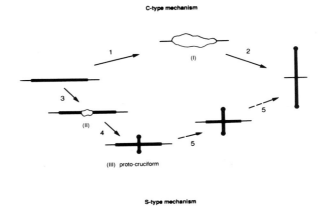

C-type mechanism

(I)

1   2

3

(II)

4   5

(III) proto-cruciform

5

S-type mechanism

Figure 2. **Two mechanisms for cruciform extrusion.** The thicker line in the species on the left is the unextruded inverted repeat, shown fully extruded in the species on the right. Two possible pathways for the formation of the cruciform are indicated. In the upper pathway the interconversion proceeds via species I (steps 1 and 2), where the entire region is unpaired. In the lower pathway the initial unpairing (step 3) is restricted to the central region (II). The four-way junction begins to assemble (step4), to form the proto-cruciform (III), which then branch migrates (step5) to the fully formed cruciform.

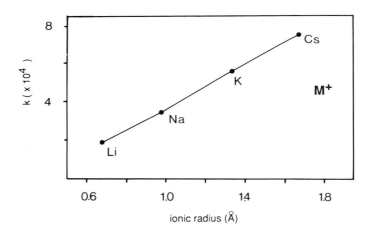

Figure 3. **Extrusion rate by an S-type cruciform depends upon the size of the cation present.** The rate constant at 37°C for the extrusion of pIRxke/vec was measured in 10 mM Tris.HCl pH 7.5 with 50 mM of the indicated Group Ia metal chloride.

4. Within the Ia and IIa metals, there is an excellent correlation between the rate of cruciform extrusion and ionic radius. This is illustrated for the Ia cations in Figure 3. The theory of ion binding to normal DNA, ie a cylindrical polyelectrolyte, is well treated using electrostatic considerations alone [Manning, 1978; Zimm & LeBret, 1983]. Counterions reduce phosphate charge by about 80% by acting like a screening cloud. This accords with the failure to localise ions in crystal structures of DNA. In constrast, tRNA has a number of high affinity binding sites for Mg^{++} [Jack et al, 1977; Holbrook et al, 1977; Quigley et al, 1978], generated by the tertiary folding of the molecule. The observed dependence of extrusion rate on ionic *size* implies a selective ion binding [Tam & Williams, 1985] in the transition state, thereby suggesting that the transition state has a structure which contains electronegative clefts capable of such selective ion binding. Therefore this is further evidence for the transition state having significant four-way junction character, just as in the scheme outlined in Figure 2.

In another ongoing study of the S-type extrusion reaction, we have constructed a series of variants of pIRbke8, in which one or two mutations (mainly A to G or *vice versa*), have been introduced into the symmetrical unit of the bke inverted repeat [A. Murchie & DMJ. Lilley, unpublished data]. Of the mutations studied, those showing the most significant alterations to extrusion rates were those in which the base changes are close to the sequence dyad. Thus the t1/2 at 37°C can change from 70 m in pIRbke8 (central sequence AGAATTCT) to less than 2 m for ATATATAT or more than 500 m for AGACGTCT. With a few special exceptions, sequence changes further from the centre of the inverted repeat only alter the rates by a factor of two or less. This suggests that only the *central region* of the inverted repeat becomes altered in the formation of the transition state. This is the just the predicted result based on the S-type mechanism shown in Figure 2. As we shall see shortly, in the C-type cruciform there is almost total insensitivity to sequence changes in the inverted repeat.

To date we have uncovered no data which seriously challenge the mechanistic models described above, and we feel increasingly confident that these are a good description of the physical processes involved.

Sequences which determine the kinetic character of cruciforms.

While we have described two mechanisms by which cruciform extrusion may occur, we have given no clue as to what determines which pathway any particular sequence takes. Comparison of the base sequences of the ColE1 and bke cruciforms (see Figure 1) gives no clue to the origins of the differences - both are 50-60% A+T, and neither has additional sequence motifs such as purine-pyrimidine alternation. At this point, however, it is worth recalling that it is the ColE1 sequence which exhibits the unusual behaviour, most cruciforms are S-type. With this in mind, there is indeed one respect in which pColIR315 is quite

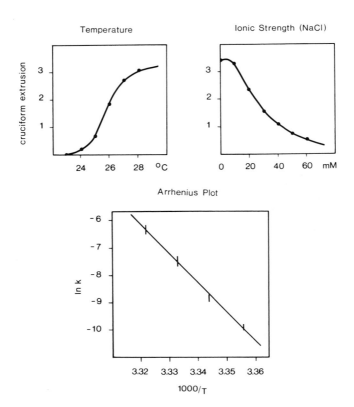

Figure 4. **Kinetic character of a bke-like cruciform surrounded by ColE1 flanking sequences.** The extrusion of pIRxke/col was studied as a function of temperature and ionic strength. The Arrhenius plot shows rate constants measured in 10 mM Tris.HCl pH 7.5, as a function of reciprocal temperature. The measured slope is -Ea/R.

abnormal. The ColE1 sequences which flank the inverted repeat are very rich in A-T base pairs. On both sides of the inverted repeat the base composition rises rapidly to over 80% A+T. We were almost forced to consider that this might be responsible for the C-type kinetic behaviour of the ColE1 cruciform. We decided to test the possibility that the A+T rich ColE1 sequences somehow influence the cruciform extrusion pathway, and confer C-type kinetics on the adjoining inverted repeat [Sullivan & Lilley, 1986].

We performed the following series of experiments:

1. We deleted the ColE1 cruciform from pColIR315, and replaced it with a new inverted sequence which was closely similar to the inverted repeat of pIRbke8. Thus we constructed a plasmid in which a bke-like cruciform resides in the context of the A+T rich ColE1 sequences. The new plasmid (pIRxke/col) exhibits typical C-type cruciform extrusion, see Figure 4, ie maximal rate at 0 mM NaCl, and Ea of 215 kcal mole^{-1}.

2. We performed the reverse experiment, by cloning oligonucleotides to generate a ColE1 inverted repeat sequence in the BamHI site of pAT153, the location at which the bke inverted repeat of pIRbke8 normally resides. As may be seen in Figure 5, the resulting plasmid (pIRCol/vec) exhibited typical S-type extrusion kinetics, ie maximal extrusion at 50 mM NaCl, and Ea of 50 kcal mole^{-1}.

These results show that the kinetic class, and therefore in all probability the mechanistic pathway, of cruciform extrusion is determined by sequences which lie outside the inverted repeat. The sequence of the inverted repeat itself seems to be of little or no importance in this selection. The A+T rich ColE1 flanking sequences appear to confer C-type extrusion kinetics on whatever sequences are placed next to them. We have termed them C-type Inducing Sequences (CIS). Further experiments have revealed the following properties of these CIS elements:

1. Only a single sequence is required. Either the right or left hand ColE1 sequences function equally well, and some other A+T rich sequences may replace them both. However, not all A+T rich sequences are active as CIS elements.

2. Where an inverted repeat is flanked by one C-type and one S-type sequence, the dominance is determined by the ionic strength, ie the kinetics are C- and S-type at 0 mM and 50 mM NaCl respectively.

3. Polarity can be unimportant. The left-hand-side 100 bp ColE1 sequence can confer C-type kinetics on inverted repeats placed at either end in different constructs.

4. The effects may be modulated over significant lengths of DNA. We have observed effects transmitted over 100 bp.

5. The C-type inducing effect may be blocked by insertion of the sequence GCCCCGGGGC between the CIS and the inverted repeat. The same sequence on the far side of the inverted repeat does not prevent C-type extrusion.

6. By use of Bal31 exonucleolysis and recloning, we have identified a region of 30 bp of very A+T rich DNA in the ColE1 left-hand side flanking sequence, which is very important for C-type induction. We have cloned a synthetic oligonucleotide of the same sequence, which is sufficient to confer C-type kinetics on an adjacent inverted repeat.

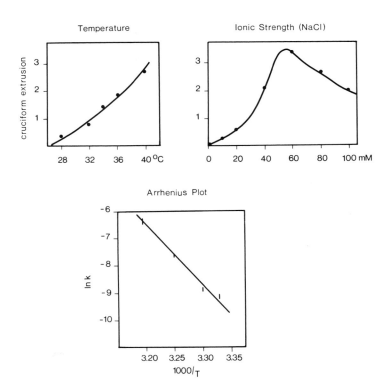

Figure 5.   Kinetic character of a ColE1-like cruciform surrounded by pIRbke8 flanking sequences. The extrusion of pIRCol/vec was studied as a function of temperature and ionic strength. The Arrhenius plot shows rate constants measured in 10 mM Tris.HCl pH 7.5, 50 mM NaCl, as a function of reciprocal temperature.

The origin of C-type induction.

How do the A+T rich CIS elements affect the entire kinetic character of cruciform extrusion at a remote location? We believe it is likely that the effects have their origin in the polymorphic and dynamic nature of A+T rich DNA [Arnott & Selsing, 1974; Klug *et al*, 1979; Mahendrasingham et al, 1983; Patel et al, 1985a,b; McClellan *et al*, 1986; Suggs &

Wagner, 1986; Lane *et al*, 1987]. This is supported by the following observations:

1. The CIS elements of ColE1 (in pColIRΔxba) are chemically reactive towards bromoacetaldehyde, glyoxal and osmium tetroxide [J. Furlong, G Gough & DMJ. Lilley, unpublished data]. The left- and right-hand side ColE1 sequences are independently reactive. Reactivity of CIS elements is dependent upon the presence of negative supercoiling, and is suppressed by raising the salt concentration.

2. The C-type induction may be completely prevented by binding distamycin [Zimmer *et al*, 1971] at 5 μM [K. Sullivan & DMJ. Lilley, unpublished data]. Distamycin-like antibiotics are known to bind to runs of successive AT base pairs in the minor groove, and thereby stabilise the double helical structure.

3. C-type inducing behaviour may be induced in normally S-type sequences (pIRbke8 for example) by inclusion of helix destabilising solvents such as 40% dimethyl formamide into the extrusion buffer [K. Sullivan & DMJ. Lilley, unpublished data].

Thus the CIS elements are reactive towards chemical probes of perturbed DNA structure, and are inactivated by binding a compound which has the effect of stabilising A+T rich DNA. Conversely, normally S-type cruciforms may be extruded at low temperature in the absence of salt if the DNA structure is weakened with organic solvents. Recently we have used a statistical mechanical approach to calculate the helix opening probabilities of the DNA sequences used in these studies [F. Schaeffer & DMJ. Lilley, unpublished data]. We find that the sequences which are experimentally observed to be CIS elements have high predicted probabilities of opening as cooperatively melting units at relatively low temperatures. In contrast, the S-type sequences are predicted to be quite stable helical structures at the same temperatures. Thus it seems possible that the mechanism of C-type induction is a striking example of telestability [Burd *et al*, 1975] effects. We cannot rule out, however, a more dynamic component in the process, whereby fluctuations arising in the CIS elements are transiently mobile [Englander *et al*, 1980; Sobell *et al*, 1983].

Perhaps the final stage of the process of C-type induction is the most difficult to understand - how the inverted repeat is actually induced to udergo cruciform extrusion *via* the C-type pathway. We will restrict ourselves to the observation that if the probability of opening along the length of the inverted repeat is finite, due to the presence in *cis* of the CIS element, then it may open as the kind of bubble depicted in Figure 2. Once completely open this would be expected to form the fully extruded cruciform rather than the reformed duplex.

Alternating $(A-T)_n$ sequences - a third kinetic class?

We have studied cruciform formation in a supercoiled plasmid pXG540, containing an inserted piece of *Xenopus lævis* aT1 gene, which includes the sequence $(A-T)_{34}$ [Greaves *et al*, 1985]. We noted that this sequence, in contrast to all other inverted repeats previously studied in this laboratory, showed no kinetic barrier to cruciform extrusion. Similar conclusions were reached by others studying synthetic $(A-T)_n$ inserts in plasmids [Haniford & Pulleyblank, 1985; Panyutin et al, 1985]. Do these results indicate a third class of cruciform kinetics, perhaps a third alternative extrusion pathway which is only available to the alternating A-T sequences?

There are no experimental data available on the kinetics of extrusion in these sequences, since it is not possible to perform kinetic experiments on a cruciform which cannot be removed, even at low temperature. We may make some speculative observations however. The possibility of a third kinetic pathway cannot be excluded. Recent studies of alternating $(A-T)_n$ segments in $linear$ DNA fragments [McClellan et al, 1986; Suggs & Wagner, 1986; Lane et al, 1987] have suggested that these tracts adopt a novel geometry which is chemically reactive, susceptible to a number of nucleases, and torsionally deformable. Thus if the ground state for the extrusion reaction is perturbed, it is possible that a different pathway becomes accessible.

There is, however, a quite different explanation for the observed kinetic behaviour of these sequences. Statistical mechanical calculations [F. Schaeffer & DMJ. Lilley, unpublished data] suggest that $(A-T)_{34}$ will undergo a transition as a cooperatively melting unit, at a temperature 10 degrees below that of the ColE1 sequences. Thus it is possible that the alternating sequence behaves as its own CIS element, and furthermore that the transition temperature at which the extrusion becomes too fast to measure (about 30°C for ColE1) is close to 0°C or lower. Thus rather than being a new kinetic class and mechanism, it is possible that the $(A-T)_n$ sequences are just a special example of a C-type cruciform.

Some concluding remarks.

Cruciform extrusion is a complex process. Despite this, it has turned out to be surprisingly amenable to kinetic analysis, and we are increasingly confident in the proposed alternative pathways described for the extrusion process. Indeed we believe this is rapidly becoming the most thoroughly understood structural transition to occur in supercoiled DNA molecules. The mechanism by which remote sequences may exert their C-type inducing effect is perhaps the most striking example of a long-range structural effect yet observed in DNA. CIS elements, though not formally participating in the extrusion process, determine the entire $character$ of a structural $transition$. While cruciform structures themselves have yet to be ascribed any significant biological role, it is clear that their study is revealing remarkable contextual effects in supercoiled DNA. The extent of these influences was unsuspected hitherto. The long term goal in these studies must be a full understanding of the kinetic and dynamic properties of specific DNA sequences placed under torsional stress in a negatively supercoiled molecule.

Acknowledgments.

We thank Professor RJP Williams FRS, Dr J Butler, Professor GS Manning, Dr F Schaeffer and Dr JA McClellan for many valuable discussions. We are grateful to the MRC, CRC, the Royal Society and the Wellcome Trust for financial support.

References.

Arnott, S. & Selsing, E (1974) Structures of poly(dA).poly(dT) and poly(dT).poly(dA).poly(dT). *J.Molec.Biol.* **88**: 509-521

Burd, JF, Wartell, RM, Dodgson, JB & Wells, RD (1975) Transmission of stability (Tele-stability) in deoxyribonucleic acid. *J.Biol.Chem.* **250**: 5109-5113

Courey, AJ & Wang, JC (1983) Cruciform formation in a negatively supercoiled DNA may be kinetically forbidden under physiological conditions. *Cell* **33**: 817-829

deMassey,B, Studier,FW, Dorgai,L, Appelbaum,F & Weisberg, RA (1984) Enzymes and the sites of genetic recombination: Studies with gene-3 endonuclease of phage T7 and with site-affinity mutants of phage λ. *Cold Spring Harbor Symp. Quant. Biol.* **49**: 715-726.

Dingwall,C, Lomonossoff,GP & Laskey,RA (1981) High sequence specificity of micrococcal nuclease. *Nucleic Acids Res.* **9**: 2659-2673.

Englander, SW, Kallenbach, NR, Heeger, AJ, Krumhansl, JA & Litwin, S (1980) Nature of the open state in long polynucleotide double helices: Possibility of soliton excitations. *Proc.Natl.Acad.Sci.USA* **77**: 7222-7226

Furlong, JC & Lilley, DMJ (1986) Highly selective chemical modification of cruciform loops by diethyl pyrocarbonate. *Nucleic Acids Res.* **14**: 3995-4007.

Gellert, M, Mizuuchi, K, O'Dea, MH, Ohmori,H & Tomizawa,J (1979) DNA gyrase and DNA supercoiling. *Cold Spring Harbor Symp. Quant, Biol.* **43**: 35-40.

Gellert, M, O'Dea,MH & Mizuuchi, K (1983) Slow cruciform transitions in palindromic DNA *Proc. Natl. Acad Sci, USA* **80**:5545 - 5549.

Gierer,A (1966) A model for DNA-protein interactions and the function of the operator. *Nature* **212**: 1480-1481.

Gough,GW (1986) *Ph.D. Thesis*, University of Dundee.

Gough,GW & Lilley,DMJ (1985) DNA bending induced by cruciform formation *Nature* **313**: 154 - 156.

Gough,GW, Sullivan,KM & Lilley,DMJ (1986) The structure of cruciforms in supercoiled DNA: probing the single-stranded character of nucleotide bases with bisulphite. *EMBO J.* **5**:191- 196.

Greaves,DR, Patient,RK & Lilley,DMJ (1985) Facile cruciform formation by an $(A-T)_{34}$ sequence from a *Xenopus* globin gene. *J. Molec. Biol.* **185**: 461-478.

Haasnoot,CAG, Hilbers, CW, van der Marel,GA, van Boom,JH, Singh,UC, Pattabiraman,N & Kollman,PA (1986). On loopfolding in nucleic acid hairpin-type structures. *J.Biomolec.Struc.&Dynamics* **3**: 843-857.

Haniford,DB & Pulleyblank,DE (1985) Transition of a cloned $d(AT)_n$-$d(AT)_n$ tract to a cruciform *in vivo Nucleic Acids Res.* **13**: 4343-4363.

Holbrook,SR, Sussman,JL, Wade Warrant,R, Church,GM & Kim,S-H (1977) RNA-ligand interactions: (I) Magnesium binding sites in yeast tRNA[Phe] *Nucleic Acids Res.* **8**: 2811 - 2820

Holliday,R (1964) A mechanism for gene conversion in fungi. *Genet. Res.* **5**: 282 - 304.

Jack,A, Ladner,JE, Rhodes,D, Brown,RS & Klug,A (1977) A crystallographic study of metal-binding to yeast phenylalanine transfer RNA. *J.Molec.Biol.* **111**: 315 - 328.

Klug,A, Jack,A, Viswamitra,MA, Kennard,O, Shakked,Z & Steitz,TA (1979) A hypothesis on a specific sequence-dependent conformation of DNA and its relation to the binding of the lac-repressor protein. *J.Molec.Biol.* **131**: 669-680.

Lane,MJ, Laplante,S, Rehfuss,RP, Borer,PN & Cantor,CR (1987) Actinomycin D facilitates transition of AT domains in molecules of sequence $(AT)_n AGCT(AT)_n$ to a DNaseI detectable alternating structure. *Nucleic Acids Res.* **15**: 839-852.

Lilley,DMJ (1980) The inverted repeat as a recognisable structural feature in supercoiled DNA molecules. *Proc.Natl.Acad.Sci. USA* **77**: 6468 - 6472.

Lilley,DMJ (1982) Dynamic, sequence-dependent DNA structure as exemplified by cruciform extrusion from inverted repeats in negatively supercoiled DNA. *Cold Spring Harbor Symp. Quant. Biol.* **47**: 101-112.

Lilley,DMJ (1983) Structural perturbation in supercoiled DNA: Hypersensitivity to modification by a single-strand-selective chemical reagent conferred by inverted repeat sequences. *Nucleic Acids Res.* **11**: 3097 - 3112.

Lilley,DMJ (1985) The kinetic properties of cruciform extrusion are determined by DNA base-sequence. *Nucleic Acids Res.* **13**: 1443 - 1465.

Lilley,DMJ (1986) Cyclic adduct formation at structural perturbations in supercoiled DNA molecules. In *The role of cyclic nucleic acid adducts in carcinogenesis and mutagenesis (IARC Publication No. 70)* (Eds B. Singer & H. Bartsch), Lyon.

Lilley,DMJ & Hallam,LR (1984) The thermodynamics of the ColE1 cruciform: comparisons between probing and topological experiments using single topoisomers *J.Molec.Biol.* **180**: 179 - 200.

Lilley,DMJ, & Kemper,B (1984) Cruciform-resolvase interactions in supercoiled DNA. *Cell* **36**: 413 - 422.

Lilley,DMJ & Markham,AF (1983) Dynamics of cruciform extrusion in supercoiled DNA: use of a synthetic inverted repeat to study conformational populations. *EMBO J.* **2**: 527 - 533.

Lilley,DMJ & Paleček,E (1984) The supercoil-stabilised cruciform of ColE1 is hyper-reactive to osmium tetroxide. *EMBO J.* **3**: 1187 - 1192.

Mahendrasingham,A, Rhodes,NH, Goodwin,DC, Nave,C, Pigram,WJ, Fuller,W, Brahms,J & Vergne,J (1983) Conformational transitions in oriented fibres of the synthetic polynucleotide poly[d(AT)].poly[d(AT)] double helix. *Nature* **301**: 535 - 537.

Manning,GS (1978) The molecular theory of polyelectrolyte solutions with applications to the electrostatic properties of polynucleotides. *Quart. Rev. Biophys.* **11**: 179 - 246.

McClellan,JA, Paleček,E & Lilley, DMJ (1986) (A-T)$_n$ tracts embedded in random sequence DNA - formation of a structure which is chemically reactive and torsionally deformable. *Nucleic Acids Res.* **14**: 9291-9309.

Mizuuchi,K, Kemper,B, Hays,J & Weisberg,RA (1982a) T4 endonuclease VII cleaves Holliday structures. *Cell* **29**: 357 - 365.

Mizuuchi,K, Mizuuchi,M & Gellert,M (1982b) Cruciform structures in palindromic DNA are favoured by DNA supercoiling. *J.Molec.Biol.* **156**: 229 - 243.

Naylor,LH, Lilley,DMJ & van de Sande,JH (1986) Stress-induced cruciform formation in a cloned (CATG)$_{10}$ sequence. *EMBO J.* **5**: 2407-2413.

Panayotatos,N & Wells,RD (1981) Cruciform structures in supercoiled DNA. *Nature* **289**: 466 - 470.

Panyutin,I, Klishko,V & Lyamichev,V (1984) Kinetics of cruciform formation and stability of cruciform structure in superhelical DNA. *J. Biomolec. Struc. & Dynamics* **1**: 1311 - 1324.

Panyutin,I, Lyamichev,V & Mirkin,S (1985) A structural transition in d(AT)$_n$.d(AT)$_n$ inserts within superhelical DNA. *J. Biomolec. Structure & Dynamics* **2**: 1221 - 1234.

Patel,DJ, Kozlowski,SA, Hare,DR, Reid,B, Ikuta,S, Lander,N & Itakura,K (1985a) Conformation, dynamics and structural transitions of the TATA box region of self-complementary d[(C-G)$_n$TATA(C-G)$_n$] duplexes in solution. *Biochemistry* **24**: 926 - 935.

Patel,DJ, Kozlowski,SA, Weiss,M & Bhatt,R (1985b) Conformation and dynamics of the Pribnow box region of the self-complementary d(CGATTATAATCG) duplex in solution. *Biochemistry* **24**: 936 - 944.

Platt.JR (1955) Possible separation of inter-twined nucleic acid chains by transfer-twist. *Proc.Natl.Acad.Sci. USA* **41** : 181 - 183.

Quigley,GJ, Teeter,MM & Rich,A (1978) Structural analysis of spermine and magnesium ion binding to yeast phenylalanine transfer RNA. *Proc.Natl.Acad.Sci. USA* **75**: 64 - 68.

Scholten,PM & Nordheim,A (1986) Diethyl pyrocarbonate: a chemical probe for DNA cruciforms. *Nucleic Acids Res.* **14**: 3981 - 3993.

Sheflin,LG & Kowalski,D (1984) Mung bean nuclease cleavage of a dA+dT-rich sequence or an inverted repeat sequence in supercoiled PM2 DNA depends on ionic environment. *Nucleic Acids Res.* **12**: 7087-7104.

Sinden,RR & Pettijohn,DE (1984) Cruciform transitions in DNA. *J. Biol. Chem.* **259**: 6593 - 6600.

Sobell,HM, Sakore,TD, Jain,SC, Bannerjee,A, Bhandary,KK, Reddy,BS & Lozansky,ED (1983) β-kinked DNA- a structure that gives rise to drug intercalation and DNA breathing - and its wider significance in determining the premelting and melting behaviour of DNA. *Cold Spring Harbor Symp. Quant. Biol.* **47**: 293 - 314.

Suggs,JW & Wagner,RW (1986) Nuclease recognition of an alternating structure in a d(AT)$_{14}$ plasmid insert. *Nucleic Acids Res.* **14**, 3703-3716.

Sullivan,KM & Lilley,DMJ (1986) A dominant influence of flanking sequences on a local structural transition in DNA. *Cell* **47**: 817-827

Sullivan,KM & Lilley,DMJ (1987) The influence of cation size and charge on the extrusion of a salt-dependent cruciform. *J.Molec.Biol* **193**: 397-404.

Symington,L & Kolodner,R (1985) Partial purification of an enzyme from *Saccharomyces cerevisiae* that cleaves Holliday junctions. *Proc. Natl. Acad. Sci. USA* **82**: 7247 - 7251.

Tam,S-K & Williams,RJP (1985) Electrostatics and biological systems. *Struct. Bonding* **63**, 103-151.

Wemmer,DE, Wand,AJ, Seeman,NC & Kallenbach,NR (1985). NMR analysis of DNA junctions: Imino proton NMR studies of individual arms and intact junction. *Biochemistry* **24**: 5745 - 5749.

West,SC & Korner,A (1985) Cleavage of cruciform DNA structures by an activity from *Saccharomyces cerevisiae*. *Proc. Natl.Acad.Sci. USA* **82**: 6445 - 6449.

Wu,H-M & Crothers,DM (1984) The locus of sequence-directed and protein induced DNA bending. *Nature* **308**: 509-513.

Zimm,BH & Le Bret,M (1983) Counter-ion condensation and system dimensionality. *J. Biomolec. Struct. Dynam.* **1**, 461-471

Zimmer,C, Reinert,KE, Luck,G, Waehnert,U, Loeber,G & Thrum,H (1981) Interaction of the oligopeptide antibiotics netropsin and distamycin A with nucleic acids. *J. Molec. Biol.* **58**: 329-338.

TORSIONAL STRESS, UNUSUAL DNA STRUCTURES, AND EUKARYOTIC GENE EXPRESSION

Charles R. Cantor, Subhasis Bondopadhyay, Shamir K. Bramachari[*], Cho-Fat Hui[1], Michael McClelland, Randall Morse[2] and Cassandra L. Smith

Department of Genetics and Development, Columbia University, New York, NY 10032, USA; [*]Molecular Biophysics Unit, Indian Institute of Science, Bangalore 560012, India; [1]Institute of Molecular Biology, Academia Sinica, Tapei 119 29, Republic of China; [2]Laboratory of Cellular and Developmental Biology, NIDDKD, NIH, Bethesda, MD 20892, USA

Torsional stress in DNA plays a significant role in regulating gene expression in prokaryotes. It has been known for more than a decade that different promoters show a differential response to the level of in vivo supercoiling in E. coli (Smith et al., 1978), and extensive studies indicate that the in vivo level of supercoiling itself is tightly regulated (Menzel and Gellert, 1983). In E. coli and other bacteria the amount of torsional stress is sufficient (Sinden et al., 1980) to favor the formation of locally untwisted alternate DNA structures. The energetic cost of forming these from the normal unconstrained DNA double helix is more than compensated for by relief of some of the excess torsional tension in the DNA (Cantor and Efstratiadis, 1984).

The effect of DNA topology and torsional tension on gene expression in eukaryotes is much less clear. A number of experiments have shown clearly that supercoiled DNA in eukaryotic cells is more actively transcribed than linear DNA. These include experiments in which both forms of DNA are transfected or injected into eukaryotic cells and the relative transcription levels compared (Harland et al., 1983; Weintraub et al., 1986), and experiments in which actively transcribed and untranscribed DNA molecules are fractionated from cells and their relative supercoiling compared (Barsoum and Berg, 1985; Luchnik et al., 1982; 1985). However, the complexity of the systems involved makes it difficult to define the level at which the DNA topology plays a critical role and the results are controversial (Petryniak and Lutter, 1987). The complexity also makes it difficult to distinguish, unequivocally, effects of DNA circularity and excess torsional stress. The problem is due, in part, to the relatively low levels of torsional stress measureable for DNA in eukakryotic cells (Sinden et al., 1980). This implies that either small amounts of stress can be localized into domains, or else that, overall, small amounts of stress can somehow have large global consequences.

Others have explored the relationship between supercoiling, chromatin structure and transcription by in vitro reconstitution systems (Ryoji and Worcel, 1984). Again, while tantalizing hints of important phenomena have emerged, it is difficult to be absolutely certain what causes and events are dominant in these systems.

A major reason for the difficulty in relating torsional stress and gene expression in eukaryotes is the relatively small amount of information available about the torsional properties of chromatin. Before the effects of supercoiling on gene expression in eukaryotes can be understood, we will need answers to questions like the following: What kinds of alternate DNA structures are favored by torsional stress, what sequences are preferred by these structures, and do proteins exist that allow class-specific recognition for these sequences and structures? How do nucleosomes interact with alternate structures: do they favor or inhibit their formation? Does chromatin restrict or facilitate the diffusion of torsional stress? In other words does chromatin structure allow torsional stress to be localized in particular regions? Are there domains in chromatin that act independently and if so, what determines the formation and elimination of these domains? What is the relationship between the timing of chromatin assembly, supercoiling, and binding of trans acting regulatory factors?

Some of these questions can be answered, in principle, by appropriately designed in vivo experiments. However, many of these will require in vitro experiments, ultimately with purified components (see for example, Hovatter and Martinson, 1987). Thus a major concern is to develop reliable in vitro systems to study chromatin assembly and specific transcription.

In this chapter we will describe progress in approaching a number of the questions posed above. All of our own work has dealt with in vitro systems using small circular DNA molecules. While the adequacy of such systems as chromatin models remains unproven, they are necessary models because only in these systems does current technology allow DNA topology to be accurately measured and controlled.

Effect of DNA sequence on alternate DNA structures induced by torsional stress

Certain types of underwound DNA structures like Z-DNA and cruciforms are well known and well characterized examples of the response of DNA to excess supercoiling in vitro and in vivo. These structures have fairly rigid (but not absolute) requirements for particular DNA sequences. Other specific types of DNA sequences are known to form torsionally induced alternate structures in vitro, and such sequences are linked to in vivo observations of nuclease hypersensitivity (Cantor and Efstratiadis, 1984). However, all these particular sequences represent a very small fraction of possible DNA sequences, and all are low complexity or symmetric structures.

We were curious whether a significant fraction of all possible DNA sequences could undergo a transition to alternate structures under sufficient torsional stress. To address this question we took advantage of the properties of form V DNA (Bramachari et al., 1987). This DNA is prepared by annealing complementary circular DNA single strands as shown in Figure 1. The preparation of form V DNA requires that the linking number be zero. Thus the superhelical density of the resulting duplex is -1 which is more than ten times the superhelical density of DNA prepared by other in vitro methods or observed in vivo. At such extreme torsional stress one can anticipate that a significant fraction of the total DNA will be forced into alternate structures. For example, if half the DNA existed in the B-form, most of the rest would have to be left-handed helix in order that the resulting writhing number of the DNA not be excessive.

We probed the structure of DNA sequences in form V by the method shown schematically in Figure 2. Various DNA restriction methylases have been shown to be sensitive to DNA structure (Vardimon and Rich, 1984; Zacharias et al., 1984). In general these enzymes appear capable of recognizing and methylating a specific sequence only when that sequence is near to a relatively unconstrained B-form. Less is known about how the enzymes will be affected by alternate DNA structures near the actual recognition sequence, but, in view of the lack of sequence dependence of methylation on these flanking regions, one can reasonably suppose that the enzymes see DNA structure only at or very near their recognition site. Thus observation of unimpaired methylation of a site in form V DNA will be good evidence that this sequence is still in the B form while reduction or prevention of methylation will indicate the presence of altered DNA structure. It will not, however, reveal the nature of that alteration.

The pattern of in vitro enzymatic methylation of pBR322 form V DNA was measured by two methods: indirectly, by the effect of that methylation on subsequent restriction nuclease cleavage, and directly, by the incorporation of 14C from labeled S-adenosylmethionine. The results indicate that, as expected, roughly half of the DNA sites inspected are

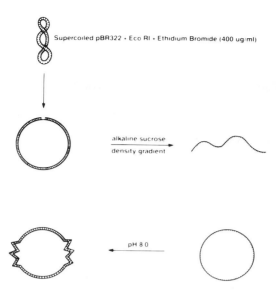

Supercoiled pBR322 + Eco RI + Ethidium Bromide (400 ug/ml)

alkaline sucrose
density gradient

pH 8 0

Figure 1.   Preparation of form V pBR322 DNA by reannealing complementary circular DNA single stranded circles, prepared from singly nicked supercoiled DNA.

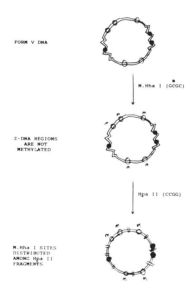

FORM V DNA

M.Hha I (GCGC)

Z-DNA REGIONS
ARE NOT
METHYLATED

Hpa II (CCGG)

M.Hha I SITES
DISTRIBUTED
AMONG Hpa II
FRAGMENTS

Figure 2.   One procedure for determining regions of altered DNA structure. M Hha I will incorporate radiolabel (m) into only those recognition sites in B-DNA.   The location of these sites is revealed by the subsequent pattern of radioactivity in Hpa II fragments.

deficient in methylation and thus present in an altered DNA structure. No clear pattern of particular sequence preferences for these altered structures was observed (Bramachari et al., 1987). For example, some, but not all, alternating purine-pyrimidine tracts were unmethylatable suggesting that they might be in the Z-form. However, many less regular DNA sequences were also under- or un-methylated. It thus appears that under sufficient torsional stress, a wide variety of different alternate structures can form, but the details of these structures are not yet known.

One result of the form V studies was unexpected. From prior analyses of the energetics of torsionally induced alternate structure formation (Stirtivant et al., 1982; Peck and Wang, 1983), most of the unfavorable energy of such structures resides in the cost of making the junction between them and neighboring stretches of B-DNA. Thus we expected to observe relatively long stretches of methylatable and unmethylatable regions in form V. However, what we actually observed was very short stretches of alternate and normal structures. For example, in the sequence shown below, starting from the Eco RI site of pBR322, alternate and normal structures appear to interchange four times in less than 40 nucleotides. Residues known to be normal are shown underlined, while residues known to be in alternate structures are shown in lower case letters.

GAATTCTCATGTTTGACagctATCATCGATaagcttTAA

It is not clear whether this mosaic of short structural domains is the result of the extremely high torsional stress, which simply overcomes the normal barriers between different structures or whether it is a consequence of the large overall amount of altered structure. For example, it is not known whether there is any effect of B-DNA length on the energetics of structure boundaries.

Effect of nucleosomes on diffusion of torsional stress

The extremely high superhelical densities of form V DNA described above can hardly be an appropriate model for DNA in chromatin in vivo, since the measured overall superhelical density in eukaryotic cells is quite low (Sinden et al., 1980). However these measurements do not rule out two possibilities: small domains of superhelical density might exist in linker regions between nucleosomes. Alternatively, the writhing energy of DNA within one or more nucleosomes might be revealed by alteration in the structure of that nucleosome or even loss of the protein components by dissociation.

In either case, the structural consequencs for the DNA will depend critically on whether the locally produced excess torsion is constrained or whether it can diffuse into the bulk chromatin. If diffusion occurs, the torsional energy will rapidly dissipate since the overall superhelical density will be very low and the energy depends on the square of the density. If diffusion is prevented, the local energy could remain quite high. Not only will the superhelical density be large, but, in addition, the energetic cost of constraining that density in a small region is known to be extremely high (Shore and Baldwin 1983; Horowitz and Wang 1983) because of the stiffness of short DNA segments. For this reason we felt that it was critical to examine whether nucleosomes can act as a barrier to the diffusion of torsional stress.

In the absence of practical direct methods for producing constrained supercoils and then watching them diffuse, we were encouraged to test diffusion indirectly. Torsional diffusion requires that either DNA rotate on the surface of the nucleosome, or that nucleosomes rotate with respect to each other by twisting or untwisting of the linker DNA between them. We can measure these rotations by examining the effect of nucleosomes on the thermal untwisting of DNA. Naked DNA is known to untwist about 0.01 degree per base pair per degree C in the range from 0 to 40 C (Depew and Wang, 1975; Pulleyblank et al. 1975). This produces a net untwisting of about four turns in a 4kB closed circular DNA, which is easily observed by gel electrophoretic analysis.

If both nucleosomal and linker DNA were free to untwist, the presence of nucleosomes should leave this thermal untwisting unaltered. If only the linker but not the core nucleosome could untwist, a DNA molecule saturated with nucleosomes should show only a quarter of the normal thermal untwisting. In practice, the presence of nucleosomes completely eliminates DNA untwisting whether or not histone H1 is present and whether or not the N-terminal histone tails are intact (Morse and Cantor, 1985;1986). These results show that adjacent nucleosomes in chromatin cannot rotate relative to each other, and thus each nucleosome represents an isolated topological domain. However, the results are valid only in the limit of low superhelical density, and more experiments are needed to determine if the rotation barrier can be overcome at higher, though still physiologically realizable, superhelical densities.

Effect of DNA topology on in vitro transcription

Our ultimate aim is to relate specific topological structures with gene activity. Since DNA topology is difficult to control, precisely, in vivo, we have sought to construct a reliable in vitro system where DNA structure and topology could be manipulated at will, and correlated with the consequences for specific transcription. In typical eukaryotic in vitro transcription systems, DNA is added to a whole cell extract, and after incubation, specific transcription can be measured, directly, by a run off assay where the template DNA is linearized, or indirectly, by annealing any RNA synthesized to an appropriate radiolabeled DNA probe and digesting the complex with S1 nuclease. The single end labeled probe is chosen so that it overlaps any anticipated transcription inititation sites. Thus the amounts and lengths of surviving radiolabeled probe fragments reveal the extent and origins of particular transcripts. The system is relatively crude because of the current lack of a completely defined set of the necessary purified transcription factors required by RNA polymerase II for appropriate transcription start sites and levels.

We chose SV40 DNA for these initial studies because of the extensive available knowledge about SV40 transcription and because of much prior work indicating possible in vivo relationships between SV40 topology and transcription, described in the introduction. When typical in vitro transcription conditions are used with a Hela cell extract (Manley et al., 1980; Manley et al., 1983), in the absence of added T antigen, closed circular SV40 DNA shows mostly specific transcription from the early early and early late start sites, and a small amount of transcription from a weak start site (also called the in vitro start site), (Ghosh and Lebowitz, 1981; Benoist and Chambon, 1981; Hansen et al., 1981). However it is difficult to correlate this transcription with DNA topology, measured by standard agarose gel electrophoresis, for two reasons. Extensive nuclease nicking and cutting occurs so that the state of the DNA after transcription is a complex mixture of species (Manley et al., 1983; Mertz, 1982; Hen et al., 1982; Sergeant et al., 1984; Jove et al., 1984). Virtually no closed circular DNA remains; instead most of the non-linearized DNA fails to enter the agarose gel, presumably because it has become highly catenated by the action of topoisomerases present in the whole cell extract. This problem is particularly severe since the high DNA concentrations needed for effective transcription in typical in vitro systems will naturally favor intermolecular catenation.

To try to circumvent the problems described above, we have systematically varied in vitro transcription conditions.

From our observations to date, we are able to eliminate unwanted nuclease and catenation reactions to a significant extent by combining several measures. The whole cell extract is fractionated to remove any residual endogenous DNA. The nucleotide concentration used for transcription is reduced by an order of magnitude when compared with conventional procedures, and the salt is also reduced considerably as shown by the detailed conditions described in the legend to Figure 3. Finally the DNA concentration is varied to find the lowest concentration at which adequate levels of transcription suitable for experimental study can still be realized.

A typical set of results are shown in Figure 3 for four levels of input DNA, either linear or supercoiled. A number of distinct advantages of our conditions are illustrated by these results. The input linear DNA remains largely intact, while the supercoiled DNA is mostly converted to a set of partially to totally relaxed circular species. At low DNA concentrations very little of the circular DNA is entrapped at the origin (approximately 2%). At the lowest DNA concentrations tested, 2.5 μg/ml, supercoiled DNA leads to almost a maximal level of transcription while linear DNA shows almost an insignificant level of transcription. This distinction is less as the DNA concentration is raised, suggesting that the preferential transcription of input circular DNA is determined by components in the whole cell extract that are present at very low concentrations.

The transcription results in Figure 3 are very encouraging since they demonstrate a clear in vitro effect of DNA topology on the level of gene expression. However, we are a long way from understanding the molecular origin of these results. For example, it is not clear which of the DNA species present after the transcription reaction were actually responsible for the transcription. While it is tempting to argue that the highly supercoiled species are responsible, they represent a minority of the DNA actually observed. Since most of the supercoiled DNA were converted to a wide

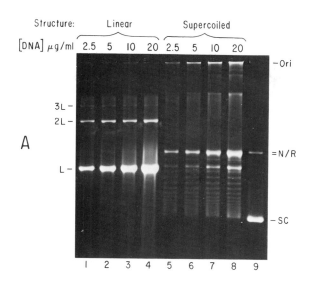

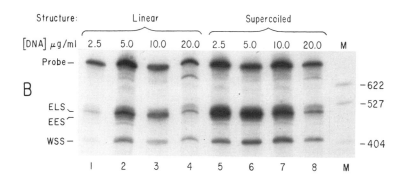

Figure 3. **In vitro** transcription of SV40 DNA as a function of DNA topology. DNA at concentrations indicated was added to a Hela whole cell extract in the presence of 12.5 uM NTPs, 5 mM creatine phosphate, 40 mM KCl, 5 mM $MgCl_2$, and incubated at 30 C for 1 hour. A third of each sample was treated with 10 mM EDTA at 0 C, deproteinized by SDS-proteinase K, and then phenol-chloroform extracted, and the resulting DNA was analyzed by electrophoresis on 1% agarose (panel A). The remaining two-thirds of each sample was used to prepare RNA which was characterized by S1 protection with a 5' labeled DNA probe, analyzed by electrophoresis on 5% polyacrylamide in the presence of 8 M urea (panel B). Symbols used are: Ori, gel origin; L, 2L, 3L, linear monomer, dimer, trimer; N/R, nicked or relaxed circular DNA; SC, highly supercoiled DNA; M, molecular weight markers; ELS, early late start sites; EES, early early start sites; Wss, weak start site.

distribution of topoisomers by the end of the transcription reaction, it would be difficult to predict which of these molecules were assembled into active transcription complexes. It is also not known at this moment whether nucleosome assembly is necessary to generate such active template structure. Experiments are needed to address these issues both by varying the superhelical density of the input DNA under conditions that lead to different transcription levels and by examining the pattern of proteins bound to the DNA under conditions that lead to different transcription levels.

What aspects of torsionally stressed DNA affect gene expression

Supercoiled DNA differs from relaxed DNA in so many respects that it is not immediately obvious on what level one should search for events significant in gene regulation. Earlier we stressed the ability of supercoiled DNA to promote altered DNA structures. One way in which this might function is the presence of altered structures which might be recognized as such by regulatory factors. However an equally plausible mechanism is simply an increased ability for DNA to change its structure in response to the binding of a protein. One example would be facilitated binding of factors to separated single strands.

In addition to the localized effects of torsional stress on DNA structure, the possibility exists for global effects of supercoiling on DNA behavior. This can take two forms. The tertiary structure of DNA is markedly affected by the level of supercoiling. Thus distant regions of DNA might be brought together in a supercoiled molecule which would have little chance of being proximal in relaxed DNA. Alternatively, the mechanical properties of supercoiled DNA, its ease of twisting and bending, will be greatly affected by the level of supercoiling. While we know relatively little about the details of these structures and properties, it is clear that the macroscopic behavior of a supercoiled DNA is very different from its relaxed counterpart.

One example of this is the very different behavior of both species in pulsed field gel electrophoresis. In this technique the mobility of a DNA molecule in a gel is determined principally by its ability to reorient in response to an electrical field that is periodically changing directions (Schwartz and Cantor, 1984; Smith et al., 1986). The greatest dependence of mobility on molecular weight, and hence the best resolution, occurs when the reorientation time of the molecule is nearly the same as the alternation time of the electrical field. These times scale roughly linearly in molecular weight for linear duplex DNAs and range from 1 second for 20 kB molecules to 1 hour for 10 Mb molecules. The critical factor that determines the molecular orientation time in a gel is presumably the length of the DNA molecule as it resides in the gel matrix.

To gain some insights into the shape and behavior of supercoiled DNA, we compared supercoiled and linear molecules by pulsed field gel electrophoresis. Typical results are shown in Figure 4. From these and more extensive studies, it is apparent that supercoils reorient about ten times faster in a gel than the corresponding linear molecules. This implies that the average supercoiled DNA configuration reduces the net length of the molecule by ten fold. Part of this condensation will be due just to the circularity of the DNA, but most of it must reflect the tertiary structure. It remains to be seen whether, in chromatin, torsional tension can also lead to condensed, perhaps specific tertiary structures that play important roles in gene expression.

Acknowledgement
    This work was supported by grants from the NIH, GM14825, CA39782, and CA23767, and from LKB Produktor, AB.

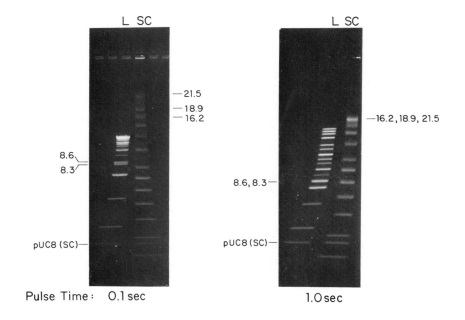

Figure 4. Pulsed field gel electrophoresis of linear (L) and supercoiled closed circular (SC) DNAs at two different pulse times. Note that the 16.2, 19.9 and 21.5 Kb supercoils and 813 and 8.6 Kb linear species are resolved only at 0.1 second pulsing. (Supercoiled pUC8 was added to all samples as an internal standard).

REFERENCES

Barsoum, J. and P. Berg (1985). Simian virus 40 minichromosomes contain torsionally strained DNA molecules. Molec. Cell. Biol. 5:3048-3057.

Benoist, C. and P. Chambon (1981). In vivo sequence requirements for the SV40 early promoter region. Nature (London) 290:304-310.

Bramachari, S.K., Y.S. Shouche, C.R. Cantor and M. McClelland (1987). Sequences that adopt non-B-DNA conformation in form V DNA as probed by enzymatic methylation. J. Mol. Biol. 193:201-212.

Cantor, C.R. and A. Efstratiadis (1984). Possible structures of homopurine-homopyrimidine S1-hypersensitive sites. Nucl. Acids Res. 12:8059-8072.

Depew, R.E. and J.C. Wang (1975). Conformational fluctuations of DNA helix. Proc. Natl. Acad. Sci. USA 72:4275-4279.

Ghosh, P.K. and P. Lebowitz (1981). Simian virus 40 early mRNA's contain multiple 5' termini upstream and downstream from a Hogness-Goldberg sequence: a shift in 5' termini during the lytic cycle is mediated by large T antigen. J. Virol. 40:224-240.

Hansen, U., D.G. Tanen, D.M. Livingston and P.A. Sharp (1981). T antigen repression of SV40 early transcription from two promoters. Cell 27:603-612.

Harland, R.M., H. Weintraub and S.L. McKnight (1983). Transcription of DNA injected into Xenopus oocytes is influenced by template topology. Nature 302:38-43.

Hen, R., P. Sassone-Corsi, J. Corden, M.P. Gaub and P. Chambon (1982). Sequences upstream from the T-A-T-A box are required in vivo and in vitro for efficient transcription from the adenovirus serotype 2 major late in promoter. Proc. Natl. Acad. Sci. USA 79:7132-7136.

Horowitz, D.S. and J. Wang (1983). Torsional rigidity of DNA and length dependence of the free energy of DNA supercoiling. J. Mol. Biol. 173:75-91.

Hovatter, K.R. and H.G. Martinson (1987). Ribonucleotide-induced helical alteration in DNA prevents nucleosome formation. Proc. Natl. Acad. Sci. USA 84:1162-1166.

Jove, R., D.E. Sperber and J.L. Manley (1984). Transcription of methylated eukaryotic viral genes in a soluble in vitro system. Nucleic Acids Res. 11:4715-4730.

Luchnik, A.N., V.V. Bakayev, I.B. Zbarskey and G.P. Georgiev (1982). Elastic torsional strain in DNA within a fraction of SV40 minichromosomes; relation to transcriptionally active chromatin. EMBO J. 1:1353-1358.

Luchnik, A.N., V.V. Bakayev, A.A. Yugai, I.B. Zbarsky and G.P. Georgiev (1985). DNAaseI-hypersensitive minichromosomes of SV40 possess an elastic torsional strain in DNA. Nucl. Acids Res. 13 1135-1149.

Manley, J.L., A. Fire, A. Cano, P.A. Sharp and M.L. Gefter (1980). DNA-dependent transcription of adenovirus genes in a soluble whole-cell extract. Proc. Natl. Acad. Sci. USA 77:3855-3859.

Manley, J.L., A. Fire, M. Samuels and P.A. Sharp (1983). In vitro transcription: whole-cell extract. Methods Enzymol. 101:568-582.

Mentzel R. and M. Gellert (1983). Regulation of the genes for E. coli DNA gyrase; homeostatic control of DNA supercoiling. Cell 34:105-113.

Mertz, J.E. (1982). Linear DNA does not form chromatin containing regularly spaced nucleosomes. Mol. Cell. Biol. 2:1608-1618.

Morse, R.H. and C.R. Cantor (1985). Nucleosome core particles suppress the thermal untwisting of core DNA and adjacent linker DNA. Proc. Natl. Acad. Sci. USA 82:4653-4657.

Morse, R.H. and C.R. Cantor (1986). Effect of trypsinization and histone H5 addition on DNA twist and topology is reconstituted minichromosomes. Nucl. Acids Res. 14:3293-3310.

Peck, L.J. and J.C. Wang (1983). Energetics of B-to-Z transition in DNA. Proc. Natl. Acad. Sci. USA 80:6206-6210.

Petryniak, B. and L.C. Lutter (1987). Topological characterization of the Simian virus 40 transcription complex. Cell 48:289-295.

Pulleybank, D.E., M. Shure, D. Tang, J. Vinograd and H.P. Vosberg (1975). Proc. Natl. Acad. Sci. USA 72:4280-4284.

Ryoji, M. and A. Worcel (1984). Chromatin assembly in Xenopus oocytes; in vivo studies. Cell 37:21-32.

Schwartz, D. and C.R. Cantor (1984). Separation of yeast chromosome-sized DNAs by pulsed field gel electrophoresis. Cell 37:67-75.

Sergeant, A., D. Bohmann, H. Zentgraf, H. Weiher and W. Keller (1984). A transcription enhancer acts in vitro over distances of hundreds of base-pairs on both circular and linear templates but not on chromatin-reconstituted DNA. J. Mol. Biol. 180:577-600.

Shore, D. and R.L. Baldwin (1983). Energetics of DNA twisting: II.Topoisomer analysis. J. Mol. Biol. 170:983-1007.

Sinden, R.R., J.O. Carlson and D.E. Pettijohn (1980). Torsional tension in the DNA double helix measured with trimethylpsoralen in living E. coli cells; analogous measurements in insect and human cells. Cell 21:773-783.

Smith, C.L., M. Kubo and F. Imamoto (1978). Promoter-specific inhibition of transcription by antibiotics which act on DNA gyrase. Nature 275:420-423.

Smith, C.L., P.W. Warburton, A. Gaal and C.R. Cantor (1986). Analysis of genome organization and rearrangements by pulsed field gradient gel electrophoresis. In, Genetic Engineering 8:(ed. J.K. Setlow and A. Hollaender), Plenum, New York, pps. 45-70.

Stirtivant, S.M., J. Klysick and R.D. Wells (1982). Energetic and structural inter-relationships between DNA supercoiling and the right-to left handed Z helix transitions in recombinant plasmids. J. Biol. Chem. 257:10159-10165.

Vardimon and A. Rich (1984).   In Z-DNA the sequence of G-C-G-C
    is neither methylated by Hha I methyltransferase nor
    cleaved by Hha I restriction endonuclease. Proc. Natl.
    Acad. Sci. USA 81:3268-3272.
Weintraub, H., P.F. Chang and K. Conrad (1986).   Expression of
    Transfected DNA Depends on DNA topology. Cell 46:115-122.
Zacharias, W., J.E. Larson, M.W. Kilpatrick and R.D. Wells
    (1984).   Hhal methylase and restriction endonulease as
    probes for B to Z DNA conformational changes in
    d(GCGC)sequences. Nucl. Acids Res. 12:7677-7692.

Development of a Model for DNA Supercoiling

Stephen C. Harvey and Robert K.-Z. Tan
Department of Biochemistry
University of Alabama at Birmingham
Birmingham, Alabama 35294

INTRODUCTION

The supercoiling of closed circular DNA in vivo is regulated by a
class of enzymes called topoisomerases, and it is known to play an
important role in gene expression. Negative supercoiling can be induced
in vitro by DNA gyrase, or by unwinding the double helix with
intercalating drugs, followed by relaxation with topoisomerase and then
removal of the intercalator. Supercoiling pressure is often used to
promote the formation of unusual DNA structures, for example, Z-DNA and
cruciforms. The study of DNA supercoiling is thus an important problem
biologically and because of its use as an agent for inducing other
structures. The topological issues surrounding DNA supercoiling have
been extensively studied (Benham, 1985; 1986; White and Bauer, 1986),
and they have been applied to the explanation of a number of events in
recombination (Wasserman et al, 1985; Cozzarelli et al, 1985; Gellert
and Nash, 1987).

In spite of the advances in understanding DNA topology, however,
little is known about the details of the three dimensional conformation
of supercoiled DNA molecules. In particular, it is not known with
certainty which of two conformations is actually the lowest energy form
of supercoiled DNA in solution, the interwound form or the toroidal
form. Most discussions of supercoiling focus on the interwound form,
because this is the shape that is observed in electron micrographs of
the molecule. The bias for the interwound form is further accentuated
by the fact that DNA models made from rubber tubing generally assume the

interwound form when they are closed into circles and subjected to supercoiling. Interestingly, the only experimental studies which purport to give information on the conformation of supercoiled DNA free in solution, the low-angle x-ray scattering studies of Brady et al. (1983; 1984) have argued that the predominant form is toroidal. It is possible that the results from electron microscopy are an artifact of laying the three-dimensional molecule down on a two-dimensional surface, because the shape of the molecule may be influenced by forces exerted during sample preparation (Vinograd and Lebowitz, 1965; Benham, 1986). It is also probable that rubber tubing is a poor model for DNA, both because it is subject to substantial gravitational effects and because the elastic properties of rubber are different from those of DNA.

There are currently two classes of models that might be used to simulate DNA supercoiling. First are the all-atom models of molecular mechanics and molecular dynamics (McCammon and Harvey, 1987). These models have the advantage of a very fine level of detail, but they are computationally very demanding. The largest nucleic acid treated by these methods so far is transfer RNA (Harvey et al., 1984; 1985; Harvey, 1986; McCammon and Harvey, 1987), but simulations on DNA have more commonly treated fewer than 25 basepairs (Levitt, 1983; Tidor et al., 1983; Singh et al., 1985; Prabhakaran and Harvey, 1985; Westhof et al., 1986). The second broad class of models are those that treat DNA as an elastic continuum (Benham, 1977; 1985; 1986; Le Bret, 1978; 1979; Selepova and Kypr, 1985; Tanaka and Takahashi, 1986; Tsuru and Wadati, 1986). These models have the advantage of mathematical rigor, but one must generally assume a conformation for the molecule, so they are not suitable for exploring conformational space and finding the minimum energy structure if it has an unusual and unexpected shape. Nor do they allow the treatment of sequence-dependent properties in a straightforward manner.

We are in the process of developing a computer-based model for DNA supercoiling. The level of detail is intermediate between the all-atom and continuum elastic models. Individual basepairs are represented as planes, and a potential energy function is defined for twisting and bending connections between neighboring basepairs. In principle, that function can be given any desired form. One can then investigate how the minimum energy conformation depends on the choice of the form for the potential energy function, and on the exact parameters used. In this paper we present a short description of the model and give some preliminary results for one particular model of DNA supercoiling.

92

DESCRIPTION OF THE MODEL

We have chosen to implement a model in which each basepair is
represented by a plane, defined by three point masses or pseudoatoms.
Figure 1 shows two basepairs from this model. The line passing through
the center of these basepairs is meant to represent the central axis of
the DNA double helix. Given the position of one basepair, the position
and orientation of the next basepair are given by specifying the
displacement along the central axis (about 3.5 Å in B–DNA), along with
three orientation angles. The first of these, the helix twist
angle, measures the rotation of successive basepairs around their common
axis. The twist angle in a ten-fold helix, for example, would be
36°. Rotations around the long axis of a basepair (corresponding
roughly to the vector connecting Cl' atoms) define the basepair roll
angle. Basepair tilt describes rotations about the basepair axis
perpendicular to the long axis. These definitions are very similar to
ones developed by Dickerson and his co-workers in describing the
crystallographic geometry of DNA (Fratini et al., 1982).

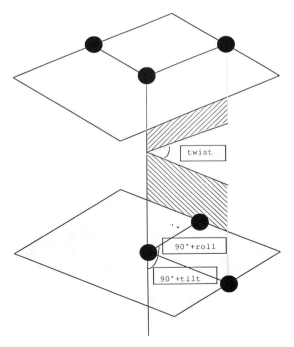

Figure 1. Details of model for two basepairs, with each basepair
represented by three pseudoatoms (black balls). See text for further
description.

For the initial form of the potential energy function, we have chosen harmonic potentials for all degrees of freedom. For example, the longitudinal stretching of the double helix is represented by the stretching of the bond connecting successive basepairs, with the energy dependence for bond stretching having the form

$$E(b) = (k_1/2)(b-b_0)^2 \qquad (1)$$

In this expression, b is the instantaneous bond length, b_0 is the ideal bond length (3.5 Å for B-DNA), and k_1 is a force constant chosen to match the longitudinal elastic modulus for B-DNA. The various angular degrees of freedom are also modeled with harmonic potentials. For example, the energy associated with deformations of the helix twist angle τ has the form

$$E(\tau) = (k_2/2)(\tau-\tau_0)^2 \qquad (2)$$

where τ_0 is the ideal value of the helix twist angle (36° in B-DNA), and k_2 is a force constant chosen to match the known torsional rigidity of B-DNA. Expressions similar to equation 2 have been used for modeling basepair roll and basepair tilt. We model the anisotropy of DNA bending by giving the force constant for tilt deformations three times the value of the force constant for roll deformations. The average value of these last two constants has been chosen to match the known elastic modulus for bending B-DNA. In addition to these expressions for deformations of the double helix, we have added harmonic potentials to maintain the geometry of individual basepair planes. The force constants associated with these latter terms are the strongest in the model, so deformations of the basepair planes are small by comparison to deformations of the helix geometry.

In order to prevent one part of the molecule from passing through another, a repulsive interaction between basepairs that are distant from one another in the covalent structure is included whenever they come into close contact with one another, i.e., less than 20Å.

The model in its present form has several advantages. It is both conceptually and mathematically simple. Harmonic potentials are generally appropriate for small deformations of elastic materials, and the quadratic form of all terms in the potential energy function lends itself to rapid and simple analysis by a variety of methods, including Newton-Raphson minimization and normal mode analysis. In addition, we believe the model has an appropriate level of detail. Variations in twist angle, basepair roll, and basepair tilt are easily analyzed, allowing an investigation of the effects of supercoiling on the local geometry of the double helix at the basepair level. The small number of pseudoatoms per basepair should make it feasible to model circular DNA molecules from a few tens of basepairs up to the size of plasmids. There is no requirement that the parameters of the model be the same for all basepairs, so it will be possible to introduce parameters representing variations in sequence. These will include both static effects such as localized bending (Marini et al, 1984; Koo et al., 1986; Hagerman, 1984; 1986; Trifonov, 1985; Tung and Harvey, 1986; see also the reviews by Hagerman and Trifonov, elsewhere in this volume), and variations in bending and twisting stiffness with sequence.

The model has two principal shortcomings in its present form. First, although the harmonic form of the potential function makes everything simple, it precludes the simulation of transitions between local minimum energy structures, such as the B to Z transition. This can be corrected in a later version by replacing the potential energy term for helix twisting with a form that has a double well. Second, the model is not double stranded, so one cannot simulate the melting of basepairs and the formation of other structures such as slipped or cruciform structures. The development of a double stranded model must wait until we have fully characterized the present model, which is intended to simulate the behavior of a continuous elastic fiber. Finally, we should mention that solvent is not treated explicitly in the model; the effects of varying solvent conditions must be mimicked by varying the parameters of the potential energy function. Although this reduces our ability to examine solvent effects, it is similar to the usual approach of molecular mechanics models of macromolecules (McCammon and Harvey, 1987) and is necessary for computational efficiency, given our current computer resources. Explicit solvent treatment must await later versions of the model and implementation of the program on a supercomputer.

RESULTS AND DISCUSSION

We have generated coordinates for closed circular models of DNA of various sizes. Those initial coordinates could be optimized by a variety of energy minimization methods (McCammon and Harvey, 1987). We have chosen molecular dynamics, because it should provide some information on the magnitude of thermal fluctuations about the minimum

energy structure, and because we wanted to get a feeling for the
computational burden for simulating transitions between conformational
states. For these initial studies, we have chosen a minimum energy tilt
angle of 3°, a typical value for B-DNA (Fratini et al., 1982). In order
to produce interesting relaxed structures, we began with a nonzero roll
angle, taking the minimum energy roll angle to be 10°.

To test the algorithm, we first took a very small closed-circular
DNA, equivalent to 50 basepairs, and brought it to its minimum energy
structure using quenched molecular dynamics at 0°K. This particular
model has a toroidal form, even though it is a relaxed circle with zero
supercoiling (Figure 2). Figure 3 shows the variation in the helix
parameters as one moves around the circle. All parameters show a
periodic behavior, with the length of the period dictated by the ten
basepair repeat of the double helix. Displacement, roll, and tilt
oscillate about their minimum energy values. The largest roll angles,
about 20°, correspond to the points of sharp bending in the projection
of the molecule that has a pentagonal form (Figure 2). The twist angle
is also oscillatory, but it is interesting to note that the twist angle
at every basepair step is smaller than the angle of 36° that was
characteristic of the starting structure and that is also the location
of the minimum energy for twisting. In coming to the minimum energy
structure, there has obviously been a net unwinding, corresponding to a
change in the total twist angle of the model. The explanation of this
phenomenon is quite simple. An examination of Figure 2 reveals that the
space curve corresponding to the helix axis has a nonzero writhe. The
quadratic form of the potential energy function for twisting prevents
any basepair from undergoing a 360° change in twist angle relative to
its neighbor, so the linking number of the molecule is unchanged during
the course of the molecular dynamics simulation. Since linking number
is conserved, any change in writhe must be offset by a corresponding
change in twist angle of the opposite sign (Benham, 1985). The positive
writhing of the model (Figure 2) is offset by the negative change in
twist (Figure 3). We have observed similar behavior in models of larger
molecules, including a 360-mer refined at 0°K and a 500-mer refined at
200°K. In the case of the larger molecules, plots of the variation of
helix parameters along the molecule differ from Figure 3 in two ways.
First, at 0°K, the amplitudes of the variations in twist, roll, and tilt
are smaller for large molecules than for the 50-mer. Second, at
elevated temperatures, the thermal noise is large enough to hide the
regular, periodic fluctuations in twist, roll, tilt, and displacement.

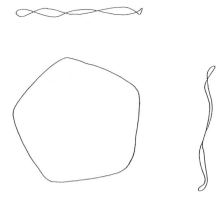

Figure 2.   Three orthogonal views of relaxed closed circular model containing fifty basepairs.   The torus about which the molecule is wrapped is most easily visualized in the front view of the model (the pentagonal form), because the inner and outer diameters of the torus correspond to inscribed and circumscribed circles in that view.

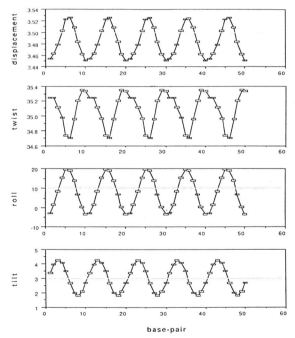

base-pair

Figure 3.   Variation of the helix parameters for the 50-mer shown in Figure 2, as one moves around the molecule.   Closing the molecule into a circle forces the helix parameters to deviate from their ideal values. The periodicities correspond to the ten basepair repeat of the double helix.

In order to test the ability of the model to refine structures that simulate supercoiling, we have performed the following computer experiment. A 360-mer was constructed with the same set of parameters as in a relaxed 360-mer, except the initial twist angle of each basepair was set at 34° relative to its neighbor. The molecule was then closed into a flat circle. Since this circle has no writhing, we can calculate an effective supercoiling density, σ, relative to the relaxed circular 360-mer by noting that

$$\sigma = \frac{\Delta Tw}{Tw}$$

On a per residue basis, ΔTw/Tw is -2°/36°, giving σ = -0.05, a value typical of supercoiling densities in vivo and in vitro. An equivalent way of calculating this number is to note that the relaxed 360-mer represents 36 turns of the double helix, while the underwound model represents only 34 turns.

It might be expected that this model could relax by simply having every basepair increase its twist angle by 2° relative to its neighbor, returning to the structure of the relaxed 360-mer. With respect to the first basepair, the second would rotate through an angle of 2°, the third through an angle of 4°,..., and basepair 360 would have to rotate through an angle of 720°. This would, of course, correspond to a relaxation of the supercoiled model by effectively changing the linking number back to 36. But the last basepair cannot make two full revolutions with respect to the first, because it is connected by a bond about which torsional rotations are opposed by a quadratic potential, so that the energy would become prohibitively high after rotation of only a few degrees. How does the model respond to this supercoiling? Figure 4 shows a sequence of backbone structures of the model when it is subjected to molecular dynamics at 300°K. In sharp contrast to the simulations of relaxed circles, very large structural changes are produced. In the final structure, at 300 ps, very pronounced writhing is observed. Although the backbone of the molecule appears to be touching itself in several places, this is not the case. Our program continuously monitors the distance of closest approach of various parts of the backbone, and no two points of this model are within 20 Å of one another. This is confirmed by a visual inspection of other views of the same structure. The structure has not yet reached equilibrium, and, when it does, it will have to be refined with a low temperature

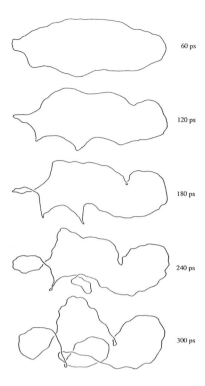

60 ps

120 ps

180 ps

240 ps

300 ps

Figure 4. Structure of a supercoiled closed circular 360-mer at
intervals of 60 ps. At t=0 (not shown), the model is a flat circle with
substantial stress in helix twist. The molecular dynamics simulation
redistributes this stress among the other degrees of freedom of the
model. The structure at 300 ps has not reached equilibrium.

molecular dynamics simulation to find the minimum energy structure. It

is not yet clear whether that structure would best be described as an

interwound form, a toroidal form, or some other class of space curve.

SUMMARY AND FUTURE DIRECTIONS

We have developed a computer-based model for the simulation of DNA

supercoiling. When fully operational, the model will allow one to use

any functional form for the energetics of deformation of DNA and to

examine how the shape of the lowest energy structure depends on the

choice of that function and on the specific parameters that are used.

Our first priority will be the implementation of a powerful energy minimizing algorithm, so that we may go directly from a starting model to the lowest energy form with a minimum of computational burden. When that lowest energy form is achieved for a particular model, we can examine the size and nature of the thermal fluctuations about that minimum using a variety of methods, including molecular dynamics, normal mode analysis, and Monte Carlo simulations (McCammon and Harvey, 1987).

We plan to carefully characterize a series of isotropic cases, in which the lowest energy roll and tilt angles are both set to 0°, and in which the stiffnesses for roll and tilt are equal. That investigation is intended to be a simulation of ordinary elastic materials. Among the questions we will ask of that model is how the choice between the interwound and toroidal forms depends on the length of the molecule and on the ratio of bend stiffness to torsional stiffness. Having examined a number of macroscopic models made up of materials other than rubber tubing, we consider it highly likely that the lowest energy forms will not be simply describable as either toroidal or interwound, except in special cases. Dr. Wilma Olson has kindly provided us with sets of coordinates from her interwound and toroidal models for supercoiled DNA (Olson et al., 1985). We will examine those conformations along with initially circular ones, to see if there are any effects of initial conformation on the final low energy structure. Any such effects would indicate errors in the algorithm.

We then hope to extend the models to include sequence dependence. Effects to be examined include variations in the lowest energy twist, roll, and tilt angles, static bends, and sequence dependent stiffness. We also plan to wind model DNAs around proteins, to see what kinds of conformational changes are produced in distant parts of the DNA molecule in response to supercoiling stresses imposed by protein binding. Finally, we will introduce non-harmonic potentials, in order to simulate transitions between right-handed and left-handed structures, and we will introduce a truly double stranded model, so that we can treat basepair melting and the formation of slipped and cruciform structures.

ACKNOWLEDGMENT

This research was supported by a grant from the National Institutes of Health (GM-34015).

REFERENCES

Benham, C.J. (1977) Proc. Natl. Acad. Sci. USA 74, 2397-2401.
Benham, C.J. (1985) Ann. Rev. Biophys. Biophys. Chem. 14, 23-45.

Benham, C.J. (1986) Comments Mol. Cell. Biophys. 4, 35-54.
Brady, G.W., Fein, D.B., Lambertson, H., Grassian, V., Foos, D., and Benham, C.J. (1983) Proc. Natl. Acad. Sci. USA 80, 741-744.
Brady, G.W., Foos, D., and Benham, C.J. (1984) Biopolymers 23, 2963-2966.
Cozzarelli, N.R., Krasnow, M.A., Gerrard, S.P., and White, J.H. (1985) Cold Spring Harbor Symp. Quant. Biol. 49, 383-399.
Fratini, A.V., Kopka, M.L., Drew, H.R., and Dickerson, R.E. (1982) J. Biol. Chem. 257, 14686-14707.
Gellert, M. and Nash, H. (1987) Nature 325, 401-404.
Hagerman, P.J. (1984) Proc. Natl. Acad. Sci. USA 81, 4632-4636.
Hagerman, P.J. (1986) Nature 321, 449-450.
Harvey, S.C., Prabhakaran, M., Mao, B., and McCammon, A. (1984) Science 223, 1189-1191.
Harvey, S.C., Prabhakaran, M., and McCammon, J.A. (1985) Biopolymers 24, 1169-1188.
Harvey, S.C. (1986) Comments Mol. Cell. Biophys. 3, 219-239.
Koo, H.-S., Wu, H.-M., and Crothers, D.M. (1986) Nature 320, 501-506.
Le Bret, M. (1978) Biopolymers 17, 1939-1955.
Le Bret, M. (1979) Biopolymers 18, 1709-1725.
Levitt, M. (1983) Cold Spring Harbor Symp. Quant. Biol. 47, 251-262.
Marini, J.C., Effron, P.N., Goodman, T.C., Singleton, C.K., Wells, R.D., Wartell, R.M., and Englund, P.T. (1984) J. Biol. Chem. 259, 8974-8979.
McCammon, J.A. and Harvey, S.C. (1987) "Dynamics of Proteins and Nucleic Acids" (Cambridge University Press, Cambridge England).
Olson, W.K., Markey, N.L., Srinivasan, A.R., Do, K.D., and Cicariello, J. (1985) in "Molecular Basis of Cancer, Part A: Macromolecular Structure, Carcinogens, and Oncogenes" (R.Rein, Ed., Alan R. Liss, New York), pp. 109-121.
Prabhakaran, M. and Harvey, S.C. (1985) J. Phys. Chem. 89, 5767-5769.
Selepova, P. and Kypr, J. (1985) Biopolymers 24, 867-882.
Singh, U.C., Weiner, S.J., and Kollman, P. (1985) Proc. Natl. Acad. Sci. USA 82, 755-759.
Tanaka, F. and Takahaski, H. (1986) J. Chem. Phys. 83, 6017-6026.
Tidor, B., Irikura, K.K., Brooks, B.R., and Karplus, M. (1983) J. Biomol. Struc. and Dynamics 1, 231-252.
Trifonov, E.N. (1985) CRC Crit. Revs. Biochem. 19, 89-106.
Tsuru, H. and Wadati, M. (1986) Biopolymers 25, 2083-2096.
Tung, C.-S. and Harvey, S.C. (1986) J. Biol. Chem. 261, 3700-3709.
Vinograd, J. and Lebowitz, J. (1965) J. Gen. Physiol. 49, 103-125.
Wasserman, S.A., Dungan, J.M., and Cozzarelli, N.R. (1985) Science 229, 711-174.
Westhof, E., Chevrier, B., Gallion, S.L., Weiner, P.K., and Levy, R.M. (1986) J. Mol. Biol. 190, 699-712.
White, J.H. and Bauer, W.R. (1986) J. Mol. Biol. 189, 329-341.

Cruciform Transitions Assayed Using a Psoralen Crosslinking Method: Applications to measurements of DNA torsional tension

David E. Pettijohn, Richard R. Sinden and Steven S. Broyles

It is now established that cruciform structures can form in supercoiled DNA at sites of inverted repeat sequences. Several different methods have been developed for detecting cruciforms, including: scission of the DNA with nucleases specific for the unpaired bases at the termini of cruciform arms (Lilley, 1980,1981; Panayotatos et al; Singleton et al,); scission with restriction enzymes (Gellert et al; Courey et al); electrophoretic resolution of DNA containing or not containing a cruciform (Mizuuchi et al; Gellert et al); electron microscopic observation (Mizuuchi et al).

We have developed procedures for assaying cruciforms that do not require separating the DNA from bound proteins and that are therefore potentially applicable to studying the cruciform structure *in vivo*. These methods rely on psoralen derivatives to crosslink the inverted repeat sequence and therefore fix it in whatever conformation it has at the time of crosslinking. A psoralen crosslink in either the linear inverted repeat sequence or within one arm of a cruciform, blocks the transition to the opposite conformational form. Since the psoralens preferentially crosslink thymidine residues on complimentary strands, the method works best when there is an AT·TA base pair sequence within the inverted repeat. An example of a suitable sequence that readily undergoes a cruciform transition and that that we we have studied most thoroughly is shown in Figure 1. This sequence is a perfectly symetric form of the lac operator selected and cloned in Sadler's laboratory (Betz et al). The inverted repeat sequence is 66 bp and it is flanked by EcoRl sites. When the sequence is in the normal linear form the EcoRl

```
                                                            *
      EcoRl_                                                 *
-GGAGCATAGAATTCCACAAATTGTTATCCGCTCACAATTCCACATGTGGAA
-CCTCGTATCTTAAGGTGTTTAACAATAGGCGAGTGTTAAGGTGTACACCTT
```

```
                                              EcoRl
               TTGTGAGCGGATAACAATTTGTGGAATTCTAATTTTTC-
               AACACTCGCCTATTGTTAAACACCTTAAGATTAAAAAG-
```

Figure 1. Sequence of the 66-bp lac operator inverted repeat.
In the plasmid pOEC12 this sequence cloned into EcoRl sites
of PMB9 is used for the studies described here. The EcoRl
sites are indicated and the *'s show the position of the
center of twofold rotational symmetry of the sequence. Some
of the thymidine residues that are preferentially crosslinked
by psoralen deriviatives are indicated by double underlining
(T).

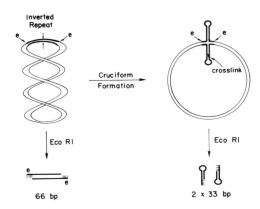

Figure 2. DNA cruciform transitions detected by a
trimethylpsoralen crosslinking assay. The lac inverted
repeat sequence (Fig. 1) cloned into supercoiled PMB9 DNA was
mixed with trimethylpsoralen and irradiated with suitable
doses of UV light to crosslink the DNA. If the negative
superhelical tension in the DNA was sufficiently high the
inverted repeat was in the cruciform and crosslinks as shown
on the right fix this structure; conversely if the torsional
tension was low the inverted repeat sequence remains in the
linear form and is fixed in that form. After excising the
inverted repeat with EcoRl endonuclease (at sites shown by e)
the 66-bp fragments released from the linear form and the 33-
bp fragments released from the cruciform are detected by gel
electrophoresis.

endonuclease cuts it out as a 66 bp fragment, but when it is in the cruciform the EcoRl sites are drawn to the base of the cruciform arms and 33 bp hairpin fragments are cut out (Fig. 2). This provides a simple procedure for testing the conformation of the sequence, as is demonstrated in Figure 3, since the two kinds of fragments can be readily distinguished by electrophoresis.

This approach also provides a method for quantitating the fractions of the inverted repeat sequence that are in the linear or cruciform. If all of the inverted repeats were crosslinked, then after cutting out the sequence with EcoRl, the relative amounts of DNA in the 33 and 66-bp fragments would directly indicate the relative amounts of the sequence in the two conformations. However, one rarely wishes to crosslink to the extent required to insert at least one crosslink in each palindromic sequence. Previously we demonstrated that the probability of crosslinking the inverted repeat sequence is the same when it is in linear form or in the cruciform (Sinden et al, 1983). Thus when only part of the inverted repeats are crosslinked, the fraction of the crosslinked inverted repeats that are cross-linked in 33-bp fragments is the fraction of the sequence that was in cruciform or: $P_{33} = F_{33}/P_{x1}$ where P_{33} is the fraction of the inverted repeat sequence existing in the cruciform, $P_{x1} = F_{33} + F_{66}$ is the fraction of the inverted repeat sequence that is crosslinked, F_{33} and F_{66} are respect-ively the fractions of the total inverted repeat sequences crosslinked in the 33-bp and 66-bp fragments.

An example showing how this analysis can be carried out is given in Figures 3 and 4. The negatively supercoiled pOEC12 DNA containing the lac operator inverted repeat was crosslinked to various extents and then the inverted repeat excised at its EcoRl sites. The DNA was linearized just before treating with EcoRl so that it lacked superhelical tension and all the uncrosslinked cruciforms converted to the linear form. Thus as shown in the first few lanes of Figure 3 where there was no crosslinking (the first lane) or less than 0.1 crosslinks per inverted repeat (the next 2 lanes) all of the DNA is excised in 66-bp fragments; however as the level of crosslinking is increased a larger fraction of the DNA is excised in 33-bp fragments showing that the crucifrom structure has been stabilized by the crosslinks. Also shown in this figure is an analysis of the amount of inverted

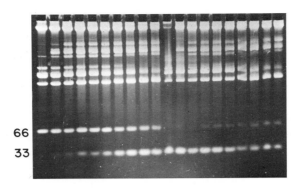

66

33

Figure 3. Electrophoretic analysis of EcoRl digests of
supercoiled pOCE12 DNA that had been crosslinked to various
extents by trimethypsoralen.   The DNA at a concentration of
100 ug/ml was equilibrated with ~0.2 ug/ml trimethylpsoralen
and irradiated at 4° C with General Electric F15T8 bulbs at
an incident light intensity of 1.6 kJ/m²/min using conditions
and readditions of trimethylpsoralen as previously described
(Sinden et al, 1980,1983).   After different periods of
irradiation samples were removed, the crosslinked DNA
purified and relaxed by linearizing with HindIII, and the
inverted repeat sequence was excised with EcoRl.   The DNA
digests were then subjected to electrophoresis on an agarose
gel.   An image of the gel stained with ethidium bromide is
shown.   The 10 lanes on the left half of the gel show from
left to right DNA samples that had increasing irradiation
times and therefore increasing numbers of psoralen
crosslinks.   The ten lanes on the right half of the gel show
the same DNA samples after heat denaturation arranged in the
same order as those on the left.   The positions of the 66 and
33-bp bands at the bottom of the gel are indicated.

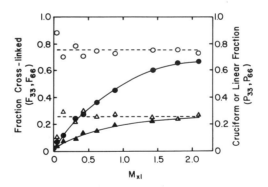

Figure 4.   Analysis of the number of crosslinked 66 and 33-bp
fragments and the calculated fraction of inverted repeats in
cruciforms.   The relative amounts of DNA in 33 and 66-bp
bands {F_{33} (●) and F_{66} (▲), respectively} were determined by
densitometry of a gel similar to that shown in Figure 3.
From these figures P_{33} (○) and P_{66} (△), the fraction of
inverted repeats in respectively cruciforms or linear forms,
were calculated as described in the text. M_{x1} is the average
number of psoralen crosslinks per inverted repeat sequence.
The DNA preparation analyzed here contained about 20% nicked
relaxed molecules.

106

repeat in the 66-bp band after heat denaturing the excised fragments. The 66-bp DNA resistant to heat denaturation is crosslinked linear DNA which allows an estimate of F_{66} as defined above. Quantitative analysis of the relative amounts of DNA excised in the two bands at different levels of DNA crosslinking is given in Figure 4. This shows that, while the absolute amount of crosslinked inverted repeats increased as expected with the level of crosslinking, the fraction P_{33} of the inverted repeats in cruciform (0.75) or in linear form P_{66} (0.25) calculated from these measurements did not vary with the amount of crosslinking. The DNA preparation used in this study contained about 20% nicked molcules so the cruciform fraction could be no higher than 80%. Thus variation in the amount of DNA bound psoralen residues over a wide range has no effect on the amount of inverted repeat in cruciform structure. This finding is significant, because it is known that intercalated psoralen molecules unwind the DNA double helix and thereby relax negative torsional strain in supercoiled DNA. Because of this factor it seemed possible that high levels of photobound psoralen might induce a conversion of cruciforms to linear forms. However the previous result suggests that this does not occur at least in the range of photobinding that has been studied. This is because the lac operator cruciform is quite stable at the low temperatures where photobinding is done, even in DNAs that have little or no torsional strain (see below and Sinden et al, 1984).

KINETICS OF CRUCIFORM TRANSITIONS

The above assay makes it possible to study kinetic parameters of cruciform formation and relaxation. In one type of experiment a nick was introduced with DNase I into a supercoiled pOEC12 plasmid having most of its lac operator sequence in cruciform. The rate of transition to the linear form was followed at different temperatures and in solutions containing different ions (Sinden et al, 1984). One such experiment at 37° C showed that the transition to linear form at low ionic strengths had a half-life of about 30 s, while the addition of 5 or 20 mM $MgCl_2$ increased the half life to about 70 and 120 s respectively (Figure 5). At 0 to 4 °C the half life for the same transition was many hours even at low ionic strengths without added Mg++. An enthalpy of

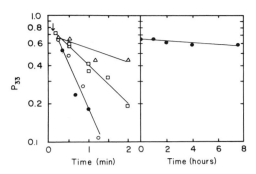

Figure 5. Transition of the cruciform to the linear form in relaxed pOEC12 plasmids. Supercoiled pOEC12 (σ = -0.06) containing about 20% nicked molecules were incubated briefly (5 s at 37°C) with DNase I in a solution containing 50mM NaCl, 5 mM MgCl$_2$, 10 mM Tris, pH 7.6 and the reaction was stopped by the addition of sodium dodecyl sulfate (SDS) and/or EDTA. The reaction was sufficient to introduce one or more nicks into >95% of the plasmids. Incubation at 37°C was continued beginning at the 10 s point indicated by the arrow above. At the times shown above samples were removed and frozen to stop the loss of cruciforms. Later the samples were thawed without warming above 4°C and psoralen crosslinking was carried out. After digesting the crosslinked plasmids with restriction endonucleases as described in Figure 3, the relative amounts of 33 and 66-bp fragments were determined by electrophoresis and the fraction (P_{33}) of the inverted repeat in cruciforms was calculated as described above. A, incubations were at 37°C in the above solution containing also: O, 10 mM EDTA; ●, 5mM EDTA and 50 ug/ml SDS; □, 50 ug/ml SDS; △, 20 mM MgCl$_2$ and 50 ug/ml SDS. B, incubation was at 0°C in the above solution containing also 5mM EDTA and 50 ug/ml SDS. (from Sinden et al, 1984 reprinted by permission).

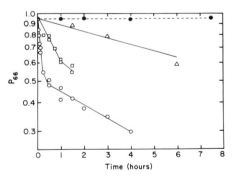

Figure 6. Transition rates from linear form to cruciform in supercoiled pOEC12 plasmids. Supercoiled pOEC12 plasmids (σ = -0.06) were isolated on CsCl gradients containing ethidium bromide, which removes cruciforms from the DNA. The CsCl and ethidium were removed at 0°C and replaced with a solution containing 50 mM NaCl, 1mM EDTA, 10 mM Tris pH 7.5 and then warmed to 37° by adding an excess of the same warm solution. At the times shown samples were removed, frozen, and later crosslinked with trimethylpsoralen to quantitate P_{66} the fraction of inverted repeats in linear form. Incubations were in the above solution plus: O, no further additions; □, 5 mM MgCl$_2$; △, 20 mM MgCl$_2$; ●, incubation at 0°C with no further additions. (from Sinden et al, 1984, reprinted by permission).

activation of 26 kcal mol^{-1} was calculated from this and other data.

The opposite transition from linear to cruciform was studied by preparing supercoiled pOEC12 DNA in the presence of ethidium bromide which positively supercoils the DNA and removes all cruciforms. After removing the ethidium at 0 °C the linear form was retained, but upon warming the DNA the cruciform transition occurred (Figure 6). The actual rates were dependent on the linking number deficit of the negatively supercoiled DNA, the ionic strength, and the temperature (Sinden et al, 1984); but the transition could go rather quickly (half life = 4.3 min) at an ionic strength of 0.06, 37°C and a physiological linking number. From the temperature dependence of the transition the enthalpy of activation was calculated to be 33.4 kcal mol^{-1}.

It has been known for sometime that the probability of a specific inverted repeat sequence existing in a cruciform is related to the specific linking number deficit of a superhelical DNA molecule (Hsieh, et al; Vologodskii, et al; Singleton et al; Sinden et al). Using the psoralen crosslinking assay, we studied the effect of temperature on the transition in pOEC12 DNA induced at different DNA linking numbers (Figure 7). The results suggested, that while the transition occurred at specific linking number deficits in the range −0.04 to −0.06 to some extent at temperatures in the range 12°C and greater, when the temperature was increased the transition occurred at lower specific linking number deficits.

ESTIMATES OF DNA TORSIONAL TENSION

Since the transition of an inverted repeat sequence to cruciform is dependent on the level of torsional strain in a supercoiled DNA molecule, it seems possible that one could exploit this dependence to estimate levels of DNA torsional tension. Indeed this approach has been developed in several studies of the levels of DNA torsional strain in prokaryotic DNAs *in vivo* (Sinden et al, 1983; Courey et al; Greaves et al; Lilley, 1986). Most methods for assaying cruciform structures require removal of the DNA from cells. This requirement raises the uncertainty that a potential cruciform structure may be lost or altered as a result of changing the environment of the inverted repeat sequence. An advantage of the psoralen cross-linking method is that most cells are permeable to psoralen derivatives and the crosslinking can be carried out *in vivo*. Thus any changes that occur after cellular disruption do not influence the analysis.

An example of the psoralen crosslinking approach to investigate the structure of inverted repeat sequences *in vivo* is described in Figure 8. The lac operator inverted repeat in pOEC12 was readily crosslinked *in vivo*, but only 66-bp fragments were detected after excission of the inverted repeat with EcoRl endonuclease Fig. 8, lane C). The inverted repeat in pOEC12 DNA purified from the same cells and deproteinized before crosslinking, yielded a high frequency of 33-bp fragments showing that the cruciform transition occurred only after the DNA was purified (lane B). Analysis of heat denatured 66-bp fragments (lane H) showed that at the highest level of psoralen photobinding *in vivo* there were on average 0.98 crosslinks per inverted repeat and these were in the linear form. Thus the inverted repeat was efficiently crosslinked *in vivo*, but only in the linear form, suggesting that the cruciform existed rarely if at all in living cells. The plasmids were crosslinked at 0 to 4°C *in vivo* to reduce the possibility (as described above) that the photobinding of psoralen reduces the number of cruciforms. Furthermore, the measured cruciform fraction P_{33} did not change when the number of psoralen crosslinks per DNA molecule was varied as in Figure 4 (Sinden et al, 1983), again indicating a lack of effect of the photobound psoralen derivativies on the cruciform transition. Also the plasmids were isolated as covalently continuous molecules before cutting out the inverted repeat to eliminate the possiblity that nicking *in vivo* relaxes cruciforms that would otherwise exist. As shown in Figure 8 lane D, the noncrosslinked plasmids isolated from bacteria at 0 to 4°C also lacked cruciforms, but the cruciform transition occurred after warming the purified plasmid. This finding also indicates that the cruciform structure does not exist in living cells. It is difficult to rule out the possibility that some unknown protein bound directly to the lac inverted repeat restrains it in the linear form, but no evidence for this type of interaction has been obtained. The lac inverted repeat is known to bind the lac repressor both *in vitro* and *in vivo* (Betz et al). However this interaction cannot influence the above results because

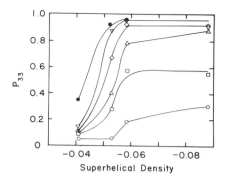

Figure 7. Effect of temperature on the DNA torsional tension-
dependent cruciform transition. pOEC12 DNA populations were
prepared having different average specific linking number
deficits shown by the superhelical densities above. These
were treated with ethidium bromide to remove preexisting
cruciforms, the ethidium removed at 0°C, replaced with 50 mM
NaCl, 1 mM EDTA, 10 mM Tris, pH 7.6 and the DNAs incubated
for 18 h at : $\bigcirc$, 0°C; $\square$, 12°C; $\triangle$, 20°; $\diamond$, 30°; $\bigtriangledown$, 37°;
●, 45°C. The fraction of the plasmids having cruciforms P_c
(same as P_{33}) was determined as in Figure 4. (from Sinden et
al reprinted by permission).

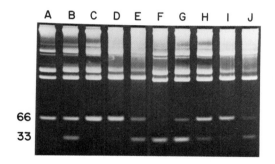

Figure 8. Electrophoresis of inverted repeat sequences in
pOEC12 DNA after psoralen crosslinking *in vitro* and *in vivo*.
Supercoiled pOEC12 DNA was isolated from bacteria grown with
chloramphenicol to amplify the number of plasmids per cell.
In different samples of the same culture the DNA was
crosslinked with trimethylpsoralen at different stages of
purification of the plasmid. The purified plasmids were
digested with a set of restriction endonucleases to relax the
DNA, then with EcoRl to excise the 33 or 66-bp inverted
repeat, and the fragments were resolved by electrophoresis.
Lanes A and B, the DNA was crosslinked after complete
purification and there were on average M_{x1} = 0.6 and 1.7
crosslinks per inverted repeat in A and B respectively. Lane
C, DNA was crosslinked *in vivo* in EDTA-permeabilized living
cells, M_{x1} = 1.0. Lanes D and E, DNA purified before
crosslinking, but kept at 0 to 4°C during all steps; lane D,
DNA incubated 30 min at 0° just before crosslinking, M_{x1} =
2.6; lane E, DNA incubated 30 min at 37° just before
crosslinking, M_{x1} = 1.2. Lanes F to J are samples of the
same DNAs as A to E, arranged in the same order, but the DNA
fragments were heat denatured before electrophoresis.

in the chloramphenicol treated cells there is a vast excess of lac inverted repeats over repressor molecules.

The most likely explanation for these findings is that the effective torsional strain in the plasmid DNA is too low *in vivo* to activate the cruciform transition. The results imply that DNA superhelical turns are restrained *in vivo* in structures (like for example nucleosomes) that restrict the equilibration of DNA superhelical turns with torsional strain in the helical winding. Other studies of the structure of prokaryotic DNA also indicate that protein-DNA complexes *in vivo* are organized to allow only a partial expression of the total potential torsional strain attributable to the DNA linking number deficit (see for reviews Pettijohn et al; Lilley).

Experiments in progress are using the cruciform transition assay to measure quantitatively the restraint of DNA superhelical turns by purified prokayotic histone-like proteins. Our results to date show that one can restrict the cruciform transition in negatively supercoiled pOEC12 plasmids by binding certain histone-like proteins. Analysis of these results shows that a reduction in DNA torsional tension after binding histone-like proteins accounts quantitatively for the reduction in transition to the cruciform. These results support the above interpretation of the results from the *in vivo* study of cruciforms.

OVERVIEW

The psoralen crosslinking procedure for assaying cruciform structures and cruciform transitions has several advantages over alternative methods. Probably the major advantage is that the approach can be applied *in vivo* since living cells are freely permeable to several psoralen derivatives. The psoralen photobinding can be performed over a wide range of temperatures and ionic conditions, permitting a flexibility of analysis not obtainable using other methods that require the activity of a nuclease or an electrophoretic analysis. Although reduced temperatures of photobinding are required in some cases to reduce the possible influence of the bound psoralens on DNA torsional tension, in many applications low levels of photobinding can be used at high temperatures without changing significantly the levels of DNA torsional tension. Because the photobinding can be carried out at temperatures near the freezing point of water, it becomes possible to study fast acting kinetic parameters of the cruciform transition; the later studies are also aided by the availability of high intensity lamps and laser sources that permit photobinding reactions to be completed after very brief irradiation times.

ACKNOWLEGDEMENT

This research was supported by a grant from the National Institutes of Health (GM 18243).

REFERENCES

Betz, J. L. and J. R. Sadler (1981). Gene **13**: 1-12.

Courey, A. J. and J. C. Wang (1983) Cell **33**: 817-829.

Gellert, M., M. H. O'Dea, and K. Mizuuchi (1983). Proc. Natl. Acad. Sci. USA **80**: 5545-5549.

Greaves, D. R., R. K. Patient, and D. M. J. Lilley (1985). J. Mol. Biol. **185**: 461-78.

Hsieh, T. and J. C. Wang (1975). Biochemistry **14**: 527-535.

Lilley, D. M. J. (1980). Proc. Natl. Acad. Sci. USA **77**: 6468-6472.

Lilley, D. M. J. (1981). Nucleic Acids Res. **9**: 1271-1289.

Lilley, D. M. J. (1986). In I. Booth and C. Higgins (Eds.), Symp. of the Society for General Microbiology. Regulation of Gene Expression, Cambridge University Press, pp 105-126.

Mizuuchi, K., M. Mizuuchi and M. Gellert (1982). J. Mol. Biol. **156**: 229-243.

Panayotatos, N. and R. D. Wells (1981). Nature (London) **289**: 466-470.

Pettijohn, D. E. (1985) N. Nanninga (Ed.) In Molecular Cytology of Escherichia coli. Structure of the Isolated Nucleoid, Academic Press, London pp. 199-227.

Sinden, R. R., J. O. Carlson and D. E. Pettijohn (1980) Cell **21**: 773-783.

Sinden, R. R., S. S. Broyles and D. E. Pettijohn (1983). Proc. Natl. Acad. Sci. USA **80**: 1797-1801.

Sinden, R. R. and D. E. Pettijohn (1984). J. Biol. Chem. **259**: 6593-6600.

Singleton, C. K. and R. D. Wells (1982). J. Biol. Chem. **257**: 6292-6295.

Vologodskii, A. V., A. N. Lukashin, V. V. Ashelevich and M. D. Kamenetskii (1979) Nucleic Acids Res. **6**: 967-982.

NMR-Distance Geometry Studies of Helical Errors and Sequence Dependent Conformations of DNA in Solution.

Dinshaw J. Patel, Lawrence Shapiro

Department of Biochemistry and Molecular Biophysics
College of Physicians and Surgeons
Columbia University,
New York, NY   10032

and

Dennis Hare

Infinity Systems
14810 2156 Avenue NW
Woodinville, Washington   98072

Key Words

Nuclear Magnetic Resonance (NMR)
Distance Geometry (DG)
DNA Solution Structures
Wobble Base Pair Mismatches
Extrahelical Purines
Homopurine·Homopyrimidine Tracts
Alternating Purine-Pyrimidine Tracts

INTRODUCTION:

The earliest information on the structure of DNA followed the successful interpretation of the X-Ray diffraction patterns of oriented high molecular weight fibers (Crick and Watson, 1954). Such studies established a right-handed 10 base pair repeat B-DNA conformation (Figure 1, top) at high humidity and a right-handed 11 base pair repeat A-DNA conformation (Figure 1, bottom) at low humidity (Langridge, et.al., 1960; Fuller, et.al., 1965; Arnott and Hukins, 1972). These studies were followed by systematic X-Ray fiber diffraction studies on synthetic polynucleotides which established polymorphism in DNA with the observed conformational variability dependent on sequence, ionic strength and humidity (reviewed in Arnott, et.al., 1983b). These results stimulated the introduction of methods that could probe DNA conformation at higher resolution.

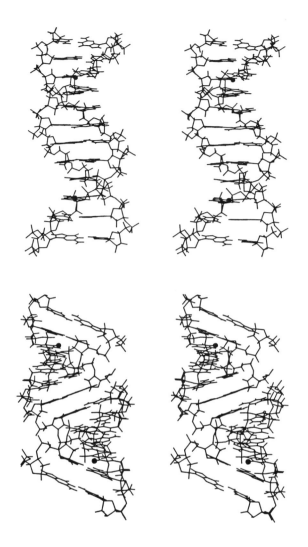

<u>Figure 1</u>. Stereo pairs of the structures of (top) B-DNA and (bottom) A-DNA as deduced from analysis of fiber diffraction data (Arnott, et.al., 1983a,b). The dodecanucleotide sequence corresponds to d(GCACGCGCGTGC).

There are several approaches that can currently provide detailed information on the structure of DNA fragments, their sequence dependent conformational variations and perturbations associated with helical errors. Single crystal X-Ray methods in the solid state (reviewed in Dickerson, et.al., 1981; Shakked and Kennard, 1985; Rich, et.al., 1984) and high resolution NMR techniques in solution (reviewed in Patel, et.al., 1982; Kearns, 1984; Wemmer and Reid, 1985; Patel, et.al., 1987a) have addressed this problem at the oligonucleotide duplex level and the complementary information so deduced has highlighted the role of sequence-structure correlations in DNA. These sequence dependent structural variations can also be detected in the DNA cleavage patterns generated by nucleases (Lomonosoff, et.al., 1981; Drew, 1984), chiral probes (Barton, 1986) and synthetic cleavage agents (Dervan, 1986; Burkhoff and Tullius, 1987).

The importance of sequence dependent structural variations in right-handed B-form DNA was established by Dickerson and Drew (1981) in their seminal analysis of the crystal structure of the self-complementary d(CGCGAATTCGCG) dodecanucleotide duplex. The observed slide, roll and twist variations at successive base pairs along the helix were explained by Calladine (1982) in terms of structural accomodations associated with relief of clashes between adjacent purine rings on partner strands. This was followed by crystal structure analysis of several other deoxyoligonucleotides all of which exhibit sequence dependent conformations in the A-DNA family of structures (reviewed in Shakked and Kennard, 1985).

The ability of NMR to probe the conformation of DNA and RNA has been greatly aided by a series of developments that hold the promise of defining structures for nucleic acids in solution to high resolution. The improvements in spectral resolution at high magnetic fields when coupled with two dimensional NMR methods has resulted in the assignment of exchangeable and nonexchangeable protons in oligonucleotide duplexes upto 2 turns of helix. Further, measurement of nuclear Overhauser enhancements (NOEs) between proton pairs separated by <5A provides approximate distance information along the length of the helix. This experimental distance information between a large number of proton pairs deduced from two dimensional NOE experiments can be combined with nucleotide bond length parameters and serve as input constraints for structure reconstruction algorithms. Both distance geometry and molecular dynamics algorithms have been used to minimize the disagreement between the refined structures and the distance bounds.

This paper reviews several applications of NMR-distance geometry methods to analyze features of the solution conformation of DNA duplexes with the emphasis on sequence dependent effects and the consequences of helical errors in the interior of duplexes.

METHODS:

Two dimensional NMR techniques (reviewed in Bax and Lerner, 1986; Wuthrich, 1986) disperse the spectral information in two frequency domains resulting in increased resolution for monitoring interactions between proximal spins. The pulse scheme for data accumulation in the two dimensional NOE experiment (NOESY) is outlined in Scheme 1. The NMR data is collected in the time domain with the free induction decays accumulated during the detection time t_2 as a function of incremental changes in the evolution time t_1. The nuclear Overhauser enhancement, which monitors the transfer of magnetization as a result of cross relaxation between adjacent pairs of spins, develops during the fixed mixing time Δ_t (Scheme 1). The time domain data sets (t_1, t_2) are then Fourier transformed to yield the two dimensional frequency domain data sets (ω_1, ω_2).

A stacked plot of the NOESY spectrum (mixing time 250 msec) of the d(CGCGAATTCGCG) duplex in 0.1M NaCl, D_2O at $25^{\circ}C$ is recorded in Figure 2. The one dimensional spectrum is on the diagonal, with off-diagonal cross peaks reflecting distance dependent NOE interactions between pairs of protons. The off-diagonal NOE patterns exhibit directionality and procedures have been developed for assignment of individual cross peaks in right-handed duplexes of known sequence in solution (Hare, et.al., 1983; Scheek, et.al., 1984; Weiss, et.al., 1984; Chazin, et.al. 1986).

The magnitude of the cross relaxation rate between spins i and j (σ_{ij}) monitored at short mixing times is proportional to the correlation time for molecular tumbling, τ_{ij}, the mixing time, Δ_t, and inversely proportional to the sixth power of the inter proton separation r_{ij}.

$$\sigma_{ij} = \frac{\tau_{ij}}{r_{ij}^6} \Delta_t$$

The volume integral for each resolved cross peak corresponding to each detectable and assignable proton-proton interaction can be measured from the NOESY contour plots as a function of mixing time.

The simple relationship between the cross relaxation rate (σ) and the interproton separation (r)

$$\frac{\sigma_{ij}}{\sigma_{kl}} = \frac{(r_{kl})^6}{(r_{ij})^6}$$

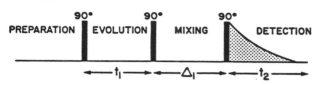

Scheme 1

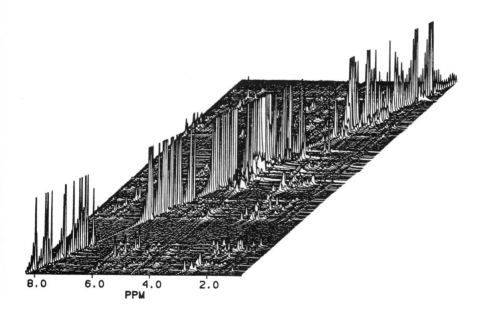

Figure 2. A stacked plot of the phase sensitive NOESY spectrum (mixing time 250 msec) of the d(CGCGAATTCGCG) duplex in 0.1M NaCl, D_2O at 25°C.

119

can then be used to estimate the interproton distances, where r_{kl} corresponds to the fixed distance yardstick and r_{ij} corresponds to the experimentally determined distance. The ratio σ_{ij}/σ_{kl} can be estimated by measuring the volume integrals of cross peaks corresponding to interactions i-j and k-l and together with the known fixed distance r_{kl} yields an estimate of the unknown distance r_{ij}.

The dependence of cross relaxation rates on rotational motion and spin-spin relaxation required the use of various distance yardsticks in order to estimate interproton distances. The thymidine H6-CH_3 distance of 3.0A was used as a yardstick for all NOEs involving CH_3 protons, the sugar H2',2'' distance of 1.85A was used as a yardstick for all NOEs involving sugar H2' and H2'' protons and the cytidine H6-H5 distance of 2.45A was used as a yardstick for all remaining NOEs between nonexchangeable protons. The thymidine imino to adenosine H2 distance of 2.8A was used as a yardstick for NOEs between imino protons and for NOEs between imino and adenosine H2 protons on neighboring base pairs.

The inter proton distances measured from the NOESY spectra were found to be dependent on the mixing time value within the initial buildup region so that it was not possible to measure an exact distance. Rather, each interproton distance was constrained by lower and upper bounds which were set to d-0.1A and d+0.1A, where d is the distance range estimated from short mixing time data. Somewhat larger bounds of d-0.3A and d+0.3A were set for weak cross peaks which were observable in NOESY spectra at longer mixing times.

The NOE based distance information can be used to reconstruct features of nucleic acid structures in solution with the major effort to data focused on distance geometry (Crippen, 1981) and molecular dynamics (Brooks, et.al., 1983) algorithms. The distance geometry approach embeds the approximate distance information along with fixed bond length parameters into cartesian space to generate a set of trial structures by selecting random distances between lower and upper bounds for each interatomic distance. Several different starting structures are then refined using nonlinear optimization techniques to minimize the disagreement between structure and distance bounds. The degree of similarity between the refined structrues is used to evaluate the extent of structural determination implicit in the data. This approach has been used to define the solution conformation of several proteins (reviewed in Wuthrich, 1986) and nucleic acids (Hare and Reid, 1986; Hare, et.al., 1986a,b; 1987a,b; reviewed in Patel, et.al., 1987a).

An alternate approach is based on molecular dynamics refinemtns which involve solution of Newton's equations of motion to determine how atomic positions change with time. The distance constraints are incorporated into the total energy function in the form of effective potentials. Molecular dynamics refinements using idealized interproton distances demonstrate a satisfactory radius of convergence to the same final refined structure independent of the starting A-DNA or B-DNA conformations (Nilsson, et.al., 1986). This combined NMR-distance geometry approach has been used to define the solution conformation of a self-complementary hexanucleotide duplex in solution (Nilsson, et.al., 1986).

The experimental distance information is restricted to base and sugar ring protons and hence structural features such as base pair overlap, glycosidic torsion angles and sugar pucker are well defined by the combination of NMR and structure reconstruction algorithms. By contrast, the poor resolution of the H5',5'' protons prevents an estimation of the torsion angles about the bonds C3'-O3'-P-O5'-C5'-C4' and this segment is merely constrained to bridge the gap between deoxyribose rings without violating any bond geometry or van der Waals constraints.

dG·dT MISMATCH

The pairing of noncomplementary bases to form mismatches is of considerable interest (see reviews by Modrich, 1987; Radman 1987) as are the resulting local and perhaps global conformational perturbations in the DNA helix. The early experimental approach has focused on the applications of NMR solution (reviewed in Patel, et.al., 1984) and single crystal X-Ray (reviewed in Kennard, 1985) techniques to elucidate structural features at and adjacent to the mismatch site in oligonucleotide duplexes of defined sequence and length. This review focuses on G·T mismatches and the interested reader is referred to recent reviews summarizing the research on other mismatches (Patel, et.al., 1987b; Kennard, 1987).

Our laboratory has undertaken systematic NMR studies on G·T mismatches by incorporating these errors in symmetry related positions in self-complementary dodecanucleotide duplexes (Patel, et.al., 1982b). This general strategy has been subsequently adopted in single crystal X-Ray studies where G·T mismatches were incorporated in self-complementary hexanucleotide (Ho, et.al., 1985), octanucleotide (Kneale, et.al., 1985; Hunter, et.al., 1986) and dodecanucleotide (Kennard, 1985) duplexes. These experimental investigations have been complemented by theoretical computations of the pairing possibilities of G·T mismatches (Keepers, et.al., 1984; Chuprina and Poltev, 1985). The energetics of the helix-coil transition of G·T mismatch containing duplexes have been probed by calorimetric (Patel, et.al., 1982b) and optical (Tibayenda, et.al., 1984; Aboul-ela, et.al., 1985) methods while hydrogen exchange kinetics have been followed by imino proton exchange at the individual base pair level at and adjacent to the mismatch site in dodecanucleotide duplexes (Pardi, et.al., 1982; Patel, et.al., 1984).

The ability to systematically assign the base and sugar protons in oligonucleotide duplexes by two dimensional NMR methods (Hare, et.al., 1983; Scheek, et.al., 1984; Weiss, et.al., 1984) has been extended to sequences containing G·T mismatch sites (Hare, et.al., 1986a; Quignard, et.al., 1987). We outline below the results of a combined NMR-distance geometry study on the self-complementary d(CGTGAATTCGCG) dodecanucleotide duplex (designated G·T 12-mer, Scheme 2) which contains G·T base pairs third base pair in from either end of the duplex (Hare, et.al., 1986a).

The two dimensional magnitude NOESY spectrum (14.0 to 7.0 ppm) of the G·T 12-mer duplex in 0.1M NaCl, H_2O at $5^{O}C$ is displayed as a contour plot in Figure 3A. The one dimensional spectrum along the diagonal exhibits imino protons between 12.7 to 13.7 ppm and 10.4 to 11.6 ppm while the base and amino protons resonate between 7.5 to 8.5 ppm. The observed NOE cross peaks between the guanosine imino and hydrogen-bonded cytidine amino protons (box B, Figure 3A) establish Watson-Crick G·C pairing at the G2·C11 and G4·C9 base pairs while the NOE cross peaks between the thymidine imino and adenosine H2 protons (box A, Figure 3A) establish Watson-Crick A·T pairing at the A5·T8 and A6·T7 base pairs in the G·T 12-mer duplex. A strong NOE is observed between the 11.6 and 10.5 imino protons of T3 and G10 at the mismatch site (box C, Figure 3A) and each of these imino protons exhibits weaker NOEs to the imino protons of flanking G2·C11 and G4·C9 base pairs (box D, Figure 3A). The relative magnitude of these NOEs is best visualized by recording a one dimensional slice as is shown in Figure 4A where the imino proton of G10 exhibits a strong NOE to the

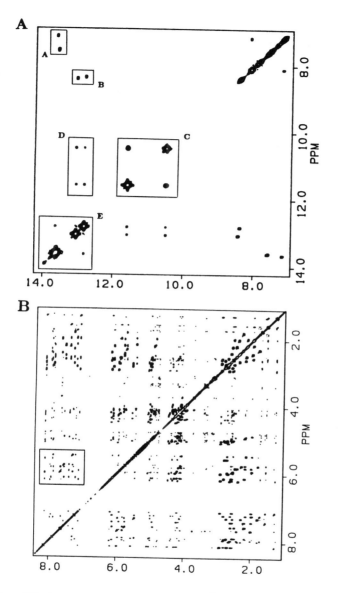

Figure 3. (A) A symmetrical contour plot (7.0 to 14.0 ppm) of the magnitude NOESY spectrum (120 msec mixing time) of the G·T 12-mer duplex in 0.1M NaCl, H$_2$0 at 5°C. (B) A symmetrical contour plot (1.0 to 8.5 ppm) of the phase sensitive NOESY spectrum (250 msec mixing time) of the G·T 12-mer duplex in 0.1M NaCl, D$_2$0 at 25°C.

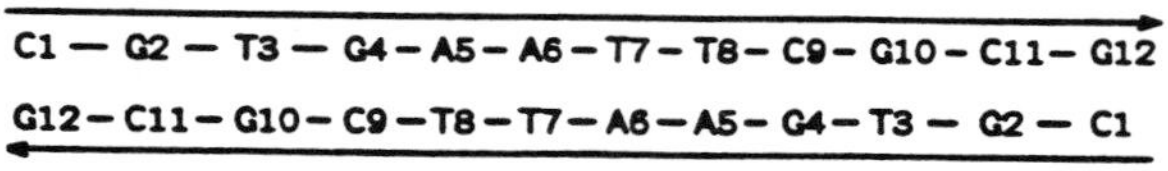

C1 — G2 — T3 — G4 — A5 — A6 — T7 — T8 — C9 — G10 — C11 — G12

G12 — C11 — G10 — C9 — T8 — T7 — A6 — A5 — G4 — T3 — G2 — C1

Scheme 2

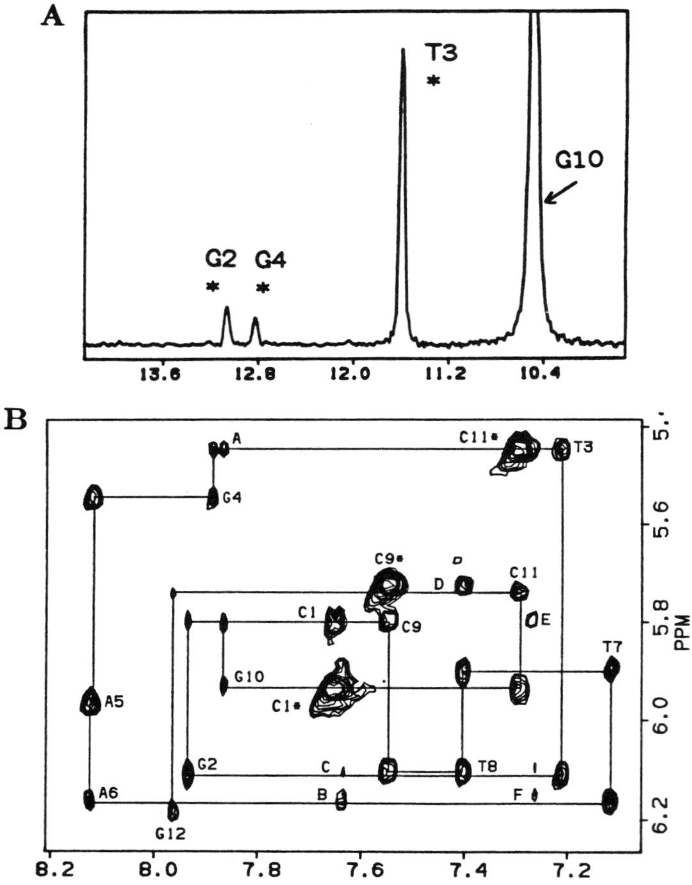

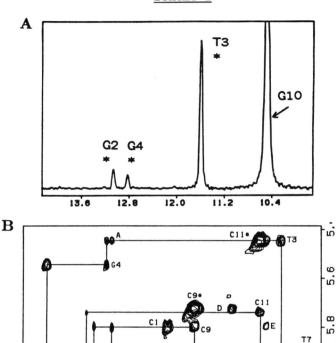

Figure 4. (A) One dimensional slice (10 to 14 ppm) establishing distance connectivities in the (G2-T3-G4)·(C9-G10-C11) segment in the NOESY spectrum (see Figure 3A) of the G·T 12-mer duplex in 0.1M NaCl, H_2O at 5°C. (B) An expanded contour plot of the NOESY spectrum (see Figure 3B) of the G·T 12-mer duplex in 0.1M NaCl, D_2O, 25°C establishing distance connectivities between base and sugar H1' protons.

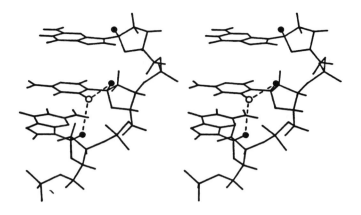

Scheme 3

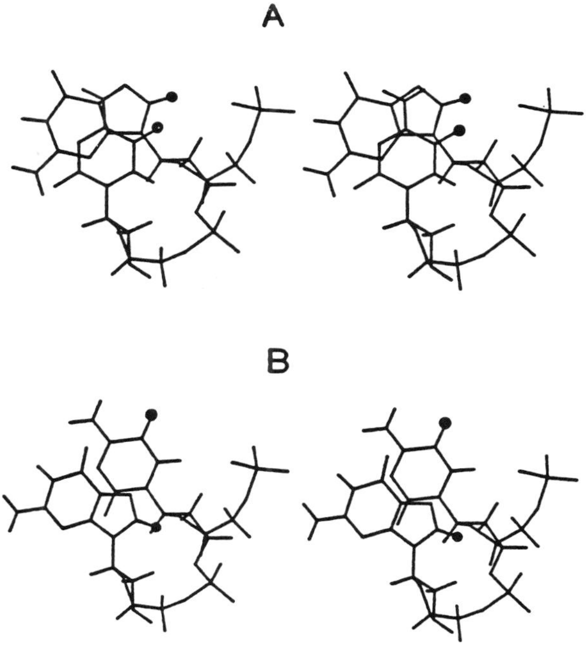

A

B

Scheme 4

imino proton of T3 within the T3·G10 mismatch pair and much weaker NOEs to the imino protons of G3 and G10 of the flanking G·C pairs in the G·T 12-mer duplex. These exchangeable proton data demonstrate that the imino protons of T3 and G10 are in close proximity and that the two bases at the mismatch site are stacked into the helix in the G·T 12-mer duplex.

The contour plot of the phase sensitve NOESY spectrum of the G·T 12-mer duplex in 0.1M NaCl, D_2O solution at $25^\circ C$ (Figure 3B) exhibits well resolved cross peaks required for assignment of the base and sugar protons along the length of the dodecanucleotide duplex. The expanded region establishing distance connectivities between the base and sugar H1' protons in the G·T 12-mer duplex is plotted in Figure 4B. Each base proton (purine H8 or pyrimidine H6) exhibits an NOE to its own and 5'-linked sugar H1' protons (Scheme 3) so that one can trace the NOE connectivities from C1 to G12 including those of the G2-T3-G4 and C9-G10-C11 segments containing the T3 and G10 bases at the mismatch site (Figure 4B). The magnitude of the base to H1' NOEs demonstrate that T3 and G10 exhibit anti glycosidic torsion angles at the T3·G10 mismatch site and the observed NOE connectivities establish that these bases at the mismatch site are stacked into the helix. The directionality of the base to sugar H1' NOE connectivities (Figure 4B) along with the observed NOEs between adjacent base protons in the base H8/H6 (3'-5') base H5/CH$_3$ direction (Scheme 4A) for the G2-T3, A6-T7, T8-T9 (cross peak D, Figure 4B) and G10-C11 (cross peak A, Figure 4B) steps demonstrate formation of a right-handed helix at and adjacent to the mismatch site in the G·T 12-mer duplex.

The NOE buildup rates of the resolved and assigned cross peaks in the NOESY spectrum of the G·T 12-mer duplex yielded 194 interproton distance constraints which were embedded into distance space and several trial structures refined by the distance geometry algorithm. Stereo pairs of one such embed and its subsequent refinement of the G·T 12-mer duplex are shown in Figure 5 (top and bottom views, respectively) with the G·T mismatch located third base pair in from either end of the dodecanucleotide duplex (Hare, et.al., 1986a). An alternate view (rotated through 90° along the helix axis) of the same G·T 12-mer embed and refinement are shown in Figure 6 (top and bottom views, respectively).

A Wobble pair is formed between T3 and G10 (Figure 7A) with the bases propeller-twisted at the mismatch site (Figure 7B) in the G·T

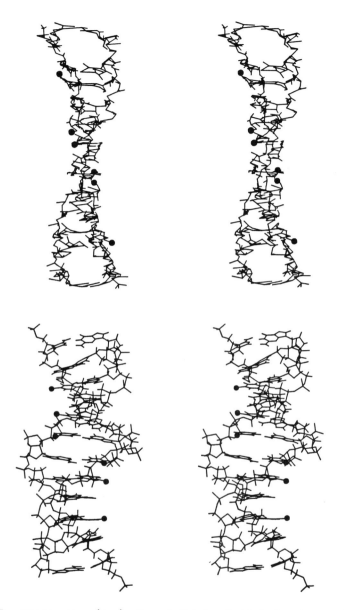

Figure 5. Stereo pairs (top) of an initial embedded structure and (bottom) of the distance geometry refined structure of G·T 12-mer duplex. The thymidine CH₃ groups are included as darkened balls for recognition purposes.

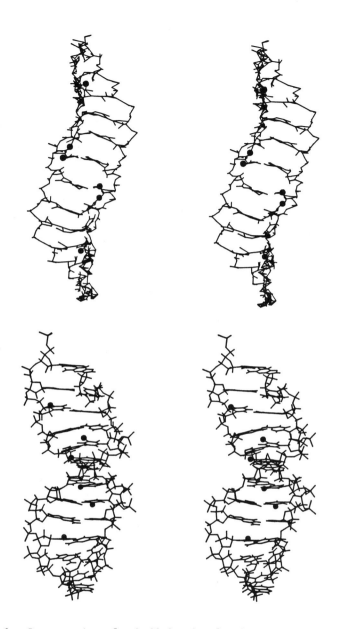

Figure 6. Stereo pairs of embedded and refined structures of the G·T 12-mer duplex rotated through 90° along the helix axis from view in Figure 5.

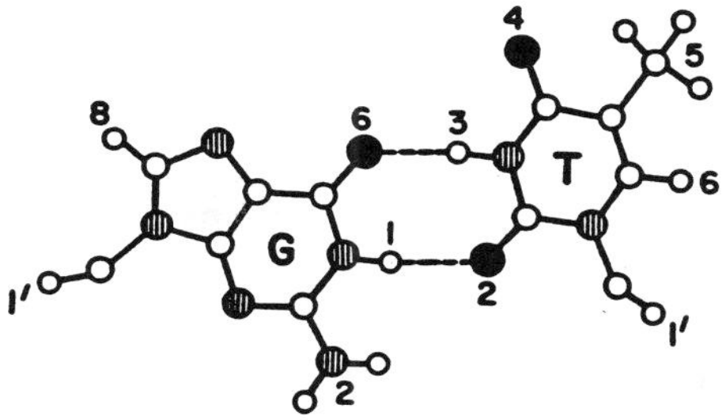

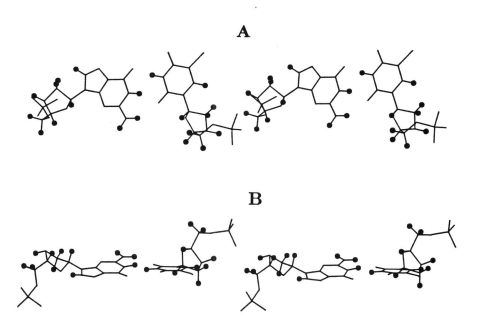

Scheme 5

A

B

Figure 7. Stereo pairs of the T3·G10 mismatch in the distance geometry refined structure of the G·T 12-mer duplex. (A) Wobble G·T pairing involving two imino-carbonyl hydrogen bonds. (B) Propeller-twisting of the Wobble G·T mismatch pair. All protons are labelled by darkened balls.

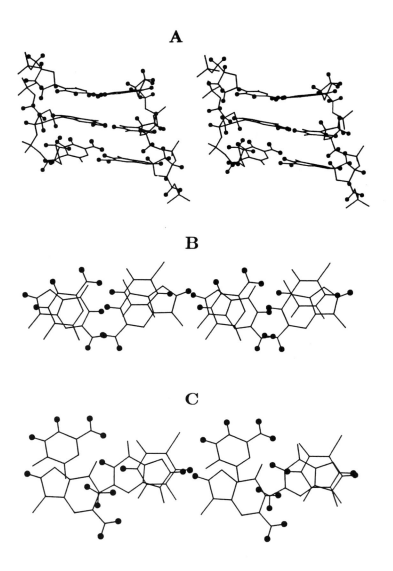

Figure 8. Stereo pairs of (A) the (G2-T3-G4)·(C9-G10-C11) segment and base stacking between (B) G2·C11 and T3·G10 pairs and (C) T3·G10 and G4·C9 pairs in the distance geometry refined G·T 12-mer duplex.

12-mer duplex. The G·T mismatch (Scheme 5) is stabilized by two imino proton-carbonyl hydrogen bonds as first proposed by Crick (1966) and our results rule out alternate pairing modes involving rare tautomeric forms (Topal and Fresco, 1976) for the G·T mismatch embeded in oligonucleotide duplexes in solution. Earlier NMR chemical shift studies had suggested formation of Wobble G·T pairs for poly(dG-dT) in solution (Early, et.al., 1978) and one dimensional NOE studies had established Wobble G·U pairing in transfer RNA (Schimmel and Redfield, 1980). The Wobble G·T pair has also been observed in the crystalline state for deoxyoligonucleotides adopting A-DNA (Kneale, et.al, 1985; Hunter, et.al., 1986), B-DNA (Kennard, 1985) and Z-DNA (Ho, et.al., 1985).

The stacking in the (G2-T3-G4)·(C9-G10-C11) segment centered about the T3·G10 mismatch site in the G·T 12-mer duplex (Figure 8A) exhibits good intrastrand overlap in the (G2-T3)·(G10-C11) purine (3'-5') pyrimidine step (Figure 8B) and partial overlap in the (T3-G4)·(C9-G10) pyrimidine (3'-5) purine step (Figure 8C).

The phosphorus spectrum of the G·T 12-mer duplex exhibits one phosphorus resonance each shifted to low and high field of the normal phosphorus spectral dispersion (Patel, et.al., 1982b). These results suggest structural perturbations in the backbone conformation of two of the eleven internucleotide phosphodiester groups in the G·T 12-mer duplex.

Overall, the NMR-distance geometry studies demonstrate that the G·T mismatch is readily incorporated into the duplex and these data along with earlier hydrogen exchange measurements (Pardi, et.al., 1982) establish that formation of a G·T mismatch results in a localized confomational perturbation.

EXTRAHELICAL ADENOSINE

The conformation of DNA helices containing extrahelical bases is of considerable interest in mechanisms of frame shift mutagenesis (Streisinger, et.al., 1966; Drake, et.al., 1983). The extrahelical base has the potential of either stacking into the duplex or looping out into solution so that addition or deletion mutants can arise due to partial misalignment of strands during replication. The repair of this helix interruption may also be governed by the conformation and stability of the extrahelical base whose orientation would depend on the type of base and the sequence of the flanking base pairs.

These issues have been addressed to date by NMR studies on extrahelical base containing oligonucleotide duplexes in aqueous solution (Patel, et.al., 1982c; Morden, et.al., 1983; Woodson and Crothers, 1987). This review focuses on the conformation of extrahelical adenosines in self-complementary tridecanucleotide duplexes. The early NMR chemical shift studies demonstrated that the extrahelical adenosine stacks into the duplex (Patel, et.al., 1982c) and faster imino proton hydrogen exchange rates were observed both adjacent to and more distant from this stacked extrahelical site (Pardi, et.al., 1982). Crystals suitable for X-Ray diffraction studies to high resolution have been grown of this extra adenosine tridecanucleotide duplex (Saper, et.al., 1986). These structural and dynamics studies have been supplemented by calorimetric (Patel, et.al., 1982c) and temperature jump (Chu and Tinoco, 1983) measurements which have probed into the thermodynamics and kinetics of the helix-coil transition, respectively. We outline below the results of a combined NMR-distance geometry study on the self-complementary d(CGCAGAGCTCGCG) tridecanucleotide duplex (designated adenosine 13-mer, Scheme 6) which contains extra adenosines between the third and fourth base pairs in from either end of the duplex (Hare, et.al., 1986b).

The distance connectivities between imino protons (12.2 to 14.0 ppm) and the amino and base protons (7.4 to 8.6 ppm) can be monitored in the magnitude NOESY spectrum of the adenosine 13-mer duplex in 0.1M NaCl, H_2O, $5^{\circ}C$. The expanded contour plot of this region establishes Watson-Crick pairing at the A6·T9 (cross peak A, Figure 9A) and G2·C12, C3·G11, G7·C8 and G5·C10 (cross peaks B, C, D and E, respectively, Figure 9A) base pairs and, in addition, exhibits NOEs between the guanosine imino protons and adenosine H2 protons on flanking residues (cross peaks, F, G and H, Figure 9A). The NOEs from the imino proton of G5 in the G5·C10 base pair to flanking residues can be readily visualized by recording a one dimensional slice through this exchangeable resonance (Figure 10A). The imino proton of G5 exhibits an NOE to the hydrogen-bonded amino proton of C10 in the G5·C10 base pair, NOEs to the imino proton of T9 and H2 proton of A6 of the flanking A6·T9 base pair in one direction and the H2 proton of extrahelical A4 in the other direction (Figure 10A). Further, we do not detect an NOE between the imino protons of C3·G11 and G5·C10 base pairs. These results establish that the extrahelical A4 stacks into

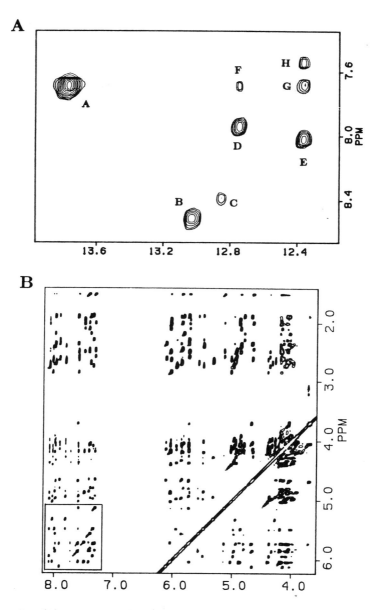

Figure 9. (A) An expanded contour plot of the magnitude NOESY
spectrum (120 msec mixing time) of the adenosine 13-mer duplex in 0.1M
NaCl, H_2O at $5^{\circ}C$ (B) An expanded contour plot of the phase sensitive
NOESY spectrum (250 msec mixing time) of the adenosine 13-mer duplex
in 0.1M NaCl, D_2O at $25^{\circ}C$.

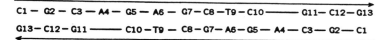

C1 — G2 — C3 — A4 — G5 — A6 — G7 — C8 — T9 — C10 ———— G11 — C12 — G13

G13 — C12 — G11 ———— C10 — T9 — C8 — G7 — A6 — G5 — A4 — C3 — G2 — C1

Scheme 6

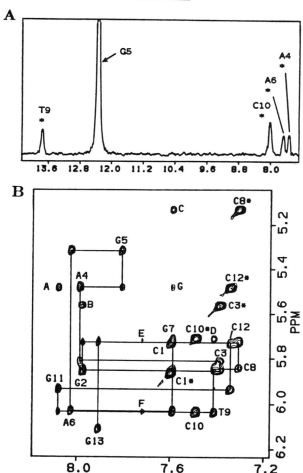

<u>Figure 10</u>. (A) One dimensional slice (7 to 14 ppm) establishing distance connectivities in the (A4-G5-A6)·(T9-C10) segment in the NOESY spectrum (see Figure 9A) of the adenosine 13-mer duplex in 0.1M NaCl, H_2O at 5°C. (B) An expanded contour plot of the NOESY spectrum (see Figure 9B) of the adenosine 13-mer duplex in 0.1M NaCl, D_2O at 25°C establishing distance connectivities between base and sugar H1' protons.

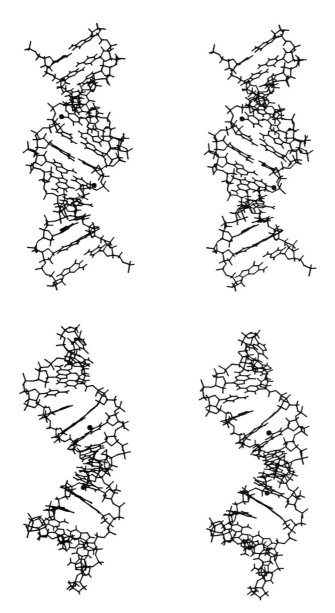

<u>Figure 11</u>. Stereo pairs of the distance geometry refined structure of
the adenosine 13-mer duplex. The top and bottom views are related by
a 90° rotation along the helix axis.

A

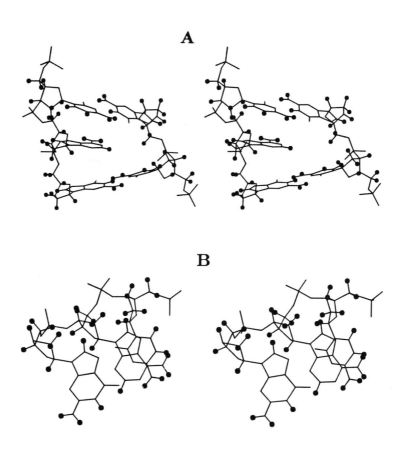

B

Figure 12. Stereo pairs of (A) the (C3-A4-G5)·(C10-G11) segment and (B) base stacking in the C3-A4-G5 segment of the distance geometry refined structure of the adenosine 13-mer duplex.

the duplex and is a spacer between flanking G·C base pairs in the
adenosine 13-mer duplex.

The phase sensitive NOESY spectrum of the adenosine 13-mer
duplex in 0.1M NaCl, D_2O at 25°C exhibits unusually well resolved
cross peaks as shown in an expanded contour plot spanning the 3.5 to
8.5 ppm and 1.5 to 6.5 ppm regions (Figure 9B). These cross peaks can
be assigned from the directionality of the NOEs between base and sugar
protons on the same and adjacent nucleotide units along the sequence.
The distance connectivities can be readily followed from residue C1 to
G13 in the expanded region relating the base protons with the sugar
H1' protons in the adenosine 13-mer duplex (Figure 10B). Most
importantly, the NOE connectivities can be monitored through the
C3-A4-G5 segment, which includes the extrahelical adenosine that lacks
a base on the partner strand. These results establish that A4 stacks
into the duplex and that the extrahelical adenosine and the rest of
the helix is right-handed in the adenosine 13-mer duplex in solution.

The distance geometry refinements were undertaken on an
experimental data set containing 216 interproton distance constraints
for the entire adenosine 13-mer duplex. Stereo pairs of two views of
one such refinement of the adenosine 13-mer duplex are shown in Figure
11 (Hare, et.al., 1986b). The extrahelical adenosine stacks into the
duplex between flanking G·C pairs with the helix axis kinked at the
extrahelical site (Figures 11 and 12A). The bases C3 and A4 are
partially stacked but no stacking is detected between A4 and G5 in the
C3-A4-G5 segment of the adenosine 13-mer duplex (Figure 12B).

A single phosphorus resonance is shifted downfield of the normal
spectral dispersion in the adenosine 13-mer duplex (Patel, et.al.,
1982c) indicative of a altered backbone at one phosphodiester linkage
in the vicinity of the stacked extrahelical residue.

It should be noted that the conformation of the extrahelical
adenosine is independent of flanking sequence since additional NMR
studies have demonstrated that the extrahelical adenosine stacks into
the duplex either when it is flanked by purines or when it is flanked
by pyrimidines (Kalnik, et.al., 1987).

HOMOPURINE·HOMOPYRIMIDINE TRACTS

There is ample evidence from physical and biological experiments
that homopurine·homopyrimidine stretches in DNA exhibit unusual
structural features (Wells, et.al., 1977; Arnott, et.al., 1983b).

Further, it has been demonstrated that the properties of one homopurine·homopyrimidine stretch can be influenced by another such stretch some distance away (Burd, et.al., 1975). Of equal interest are the conformation of DNA at junctions of (purine)$_n$ and (pyrimidine)$_n$ tracts on the same strand of the duplex which results in either a purine-pyrimidine step or a pyrimidine-purine step depending on the sequence.

A heteronomous DNA model in which partner strands adopt A and B helical DNA conformations has been put forward for poly(deoyxpurine)·poly(deoxypyrimidine) in the fiber state (Arnott, et.al., 1983a). The crystal structure of d(GGGGCCCC) demonstrates stacking of adjacent guanosines but not adjacent cytidines in the octanucleotide duplex and these features can be used to generate a model of poly(dG)·poly(dC) (McCall, et.al., 1985). A small structural discontinuity was detected at the central G-C step in the crystal structure of the d(GGGGCCCC) duplex.

Single-strand nuclease hypersensitivity has been detected within (purine)$_n$·(pyrimidine)$_n$ stretches when embedded in superhelical closed circular DNA (Schon, et.al., 1983; Nickol and Felsenfeld, 1983). The structural basis for this conformational change remains to be demonstrated (Pulleyblank, et.al., 1985; Evans and Efstratiadis, 1986) but it is faciliated by low pH (Lyamichev, et.al., 1986).

Recent studies on the anomalous gel mobility of kinetoplast DNA (Marini, et.al., 1982) have defined a bending locus with (CA$_5$T) tracts symmetrically distributed about it at a repeat interval of 10 base pairs (Wu and Crothers, 1984). Additional studies have emphasized the importance for bending of (A)$_n$·(T)$_n$ stretches in phase with the helical repeat (Hagerman, 1986; Koo, et.al., 1986; Ulanovsky, et.al., 1986). The bent DNA can be visualized by electron microscopy (Griffith, et.al., 1986) and its structural features probed by the unusual footprinting patterns detected for cleavage by hydroxyl radicals in (A)$_n$·(T)$_n$ stretches (Burkhoff and Tullius, 1987).

The question of unusual DNA conformations within and at the border of homopurine·homopyrimidine stretches can be approached by NMR studies on dodecanucleotide duplexes containing these segments. Our strategy was to study related sequences containing a specific difference at a single central site in the helix. Thus, NMR distance geometry studies have been undertaken on the self-complementary d(GGAAAGCTTTCC) duplex (Pur$_6$-Pyr$_6$ 12-mer, Scheme 7) which contains a

G1—G2—A3—A4—A5—G6—C7—T8—T9—T10—C11—C12

C12—C11—T10—T9—T8—C7—G6—A5—A4—A3—G2—G1

Scheme 7

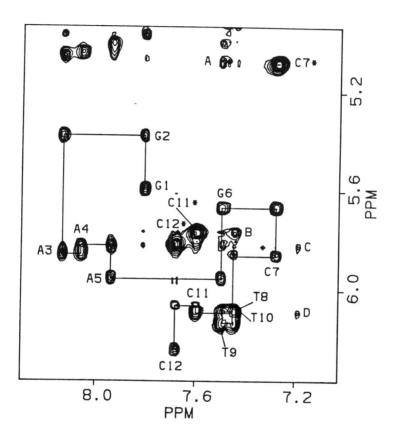

Figure 13. An expanded contour plot of the phase sensitive NOESY spectrum (250 msec mixing time) of the Pur_6-Pyr_6 12-mer duplex in 0.1M NaCl, D_2O, $25^\circ C$ establishing distance connectivities between base and sugar H1' protons.

139

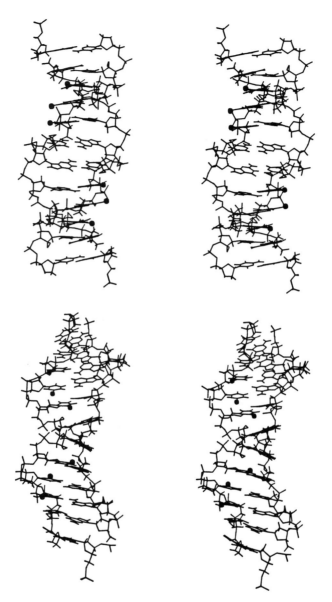

Figure 14. Stereo pairs of the distance geometry refined structure of the Pur_6-Pyr_6 12-mer duplex. The top and bottom views are related by a $90°$ rotation along the helix axis.

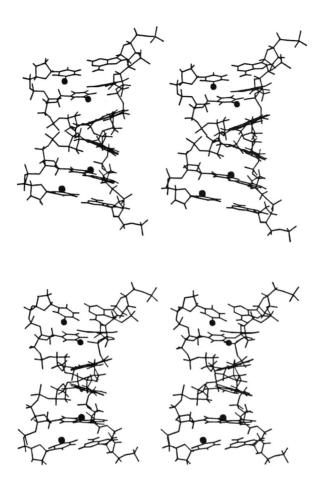

Figure 15. Stereo pairs of the central hexanucleotide segment of the Pur_6-Pyr_6 12-mer duplex as determined in separate distance geometry refinements.

central G-C step and the related self-complementary d(CCTTTCGAAAGG) duplex (Pur_6-Pur_6 12-mer, Scheme 8) which contains a central C-G step (Hare, et.al., 1987a). Such a comparison should permit an estimation of the conformational characteristics of $(purine)_n \cdot (pyrimidine)_n$ stretches and also the conformation at purine(3'-5')pyrimidine and pyrimidine(3'-5')purine junctions of such tracts.

The protons in the Pur_6-Pyr_6 12-mer duplex (Scheme 7) have been assigned following an analysis of the NOESY spectra in H_2O and D_2O solution. There was some question relating to spectral resolution at the A3-A4-A5 and T8-T9-T10 stretches in this dodecanucleotide duplex. The expanded region establishing distance connectivities between base and sugar H1' protons in the NOESY spectrum of the Pur_6-Pyr_6 12-mer duplex in D_2O is plotted in Figure 13. The H8 protons of purines A3, A4 and A5 are well resolved but the H6 protons of T8 and T10 overlap resulting in superpositioned cross peaks in the d(GGAAAGCTTTCC) duplex. Another feature of interest is the weak cross peak (designated A, Figure 13) between the H8 of G6 and the H5 of C7 in the central G6-C7 step indicative of these major groove protons being further apart in the Pur_6-Pyr_6 12-mer duplex compared to the same step in regular helices.

The resolved and assigned NOE cross peaks yielded a total of 222 experimental distances which served as input parameters for distance geometry refinements. Two views of one refined structure of the Pur_6-Pyr_6 12-mer is plotted in Figure 14 (Hare, et.al., 1987a). There is a striking discontinuity at the central G6·C7 step (Figure 14, lower view) where adjacent G·C base pairs in the center are overwound and exhibit a large roll angle opening into the major groove. This feature is also observed in all the refinements as shown in a comparison of the central AAGCTT hexanucleotide segments of two different refinements of the Pur_6-Pyr_6 12-mer in Figure 15. The separation between the H8 of G6 and the H5 of C7 varies between 5.3 and 5.9A in the various distance geometry refinements compared to a separation of 3.9A in B-DNA. There is a pronounced kink at the central G-C step and a shallow major groove in the center of the duplex relative to the same segment in classical B form DNA. The kink detected at the central G-C step in the Pur_6-Pyr_6 12-mer duplex appears to be rapidly damped out on moving away from this central step (Figure 14) so that the perturbation is localized and does not appear to propagate along the dodecanucleotide duplex.

142

C1— C2—T3—T4—T5—C6— G7— A8— A9— A10—G11—G12
G12-G11-A10-A9—A8—G7—C6—T5—T4—T3—C2—C1

Scheme 8

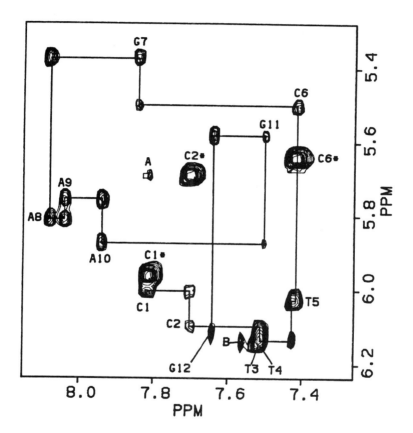

Figure 16. An expanded contour plot of the phase sensitive NOESY spectrum (250 msec mixing time) of the Pyr$_6$-Pur$_6$ 12-mer duplex in 0.1M NaCl, D$_2$0, 25°C establishing distance connectivities between base and sugar H1' protons.

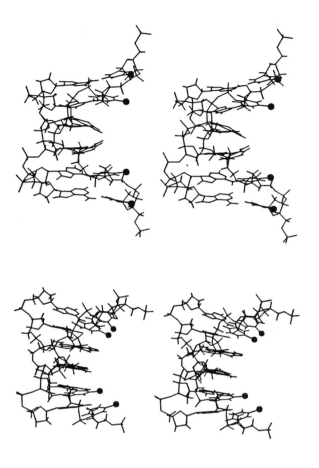

Figure 17. Stere pairs of the central hexanucleotide segment of the Pyr$_6$-Pur$_6$ 12-mer duplex as determined in separate distance geometry refinements.

The above result takes on additional significance in light of recent studies that have detected S1 nuclease sensitive sites at the junction of purine stretches and a TATA box of the general sequence -(Pur)$_n$-T-A-T-A- in the adenovirus genome (Kilpatrick, et.al., 1986). These workers suggested that the DNA deformation reflects the homopurine·homopyrimidine nature of this region and fits in well with our studies since the S1 hypersensitive sites include the purine-pyrimidine step in the (Pur)$_n$-T-A-T-A segment.

A parallel NMR-distance geometry study was undertaken on the Pyr$_6$-Pur$_6$ 12-mer duplex (Scheme 8) with the quality of the data sets evaluated from the expanded region establishing distance connectivities between the base and sugar H1' protons for this dodecanucleotide helix (Figure 16). The H8 protons of A8, A9 and A10 are well resolved but the H6 protons of T3 and T4 are superpositioned in the d(CCTTTCGAAAGG) duplex. Overall, the resolution of the cross peaks in the pyrimidine rich segment were poor resulting in fewer interproton distances estimated for this portion of the Pyr$_6$-Pur$_6$ 12-mer duplex.

The distance geometry refinements were undertaken on 184 inter proton distance constraints with the different refined structures exhibiting significant variability for the Pyr$_6$-Pur$_6$ 12-mer duplex (Hare, et.al., 1987a). This reflects an insufficient number of distance constraints in the T-T-T segment of the Pyr$_6$-Pur$_6$ 12-mer duplex. Two distance geometry refinements of the central TTCGAA hexanucleotide segment of the Pyr$_6$-Pur$_6$ 12-mer duplex are plotted in Figure 17. Despite the differences between these two refinements, one notes the absence of a kink at the central C-G step and that the major groove was not shallow in the center of the Pyr$_6$-Pur$_6$ 12-mer duplex (Figure 17) as was observed for the center of the Pur$_6$-Pyr$_6$ 12-mer duplex (Figure 15).

Overall, these results on the homopurine·homopyrimidine oligonucleotides demonstrate the formation of a structural perturbation at the central G-C step in the Pur$_6$-Pyr$_6$ 12-mer duplex which is absent at the corresponding central C-G step in the Pyr$_6$-Pur$_6$ 12-mer duplex.

ALTERNATING PURINE-PYRIMIDINE TRACTS:

Structural transitions for alternating (dC-dG)$_n$ sequences as a function of a ionic strength (Pohl and Jovin, 1972) have been assigned to conformational interconversions between right- and left-handed DNAs in solution (reviewed by Rich, et.al., 1984; Wells, et.al., 1983). Research in this area was stimulated by the discovery of the Z-DNA structure (Wang, et.al., 1979), the dinucleotide repeat features of which were anticipated by earlier NMR studies in solution (Patel, et.al., 1979).

NMR studies have been reported for $(dC-dG)_n$ oligonucleotides in both low salt (Patel, 1976; Cheng, et.al., 1984; Orbons, 1987) and high salt (Patel, et.al., 1972;1982d; Feigon, et.al., 1984; Giessner-Prettre, et.al., 1984; Orbons, 1987), characteristic of right-handed mononucleotide repeat and left-handed dinucleotide repeat DNA helices respectively. These studies have been extended to $(dC-dG)_n$ oligonucleotides containing A·T base pairs which retain or break pyrimidine-purine alternation (Feigon, et.al., 1985). The trinucleotide (GTG)·(CAC) step is a common feature of protein recognition sites on prokaryotic and eucaryotic DNA (Donlan, et.al.. 1986). It has been demonstrated that the thymidine imino protons in such G-T-G steps exhibit fast hydrogen exchange rates (Lu, et.al., 1983) probably reflecting conformational differences at such steps in control regions of DNA.

A systematic NMR-distance geometry study has been initiated to investigate $(dC-dG)_n$ duplexes containing either G-T-G or G-A-G steps in the interior of the helix. These studies have focused on the self-complementary d(CGCACGCGCGTGCG) duplex (GTG 14-mer, Scheme 9) which contains (GTG)·(CAC) steps and the self-complementary d(CGCTCGCGCGAGCG) duplex (GAG 14-mer, Scheme 10) which contains (GAG)·(CTC) steps (Hare, et.al., 1987b). The cross peaks detected in the NOESY spectra of the GTG 14-mer and GAG 14-mer duplexes in 0.1M NaCl, H_2O 5°C establish that all the G·C and A·T base pairs form Watson-Crick hydrogen bonds for both duplexes in low salt solution.

The NOESY spectrum of the GTG 14-mer duplex (Scheme 9) in 0.1M NaCl, D_2O, 25°C is unusually well resolved for an alternating purine-pyrimidine sequence and this is highlighted in the expanded region establishing distance connectivities between base and sugar H1' protons (Figure 18),. The chain can be readily traced from C1 to G14 with assignment of the connectivities aided by the observation of resolved guanosine H8 (3'-5') pyrimidine H6 cross peaks (A to E, Figure 18) for the five G-C steps in the GTG 14-mer duplex.

146

C1−G2−C3−A4−C5−G6−C7−G8−C9−G10-T11−G12−C13−G14

C14-C13-G12-T11-G10-C9−G8−C7−G6−C5−A4−C3−G2−C1

Scheme 9

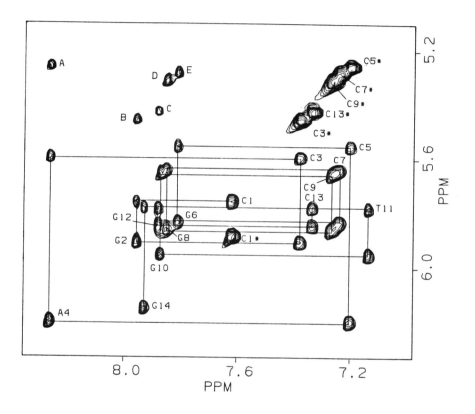

Figure 18. An expanded contour plot of the phase sensitive NOESY spectrum (250 msec mixing time) of the GTG 14-mer duplex in 0.1M NaCl, D_2O, 25°C establishing distance connectivities between base and sugar H1' protons.

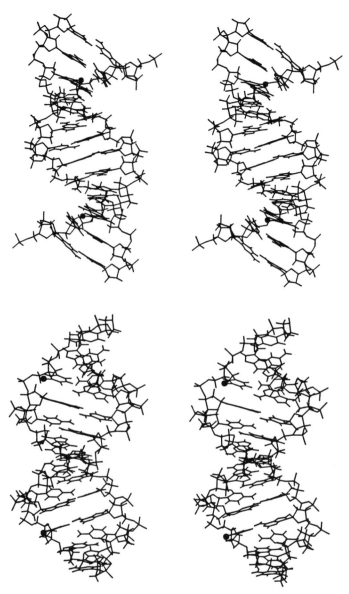

Figure 19. Stereo pairs of the distance geometry refined structure of the twelve non-terminal base pairs of the GTG 14-mer duplex. The top and bottom views are related by a 90° rotation along the helix axis.

A

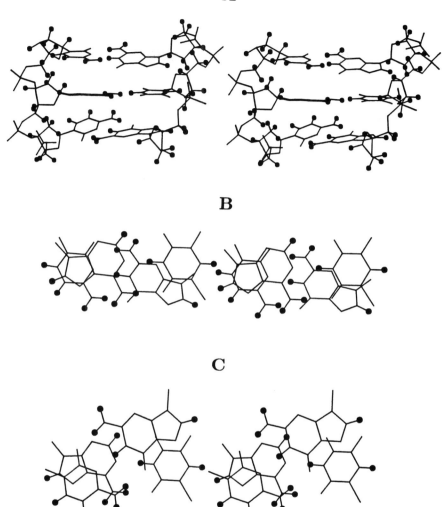

B

C

<u>Figure 20</u>. Stereo pairs of (A) the (C3-A4-C5)·(G10-T11-G12) segment
and base stacking between (B) C3·G12 and A4·T11 pairs and (C) A4·T11
and C5·G10 pairs in the distance geometry refined structure of the GTG
14-mer duplex.

149

The distance geometry refinement was undertaken on a set of 188 experimental distance constraints for the twelve non-terminal base pair segment of the GTG 14-mer duplex. Two views of one refined structure are plotted in Figure 19 (Hare, et.al., 1987b) and exhibit distinct differences from the 10 base pair repeat B-DNA (Figure 1, top) and 11 base pair repeat A-DNA (Figure 1, bottom) structures of the same sequence. It should be noted that the base pairs are tilted and displaced from the helix axis in the GTG 14-mer duplex (Figure 19). A close-up view of the (GTG)·(CAC) segment of the GTG 14-mer duplex is plotted in Figure 20A. The bases are stacked in the C-A-C segment but no stacking is observed between bases in the G-T-G segment of the GTG 14-mer duplex (Figures 20B and 20C).

The NOESY spectrum of the GAG 14-mer duplex (Scheme 10) in 0.1M NaCl, D_2O, $25^{\circ}C$ exhibits well resolved cross peaks as shown in the expanded base to sugar H1' region plotted in Figure 21. The cross peaks between base and sugar H1' protons can be readily assigned as can those between adjacent base protons in the four G-C steps (cross peaks A to D, Figure 21) in the GAG 14-mer duplex.

The distance geometry refinements were undertaken on the 184 inter proton distance constraints available for the entire GAG 14-mer duplex. Two views of one refined structure are plottted in Figure 22 (Hare, et.al., 1987b) with the GAG 14-mer duplex also exhibiting displacement and tilting of the base pairs relative to the helix axis. A close-up view of the (GAG)·(CTC) segment of the GAG 14-mer duplex is plotted in Figure 23A. The bases are partially stacked in the G10-A11-G12 segment (Figures 23B and 23C) with the most pronounced distortion reflected in a large roll angle opening into the minor groove for the (C3-T4)·(A11-G12) step (Figures 23A and 23C).

Recent analysis of X-Ray diffraction patterns of DNA fibers have demonstrated conformational polymorphism within the B-DNA and A-DNA classes of structures (Arnott, et.al., 1983b). Thus, helical parameters like rise per residue, twist between adjacent base pairs, base tilt and displacement from the helix axis vary significantly within each family. Despite this polymorphism, the conformation of the NMR-distance geometry structures of the GTG 14-mer (Figure 19) and the GAG 14-mer (Figure 22) duplexes in solution were unexpected and need additional substantiation. Unfortunately, single X-Ray studies on related sequences demonstrate formation of left-handed Z-DNA helices (reviewed in Rich, et.al., 1985) rather than the right-handed

C1—G2—C3—T4—C5—G6—C7—G8—C9—G10-A11—G12—C13—G14
C14-C13-G12-A11-G10-C9—G8—C7—G6—C5—T4—C3—G2—C1

Scheme 10

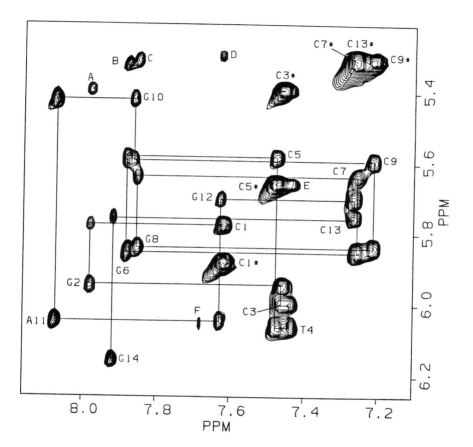

Figure 21. An expanded contour plot of the phase sensitive NOESY spectrum (250 msec mixing time) of the GAG 14-mer duplex in 0.1M NaCl, D_2O, 25°C establishing distance connectivities between base and sugar H1' protons.

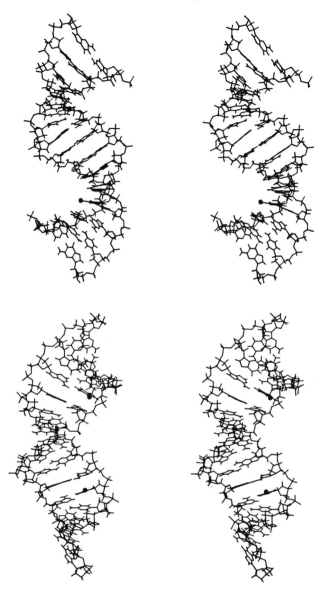

<u>Figure 22</u>. Stereo pairs of the distance geometry refined structure of the GAG 14-mer duplex. The top and bottom views are related by a 90° rotation along the helix axis.

152

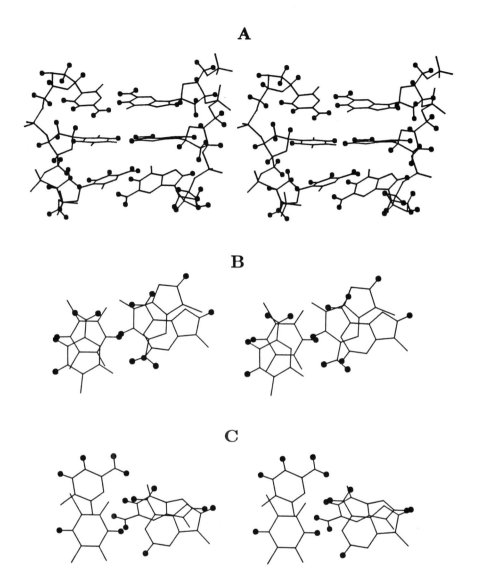

Figure 23. Stereo pairs of (A) the (C3-T4-C5)·(G10-A11-G12) segment and base stacking between (B) C3·G12 and T4·A11 pairs and (C) T4·A11 and C5·G10 pairs in the distance geometry refined structure of the GAG 14-mer duplex.

conformations observed in low salt solution. However, considerable
bending and twisting of the helix was detected during molecular
dynamics simulations of DNA (Levitt, 1983) and the GTG 14-mer and GAG
14-mer structures (Figure 19 and 22, respectively) bear a resemblance
to those conformations computed along the molecular dynamics
trajectory.

CURRENT LIMITATIONS AND FUTURE PROSPECTS

The conformation of oligonucleotide duplexes based on the
combined NMR-distance geometry approach outlined in this review should
be viewed as low resolution structures. The distance data sets are
approximate due to the two spin approximation used for their
estimation and incomplete due to partial spectral overlap of cross
peaks in certain regions of the NOESY spectra. Further, the NOE based
distance ruler is short range (<5A) and there is potential for
distance errors to propagate over longer distances resulting in a loss
of resolution. The distance geometry refined structures have not been
subjected at this stage to energy minimization procedures.

The simple two spin approximation used to estimate inter proton
distances from the volume integrals of the observed NOE cross peaks in
the two dimensional NOESY spectra is only correct at short mixing
times in the limit that Δ_t approaches zero (Olejniczak, et.al.,
1986). This precludes measurement of longer distances approaching 5A
associated with the weaker NOE cross peaks and limits the total number
of distance constraints used as input parameters in the structure
reconstruction algorithms. More recent research has taken multispin
effects including spin diffusion into account (Keepers and James,
1984; Olejniczak, et.al., 1986) and this has the potential of
providing additional distance constraints.

It should also be possible to estimate backbone torsion angles
through measurement of three bond proton-proton, proton-carbon and
proton-phosphorus coupling constants in the sugar phosphate backbone
(Karplus, 1959). Progress in this area is beginning to emerge
(Chazin, et.al., 1986; Rinkel, 1987) and should provide additional
constraints for the reconstruction algorithms.

Research has also been underway to calculate the intensity of
the cross peaks in time dependent NOESY data sets based on a multispin
relaxation analysis for a given conformation. This approach has been
applied to compute the time dependent NOESY cross peaks for several

canonical models of a $(dA-dT)_5$ decanucleotide duplex and the results compared with the experimental NOESY spectra (Suzuki, et.al., 1986). It was suggested that the computed structure based on right-handed wrinkled D-DNA was in closest agreement with the experimental data.

The accuracy of distance geometry refinements could be substantially improved by refinement techniques that can iteratively minimize the difference between experimental NOESY spectra and NOESY spectra simulated (including multispin effects) from the structure being refined. A back calculation refinement technique would refine structures against actual data rather than their interpretation into distance bounds. Such programs are currently under development and will dramatically improve the resolution of the NMR-distance geometry refined structures of oligonucleotide duplexes in solution.

It would be of interest to extend the above results to longer DNA duplexes but the problem of spectral overlap and loss of resolution associated with the larger number of protons and increased line widths must first be overcome. This should be possible through the introduction of multiple quantum (Muller, et.al., 1986) and spectral editing techniques (Griffey, et.al., 1985; Otting, et.al., 1986) which permit the spectroscopist to focus on the interactions of interest. It is now possible to introduce ^{15}N (Gao and Jones, 1987) and ^{13}C labels through direct synthesis into oligonucleotide duplexes and heteronuclear two dimensional editing techniques should permit the selective observation of protons attached to ^{13}C and ^{15}N in labelled nucleic acids.

There are several challenging nucleic acid structural problems facing the NMR spectroscopist in the future. These include the conformation of nuclease hypersensitive sites in $(deoxypurine)_n \cdot (deoxypyrimidine)_n$ tracts, bending sequences located at $(dA)_n \cdot (dT)_n$ tracts repeated in phase with the helical repeat and junctions between right-handed and left-handed duplexes. Higher order structures such as helical knots, splice sites and three- and four-stranded immobile and semimobile junctions are also of interest. Structural studies on these systems can be complemented by NMR hydrogen exchange studies aimed at monitoring the dynamics of transient helix opening at the individual base pair level.

REFERENCES

Aboul-ela, F., D. Koh., I. Tinoco, Jr., and F.H. Martin (1985). Base-base mismatches: Thermodynamics of double helix formation for d(CA$_3$XA$_3$G)·d(CT$_3$YT$_3$G) where X, Y = A, C, G, T. Nucleic Acids Research. 13: 4811-4824.

Arnott, S., R. Chandrasekaran, I.H. Hall and L.C. Puigjaner (1983a). Heteronomous DNA. Nucleic Acids Research. 11: 4141-4155.

Arnott, S., R. Chandrasekaran, I.H. Hall, L.C. Puigjaner, J.K. Walker and M. Wang (1983b). DNA Secondary Structures: Helices, Wrinkles and Junctions. Cold Spring Harbor Symp. Quart. Biol. 47: 53-65.

Arnott, S. and D.W. Hukins (1972). Optimized parameters for A-DNA and B-DNA. Biochem. Biophys. Res. Commun. 47: 1504-1509.

Barton, J.K. (1986). Metals and DNA: Molecular left-handed complements. Science. 233: 727-734.

Bax, A. and L. Lerner (1986). Two-dimensional NMR spectroscopy. Science. 232: 960-967.

Brooks, B.R., R.E. Bruccoleri, R.E. Olafson, D.J. States, S. Swaminathan and M. Karplus (1983). CHARMM. A program for macromolecular energy, minimization and dynamics calculations. J. Comput. Chem. 4: 187-217.

Burd, J.F., J.E. Larson and R.D. Wells (1975). Further studies on telestability of DNA. J. Biol. Chem. 250: 6002-6007.

Burkhoff, A.M. and T.D. Tullius (1987). The unusual conformation adopted by the adenine tracts in kinetoplast DNA. Cell. 48: 935-943.

Calladine, C.R. (1982). Mechanics of sequence-dependent stacking of bases in B-DNA. J. Mol. Biol. 161: 343-352.

Chazin, W.J., K. Wuthrich, S. Hyberts, M. Rance, W.A. Denny and W. Leupin (1986). Proton NMR assignments for d(GCATTAATGC) using experimental refinements of established procedures. J. Mol. Biol. 190: 439-453.

Cheng, D.M., L.S. Kan, D. Frechet, P.O.P. Ts'o, S. Uesuji, T. Shida and M. Ikehara (1984). Proton and phosphorus NMR studies on the conformation of d(CGCG)$_2$ and d(CGCGCG)$_2$ short helices in the B-conformation. Biopolymers. 23: 775-795.

Chu, Y. G. and I. Tinoco, Jr. (1985). Temperature jump kinetics of a double helix containing a G·T base pair and a double helix containing an extra adenine. Biopolymers. 22: 1235-1246.

Chuprina, V.P. and V.I. Poltev (1985). Alterations of the DNA double helix conformation upon incorporation of mispairs as revealed by energy computations and pathways of point mutations. Nucleic Acids Research. 13: 141-154.

Crick, F.H. (1986). Codon-anticodon pairing: The Wobble hypothesis. J. Mol. Biol. 19: 548-555.

Crick, F.H. and J.D. Watson (1954). The complementary structure of DNA. Proc. R. Soc. London. Ser A. 223: 80-81.

Crippen, G.M (1981). Distance geometry and conformation calculations. John Wiley, New York.

Dervan, P.B. (1986). Design of sequence-specific-DNA-binding molecules. Science. 232: 464-471.

Dickerson, R.E. and H.R. Drew (1981). Structure of a B-DNA dodecamer II. Influence of base sequence on helix structure. J. Mol. Biol. 149: 761-786.

Dickerson, R.E., H.R. Drew and B. Conner (1981). Single crystal X-Ray analysis of the A-, B- and Z-helices. In R.H. Sarma (Ed.) Biomolecular Stereodynamics I. Adenine Press, New York, 1-34.

Donlan, M., S. Cheung and P. Lu (1986). NMR studies of an SV40 enhancer core DNA sequence. Biophys. J. 49: 26-29.

Drake, J.W., B.W. Glickman and L.S. Ripley (1983). Updating the theory of mutation. American Scientist. 71: 621-630.

Drew, H.R. (1984). Structural specificities of five commonly used DNA nucleases. J. Mol. Biol. 176: 535-557.

Early, T.A., J. Olmsted, D.R. Kearns and A.G. Lezius (1978). Base pairing structure in poly(dG-dT) double helix: Wobble base pairs: Nucleic Acids Research. 5: 1955-1970.

Evans, T. and A. Efstratiadis (1986). Sequence-dependent S1 nuclease-hypersensitivity of a heteronomous and bent DNA duplex. J. Biol. Chem. 261: 14771-14780.

Feigon, J., A.H. Wang, G.A. van der Marel, J.H. van Boom and A. Rich (1984). A one- and two-dimensional NMR study of the B to Z transition in $(m^5dC-dG)_3$ in methanolic solution. Nucleic Acids Research. 12: 1243-1263.

Feigon, J., A.H. Wang, G. van der Marel, J.H. van Boom and A. Rich (1985). Z-DNA forms without an alternating purine-pyrimidine sequence in solution. Science. 230: 82-84.

Fuller, W., M.H. Wilkins, H.R. Wilson, L.D. Hamilton and S. Arnott (1965). The molecular configuration of DNA. IV. X-Ray diffraction study of the A-form. J. Mol. Biol. 12: 60-80.

Gao, X. and R.A. Jones (1987). Nitrogen-15-labelled Oligodeoxyoligonucleotides. Characterization by [15]NMR of d(CGTACG) containing [15]N6- or [15]N-1 labelled deoxyadenosine. J. Am. Chem. Soc. 109: 3169-3171.

Giessner-Prettre, C., B. Pullman, S. Tran-Dinh, J.M. Neumann, T. Huynh-Dinh and J. Igolen (1984). Proton NMR study of the B to Z transition of $d(CGm^5CG)_2$ and $d(CGm^5CGCG)_2$: Theory and experiment. Nucleic Acids Research. 12: 3271-3281.

Griffey, R.H., A.G. Redfield, R.E. Loomis, and F.W. Dahlquist. (1985). NMR observation and dynamics of specific amide protons in T4 lysozyme. Biochemistry. 24: 817-822.

Griffith, J., M. Bleyman, C.A. Rauch, P.A. Kitchin and P.I. Englund (1986). Visualization of the bent helix in kinetoplast DNA by electron microscopy. Cell. 46: 717-724.

Hagerman, P.J. (1986). Sequence-directed curvature of DNA. Nature. 321: 449-450.

Hare, D.R., and B.R. Reid (1986). Three-dimensional structure of a DNA hairpin in solution: Two dimensional NMR and distance geometry calculations on d(CGCGTTTTCGCG). Biochemistry. 25: 5341-5350.

Hare, D., L. Shapiro and D.J. Patel (1986a). Wobble dG·dT pairing in right-handed DNA: solution conformation of the d(CGTGAATTCGCG) duplex deduced from distance geometry analysis of nuclear Overhauser effect spectra. Biochemistry. 25: 7445-7456.

Hare, D., L. Shapiro and D.J. Patel (1986b). Extrahelical adenosine stacks into the d(CGCAGAGCTCGCG) duplex deduced from distance geometry analysis of nuclear Overhauser effect spectra. Biochemistry. 25: 7456-7464.

Hare, D.R., L. Shapiro, M. Zagorski and D.J. Patel (1987a). Solution conformation of homopyrimidine·homopurine dodecanucleotides containing central C-G and G-C steps: Distance geometry analysis of NOESY spectra. to be submitted.

Hare, D., L. Shapiro, M. Zagorski and D.J. Patel (1987b). Solution conformation of fully- and partially- alternating pyrimidine-purine tetradecanucleotide duplexes: Distance geometry analysis of NOESY spectra. to be submitted.

Hare, D.R., D.E. Wemmer, S.H. Chou, G. Drobny and B.R. Reid (1983). Assignment of nonexchangeable proton resonances of d(CGCGAATTCGCG) using two dimensional NMR methods. J. Mol. Biol. 171: 319-336.

Ho, P.S., C.A. Frederick, G.J. Quigley, G.A. van der Marel, J.H. van Boom, A.H. Wang and A. Rich (1985). G·T Wobble base pairing in Z-DNA at 1.0A resolution: the crystal structure of d(CGCGTG). The EMBO J. 4: 3617-3623.

Hunter, W.N., G. Kneale, T. Brown, D. Rabinovich and O. Kennard (1986). Refined crystal structure of an octanucleotide duplex with G·T mismatch base pairs. J. Mol. Biol. 190: 605-618.

Kalnik, M., D. Norman, B. Li, P. Swann and D.J. Patel (1987). Extra helical thymidines in DNA helices. Structural studies as a function of flanking sequence. J. Biol. Chem. to be submitted.

Karplus, M. (1959). Contact electron-spin coupling of nuclear magnetic moments. J. Chem. Phys. 30: 11-15.

Kearns, D.R. (1984). NMR studies of conformational state and dynamics of DNA. CRC Reviews in Biochemistry. 15: 237-290.

Keepers, J.W. and T.L. James (1984). A theoretical study of distance determination from NMR. Two dimensional NOESY spectra. J. Magn. Reson. 57: 404-426.

Keepers, J.W., P. Schmidt, T.L. James and P.A. Kollman (1984). Molecular mechanics studies of mismatched base analogs and extrahelical adenosine analog. Biopolymers. 23: 2901-2929.

Kennard, O. (1985). Structural studies of DNA fragments. The G·T Wobble base pair in A, B, and Z-DNA; the G·A base in B-DNA. J. Biomol. Str. Dyn. 3: 205-226.

Kennard, O. (1987). Structure of Base Pair Mismatches in DNA. In F. Eckstein and D.M. Lilley (Eds). Nucleic Acids and Molecular Biology, Vol. 1, Springer-Verlag, Berlin.

Kilpatrick, M.W., A. Torri, D.S. Kang, J.A. Engler and R.D. Wells (1986). Unusual DNA structures in the Adenovirus genome. J. Biol. Chem. 261: 11350-11354.

Kneale, G., T. Brown, O. Kennard and D. Rabinovich (1985). G·T base pair in a DNA helix: the crystal structure of d(GGGGTCCC). J. Mol. Biol. 186: 805-814.

Koo, H.S., H. Wu and D.M. Crothers (1986). DNA binding at adenine·thymidine tracts. Nature. 320: 501-506.

Langridge, R., D.A. Marvin, W.E. Seeds, H.R. Wilson, C.W. Hooper, M.H. Wilkins and L.D. Hamilton (1960). The molecular configuration of DNA. II. Molecular models and their fourier transforms. J. Mol. Biol. 2: 38-60.

Levitt, M. (1983). Computer Simulation of DNA Dynamics. Cold Spring Harbor Symposium Quart. Biol. 47: 251-262.

Lomonossoff, G.P., P.J. Butler and A. Klug (1981). Sequence dependent variation in the conformation of DNA. J. Mol. Biol. 149: 745-760.

Lu, P.S. Cheung and K. Arndt (1983). Possible molecular detent in the DNA structure at regulatory sequences. J. Biomol. Str. and Dyn. 1: 509-521.

Lyamichev, V.I., S.M. Mirkin and M.D. Frank-Kamenstskii (1986). Structure of homopurine·homopyrimidine tract in superhelical DNA. J. Biomol. Str. and Dyn. 3: 667-669.

Marini, J.C., S.D. Levene, D.M. Crothers and P.T. Englund (1982). Bent helical structure in kinetoplast DNA. Proc. Natl. Acad. Scs. USA. 79: 7664-7668.

McCall, M., T. Brown and O. Kennard (1985). The crystal structure of d(GGGGCCCC). A model for poly(dG)·poly(dC). J. Mol. Biol. 183: 385-396.

Modrich, P. (1987). DNA mismatch correction. Ann. Revs. Biochem. in press.

Morden, K.M., Y.G. Chu, F.H. Martin and I. Tinoco, Jr. (1983). Unpaired cytosine in the d(CA$_3$CA$_3$G)·d(CT$_6$G) duplex is outside the helix. Biochemistry. 22: 5557-5563.

Muller, N., R. Ernst and K. Wuthrich (1986). Multiple-quantum-filtered two dimensional correlated NMR spectroscopy of proteins. J. Am. Chem. Soc. 108: 6482-6492.

Nickol, J.M. and G. Felsenfeld (1983). DNA conformation at the 5' end of the chicken adult β-globin gene. Cell. 35: 467-477.

Nilsson, L., G.M. Clore, A.M. Gronenborn, A.T. Brunger and M. Karplus (1986). Structure refinements by molecular dynamics with NOE interproton distance restraints: Application to d(CGTACG). J. Mol. Biol. 188: 455-475.

Olejniczak, E.T., R.T. Gampe, Jr. and S.W. Fesik (1986). Accounting for spin diffusion in the analysis of 2D NOE data. J. Magn. Reson. 67: 28-41.

Orbons, L.P. (1987). DNA Polymorphism in solution. NMR studies of B, Z and Hairpin forms of DNA oligonucleotides. Ph.D. Thesis. University of Leiden, The Netherlands.

Otting, G., H. Senn, G. Wagner and K. Wuthrich (1986). Editing of two dimensional proton spectra using X-half filters. Combined use of residue-selective ^{15}N labelling of proteins. J. Magn. Reson. 70: 500-505.

Pardi, A., K.M. Morden, D.J. Patel and I. Tinoco, Jr. (1982). Kinetics for exchange of the imino protons in the d(CGCGAATTCGCG) double helix and in two similar helices that contain a G·T base pair and an extra adenine. Biochemistry. 21: 6567-6574.

Patel, D.J. (1976). Proton and phosphorus NMR studies of d(CGCG) and d(CGCGCG) duplexes in solution. Helix-coil transition and complex formation with actinomycin D. Biopolymers. 15: 533-558.

Patel, D.J., L.L. Canuel and F.M. Pohl (1979). Alternating B-DNA conformation for the oligo(dG-dC) duplex in high salt solution. Proc. Natl. Acad. Scs. USA. 76: 2508-2511.

Patel, D.J., S.A. Kozlowski, L.A., Marky, J.A. Rice, C. Broka, J. Dallas, K. Itakura and K.J. Breslauer (1982b). Structure, dynamics and energetics of deoxyguanosine·thymidine Wobble base pair formation in the self-complementary d(CGTGAATTCGCG) duplex in solution. Biochemistry. 21: 437-444.

Patel, D.J., S.A. Kozlowski, L.A. Marky, J.A. Rice, C. Broka, K. Itakura and K.J. Breslauer (1982c). Extra adenosine stacks into the self-complementary d(CGCAGAATTCGCG) duplex in solution. Biochemistry. 21: 445-451.

Patel, D.J., S.A. Kozlowski, A. Nordheim and A. Rich (1982d). Right-handed and left-handed DNA conformation: Studies of B- and Z-DNA using proton nuclear Overhauser effect and phosphorus NMR. Proc. Natl. Acad. Scs. USA. 79: 1413-1417.

Patel, D.J., S.A. Kozlowski, S. Ikuta and K. Itakura (1984). Dynamics of DNA duplexes containing internal GT, GA, AC and TC pairs: Hydrogen exchange at and adjacent to the mismatch site. Fed. Proc. Fed. Am. Soc. Exp. Biol. 43: 2663-2670.

Patel, D.J., A. Pardi and K. Itakura (1982a). DNA Conformation, Dynamics and Interactions in Solution. Science 216: 581-590.

Patel, D.J., L. Shapiro and D. Hare (1987a). Nuclear magnetic resonance and distance geometry studies of DNA structures in solution. Ann. Revs. Biophys. Biophys. Chem. 16: 423-453.

Patel, D.J., L. Shapiro and D. Hare (1987b). Conformation of DNA
    mismatches in solution. In F. Eckstein and D.M. Lilley (Eds),
    Nucleic Acids and Molecular Biology, Vol 1, Springer-Verlag,
    Berlin.
Pohl, F.M. and T.M. Jovin (1972). Salt-induced cooperative
    conformation change of a synthetic DNA: equilibrium and kinetic
    studies with poly(dG-dC). J. Mol. Biol. 67: 375-396.
Pulleyblank, D.E., D.B. Haniford and A.R. Morgan (1985). A structural
    basis for S1 nuclease sensitivity of double-stranded DNA. Cell.
    42: 271-280.
Quignard, E., G.V. Fazakerley, G. van der Marel, J.H. van Boom,.
    and W. Guschlbauer (1987). Comparison of the conformation of an
    oligonucleotide containing a central G·T base pair with the
    non-mismatch sequence by proton NMR. Nucleic Acids Research. 15:
    3397-3409.
Radman, M. (1987). Mismatch repair in E. coli Ann. Revs. Genetics in
    press.
Rich, A., A. Nordheim and A.H. Wang (1984). The chemistry and biology
    of left-handed Z-DNA. Ann. Revs. Biochem. 53: 791-846.
Rinkel, L.J. (1987). NMR studies of DNA structures in solution:
    Influence of the bases on the conformational behavior. Ph.D.
    Thesis. University of Leiden, The Netherlands.
Saper, M.A., H. Eldar, K. Mizuchi, J. Nickol, E. Appella and J.L.
    Sussman (1986). Crystallization of a DNA tridecamer
    d(CGCAGAATTCGCG). J. Mol. Biol. 118: 111-113.
Scheek, R.M., R. Boelens, N. Russo, J.H. van Boom and R. Kaptein
    (1984). Sequential resonance assignments in proton NMR spectra
    of oligonucleotides by two dimensional NMR spectroscopy.
    Biochemistry. 23: 1371-1376.
Schimmel, P.R. and A.G. Redfield (1980). Transfer RNA in solution:
    selected Topics, Ann. Revs. Biophys. Bioeng. 9: 181-221.
Schon, E., T. Evans, J. Walsh and A. Efstratiadis (1983).
    Conformation of promotor DNA: Fine mapping of S1-hypersensitive
    sites. Cell. 35: 837-848.
Shakked, Z. and O. Kennard (1985). The A form of DNA. In Biological
    Molecules and Assemblies. Vol II. Nucleic Acids and Interactive
    Proteins (Eds. M. Jurnak and A. McPherson). John Wiley, New
    York, 1-36.
Streisinger, G., Y. Okada, J. Emrich, J. Newton, A. Tsugita, E.
    Terzaghi and M. Inouye (1966). Frame shift mutations and the
    genetic code. Cold Spring Harbor Symposium Quantitative Biology.
    31: 77-84.
Suzuki, E., N. Pattabiraman, G. Zon and T.L. James (1986). Solution
    structure of [d(A-T)$_5$] via complete relaxation matrix analysis
    of NOESY spectra and molecular mechanics calculations: Evidence
    for a hydration tunnel. Biochemistry. 25: 6854-6865.
Tibanyenda, N., S.H. Debruin, C.A. Haasnoot, G.A. van der Marel, J.H.
    van Boom and C.W. Hilbers (1984). The effect of single base pair
    mismatches on duplex stability. Eur. J. Biochem. 139: 19-27.
Topal, M.D., and J.R. Fresco (1976). Complementary base pairing and
    the origin of substitution mutations. Nature. 263: 285-293.
Ulanovsky, L., M. Bodner, E.N. Trifonov and M. Choder (1986). Curved
    DNA: Design, synthesis and circularization. Proc. Natl. Acad.
    Scs. USA. 83: 862-866.
Wang, A.H., G.J. Quigley, F.J. Kolpak, J.L. Crawford, J.H. van Boom,
    G. van der Marel and A. Rich (1979). Molecular structure of a
    left-handed double helical DNA fragment at atomic resolution.
    Nature. 282: 680-686.

Weiss, M.A., D.J. Patel, R.T. Sauer and M. Karplus (1984). Two
    dimensional proton NMR study of the operator site O_L1: A
    sequential assignment strategy and its application. Proc. Natl.
    Acad. Scs. (USA) 81: 130-134.
Wells, R.D., R.W. Blakesley, J.F. Burd, H.W. Chan, J.B. Dodgson, S.C.
    Hardies, G.T. Horn, K.F. Jensen, J. Larson, I.F. Nes, E. Selsing
    and R.M. Wartell (1977). The role of DNA structure in gene
    regulation. Crit. Revs. Biochem. 4: 305-340.
Wells, R.D., R. Brennan, K.A. Chapman, T.C. Goodman, P.A. Hart, W.
    Hillen, D.R. Kellog, M.W. Kilpatrick, R.D. Klein, J. Klysik,
    P.F. Lambert, J.E. Larson, J.J. Miglietta, S.K. Neuendorf, T.F.
    O'Connor, C.K. Singleton, S.T. Stirdivant, C.M. Veneziale, R.M.
    Wartell and W. Zacharias (1983). Left-handed DNA Helices,
    Supercoiling and the B-Z junction. Cold Spring Harbor Symp.
    Quant. Biol. 47: 77-84.
Wemmer, D.E. and B.R. Reid (1985). High resolution NMR studies of
    nucleic acids and proteins. Ann. Revs. Phys. Chem. 36: 105-137.
Woodson, S.A. and D.M. Crothers (1987). Proton NMR Studies on
    bulge-containing DNA oligonucleotides from a mutational hot-spot
    sequence. Biochemistry. 26: 904-912.
Wu, H. and D.M. Crothers (1984). The locus of sequence-directed and
    protein-induced DNA bending. Nature. 308: 509-513.
Wuthrich, K. (1986). NMR of Proteins and Nucleic Acids. John Wiley,
    New York.

ACKNOWLEDGEMENTS

The research was supported by the National Institutes of Health
Grants GM35620 (SBIR) to D.H. and GM34504 to D.P. The NMR
spectrometers were purchased from funds donated by the Robert Woods
Johnson Jr. Trust and the Matheson Trust towards setting up an NMR
Center in the Basic Medical Sciences at Columbia University.

Hyperreactivity of the B-Z Junctions Probed by Two Aromatic Chemical
Carcinogens, 2-N,N-Acetoxyacetylaminofluorene and 3-N,N-Acetoxyacetyl-
Amino-4,6-Dimethyldipyrido[1,2-a:3',2'-d] Imidazole.

Laurent Marrot, Annie Schwartz, Eric Hebert,
Germain Saint-Ruf and Marc Leng

Centre de Biophysique Moléculaire, C.N.R.S.,
1A, Avenue de la Recherche Scientifique
45071 Orléans cedex 2, France

Abstract

The reaction between DNAs in various conformations and two isosteric
chemical carcinogens 2-N,N-acetoxyacetylaminofluorene and 3-N,N-acetoxya-
cetylamino-4,6-dimethyldipyrido[1,2-a:3',2'-d] imidazole has been studied.
The modification of DNA has been analysed at the nucleotide level by means
of the 3'-5'exonuclease activity of T4 DNA polymerase. Both carcinogens
bind covalently to $(dC-dG)_{16}$ and $(dG-dT)_{15}$ sequences inserted in closed
circular plasmids when the inserts are in the B form ; they do not bind to
these inserts when they are in the Z form. The reactivity of guanine
residues at the B-Z junction depends upon the superhelical density of the
plasmids and upon the base sequence at the junction. A strong hyperreactivity is observed on the 3' side of the $(dC-dG)_{16}$ in pLP32, the insert
being in the Z form. It is concluded that the reactivity of guanine
residues with both carcinogens depends upon the DNA conformation. The non-
reactivity of Z-DNA and the hyperreactivity of some sequences under topological stress might have some importance in chemical carcinogenesis.

Introduction

It is known that ultimate carcinogens bind covalently to DNA and it is
generally thought that this chemical modification is a critical event in
mutation and in the initiation stages of tumorigenesis (Singer et al.).
Sequence-dependent effects of the conformation and dynamic properties are
important factors in determining the reactivities of chemical carcinogens
with nucleic acids. For example, the reactivity of the well-known carcinogen 2-N,N-acetoxyacetylaminofluorene (N-AcO-AAF) and B-DNA depends upon
the sequence of DNA (Fuchs). All the guanines are modified but the distribution is non-uniform, the probability of a guanine to be modified varying
between 1 and 50 in a relative scale while the related chemical carcinogen
N-hydroxyaminofluorene reacts equally well with all the guanine residues
in B-DNA (Bichara et al.). On the other hand, N-hydroxyaminofluorene does
not react with guanine residues in $d(CG)_n$ sequences in the Z conformation
(Rio et al. 1983, 1986).

Initially, the purpose of our work was to determine the importance of Z-
DNA versus B-DNA and consequently the importance of B-Z junctions in the

reaction of nucleic acids and N-AcO-AAF. It had been shown that synthetic polynucleotides such as poly(dG-dC).poly(dG-dC) (B form) and poly(dG-br^{5}dC).poly(dG-br^{5}dC) or poly(dG-m^{5}dC).poly(dG-m^{5}dC) (Z form) reacted with N-AcO-AAF (Spodheim-Maurizot et al., Santella et al. 1982, Rio et al. 1983). The question was to know whether guanines in other sequences able to adopt the Z form could also be modified by N-AcO-AAF. We have undertaken a study of (dC-dG)$_{16}$ and (dG-dT)$_{15}$ sequences inserted in closed circular DNAs. These sequences are in B or in Z form depending upon the supercoiling of the plasmids. Several studies have established that the free energy of plasmid negative supercoiling is one of the most important factors in inducing the B-Z transition (general review, Rich et al.). In addition, it has seemed to us interesting to compare the reactivity of N-AcO-AAF to that of 3-N,N-acetoxyacetylamino-4,6-dimethyldipyrido [1,2-a:3',2'-d] imidazole (N-AcO-AGlu-P-3), a supposed metabolite of the carcinogen 3-amino-4,6-dimethyldipyrido[1,2-a:3',2'-d] imidazole formed in the charred parts of cooked foods, in the pyrolysis of proteins and amino acids (Sugimura et al., Hashimoto et al., Commoner et al., Saint-Ruf et al.). N-AcO-AAF and N-AcO-AGlu-P-3 show significant structural similarities and can be considered as isosteric amines (figure 1). Both compounds react preferentially on the C(8) of guanine residues and the distortions induced by both adducts in nucleic acids present several similarities (Hebert et al., Loukakou et al., Daune et al. 1981, Grunberger et al.).

We here show that both carcinogens react with (dC-dG)$_{16}$ and (dG-dT)$_{15}$ inserts in the B form. They do not react with the inserts in the Z form although both carcinogens react with the linear synthetic polynucleotide poly(dG-br^{5}dC).poly(dG-br^{5}dC) in the Z form. Analysis at the nucleotide level shows that the reactivity of the junctions depends upon their sequence and upon the superhelical density of the plasmid. A striking result is the large hyperreactivity of the junction on the 3' side of the (dC-dG)$_{16}$ insert in native supercoiled pLP32.

Materials and Methods

Two plasmids have been used : pLP32 which is a pBR322 derivative containing a (dC-dG)$_{16}$ segment at the single BamH I site (a gift of Dr. J.C. Wang), pCM4 which is also a pBR322 derivative containing a 300 base pairs segment of a repeat DNA from Cebus appela DNA inserted at the Pst I site (a gift of Dr. B. Malfoy). The plasmids were relaxed or negatively supercoiled by treatment with topoisomerase I from chicken erythrocytes in the presence of ethidium bromide (Nordheim et al. 1982, Camerini-Otero et al.).

The nucleic acids were modified by N-AcO-AAF and by N-AcO-AGlu-P-3 as described. The amounts of covalently bound carcinogens (named AAF and AGlu-P-3, respectively) were determined by ultra-violet absorption (Fuchs et al., Loukakou et al.).

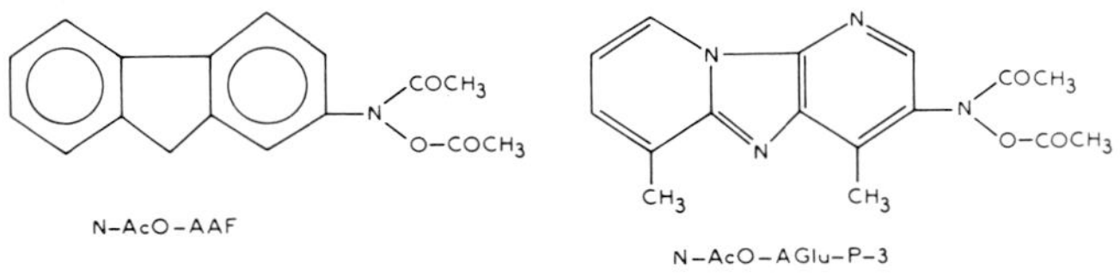

N-AcO-AAF

N-AcO-AGlu-P-3

Figure 1 - Formulae of N-AAF and N-AGlu-P-3.

The DNA binding spectra of the carcinogens were done as previously described (Marrot et al.).   Briefly, after digestion with the appropriate
restriction enzymes, carcinogen-modified DNAs were labeled at the 5' end
with polynucleotide kinase and[γ-^{32}P] ATP, then a second restriction cut
was made and the fragments were isolated from 1,5 % agarose gel by elec-
troelution. The labeled fragments were digested using the 3'-5' exonuclease activity of T4 DNA polymerase according to the procedure of Fuchs.
The modified sites were identified by electrophoresis through polyacrylamide gels in 7 M urea. Chemical degradation sequencing reactions were also
performed according the procedure of Maxam and Gilbert.

Results

We have studied the reactivity of N-AcO-AAF or N-AcO-AGlu-P-3 with the two
plasmids pLP32 and pCM4 (pLP32 contains the (dC-dG)$_{16}$ insert and pCM4
contains the (dG-dT)$_{15}$ insert), the plasmids being either relaxed or negatively supercoiled. The analysis of AAF and AGlu-P-3 modification spectra
at the nucleotide level was made possible following the observation that

the 3'-5' exonuclease activity of T4 DNA
polymerase was blocked in the vicinity of
an AAF or an AGlu-P-3 adduct (Fuchs, Marrot
et al.).

The scheme of the experiment was as follows. After reaction of the   carcinogens
with the plasmids, restriction fragments
radio-actively labeled at the 5' end   were
digested with T4 DNA polymerase under conditions in which the enzyme functions as a
3'-5' exonuclease. The resulting fragments
were resolved on a Maxam-Gilbert sequencing
gel. The results are shown in figures 2 and
3 and can be summarized in three points.

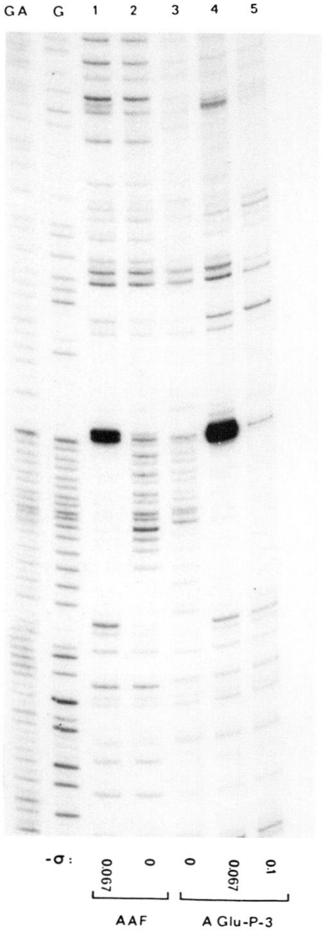

Figure 2 - Autoradiogram of the   lower
strand of pLP32 modified by N-AcO-AAF and
N-AcO-AGlu-P-3, respectively. The digestion
products of the (185-524) EcoRV-Sau96 I
restriction fragments were separated by
electrophoresis through a polyacrylamide
gel. Lane (1), AAF modified native supercoiled pLP32, lane (2) AAF modified relaxed
pLP32, lane (3) AGlu-P-3 modified relaxed
pLP32, lane (4) AGlu-P-3 modified native
supercoiled pLP32, lane (5) AGlu-P-3 modified highly negatively supercoiled pLP32.
The samples are carcinogen-modified   at
about 0.3 %. The lanes on the left-hand
side are the Maxam-Gilbert sequencing reactions (G, G+A) of the same unmodified fragment.

165

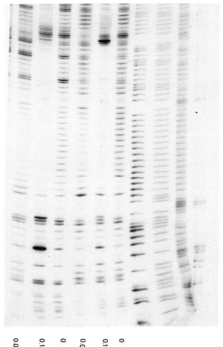

```
      1   2   3   4   5   6   G   GA  TC  C
```

-σ :   0.067   0.1   0   0.067   0.1   0

A Glu-P.3   A A F

Figure 3 - Autoradiogram of the upper strand of pCM4 modified by N-AcO-AAF and N-AcO-AGlu-P-3, respectively. The digestion products of (3589-3660) HinpI-HpaII restriction fragments were separated by electrophoresis as in figure 2. Lane (1), AGlu-P-3 modified native pCM4, lane (2) AGlu-P-3 modified highly negatively supercoiled pCM4, lane (3), AGlu-P-3 relaxed pCM4, lane (4), AAF modified native pCM4, lane (5) AAF modified highly negatively supercoiled pCM4, lane (6) AAF modified relaxed pCM4. The samples are carcinogens modified at about 0.3 %. The lanes on the right-hand side are the Maxam-Gilbert sequencing re-actions of the unmodified fragment.

1) In the relaxed plasmids, all the guanine residues are modified (lanes 2 and 3, figure 2 ; lanes 3 and 6, figure 3). There is a constant shift of about three nucleotides towards the higher molecular weights in the enzyme-generated fragments as compared to the chemically generated guanine cleavage fragments which is due on one hand to the destruction of guanine residue in Maxam-Gilbert reaction and on the other hand to the presence of the bound carcinogen and the 3'-OH generated by the enzyme. It is important to note that if one assumes that the band intensity on the gel corresponds to the binding site frequency, all the guanines are not equally modified (figure 4). Moreover, the guanines which are more (or less) reactive with N-AcO-AAF are not the ones which are more (or less) reactive with N-AcO-AGlu-P-3.

2) In negatively supercoiled plasmids at native superhelical density (pLP32) or at a superhelical density $\sigma = - 0.1$ (pCM4), the guanines in the $(dC-dG)_{16}$ insert (lanes 1 and 4, figure 2) and in the $(dG-dT)_{15}$ insert (lanes 2 and 5, figure 3) are not modified. Since the inserts are in the Z conformation (Peck et al. 1982, Malfoy et al.) we conclude that Z-DNA does not react with the two carcinogens. In pCM4, the $(dG-dT)_{15}$ sequence is followed on the 3' side by a $d[(GT)_2 \ G_2 \ (GT)_4 \ G_2 \ (GT)_2]$ sequence. These sequences are separated by four bases (ACTC). The guanines in the total sequence (55 base pairs long) are not modified by the carcinogens.

166

AAF

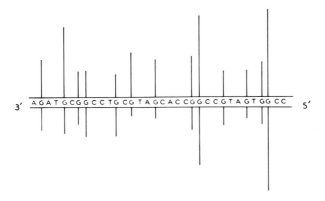

3′ AGATGCGGCCTGCGTAGCACCGGCCGTAGTGGCC 5′

AGlu−P−3   AAF

T °C: 16   23   16   23

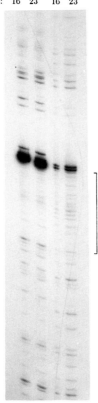

:
C
G
C
G
:

Figure 4 - Distribution of AGlu-P-3 and AAF adducts along pLP32 on 5' side of the (dC-dG)$_{16}$ insert. The bars represent the reactivity index at a given guanine residue as determined by densitometry of the autoradiogram of the digests (figure 2). The reactivity index at a given residue is the ratio of the area of the peak corresponding to this residue over the sum of the areas of all the peaks.

Figure 5 - Autoradiogram of the lower strand of pLP32 modified by N-AcO-AAF or N-AcO-AGlu-P-3. The reaction between the carcinogens and pLP32 has been done at 16°C or 23°C during 40 minutes, the carcinogens being added to negatively supercoiled DNA as explained in the text.

167

3) A striking result is that the reactivity of both carcinogens and the guanine residues on the 3' side of the $(dC-dG)_{16}$ insert depends upon the degree of negative superhelical density of pLP32. The guanines are much more reactive in native supercoiled pLP32 than in relaxed pLP32 (compare lanes 1 and 2 or lanes 3 and 4 in figure 2). Their reactivities are about 20 and 40 times larger with N-AcO-AAF and with N-AcO-Glu-P-3, respectively than the more reactive guanines on each side of the insert. On the other hand, this junction is much less but still reactive with highly negatively supercoiled pLP32 and N-AcO-AGlu-P-3 (lanes 4 and 5, figure 2).

Reaction with the $(dC-dG)_{16}$ insert during the B-Z transition

The conformation of the junction on the 3' side of the $(dC-dG)_{16}$ insert in pLP32 is unknown. We wanted to know whether the conformation of this junction could present some similarity with a transient conformation adopted by the insert during the B-Z transition. The rate of the B-Z transition of the insert depends upon the superhelical density of the plasmid. At $20°C$ and at native superhelical density, the half-time of the transition is about 1 hour (Peck et al. 1986, Pohl). It is thus possible to study the reactivity of the carcinogens and pLP32, the insert being in the B form, the Z form or transiting from the B form to the Z form. The experiment was carried out by first relaxing the plasmid by addition of ethidium bromide and then ethidium bromide was removed by filtration on a short dowex 50 column in the cold (Peck et al. 1986). The reaction between the carcinogens and the plasmid was done at two temperatures ($16°C$ and $23°C$) during 40 minutes. The samples were then treated with T4 DNA polymerase and the resulting fragments were analysed by gel electrophoresis. As shown in figure 5, the guanines in the insert are modified as the other guanines in the B regions. However the two guanines on the 3' side of the insert are still hyperreactive.

Discussion

In this work, we have compared the reactivity of two isosteric chemical carcinogens, N-AcO-AAF and N-AcO-AGlu-P-3 with nucleic acids in various conformations.

Both carcinogens react with B-DNA. Analysis at the nucleotide level shows that all the guanines are modified, but they are not equally reactive. Moreover, the AAF modification spectrum differs from the AGlu-P-3 modification spectrum.

In a recent work, it has been proposed that the reaction of a guanine residue and N-AcO-AAF does not occur in DNA being in the classical B conformation (named G) because the phosphate group of DNA and the acetyl group of the carcinogen prevent access to the C(8) of guanine. The reaction occurs with a transient conformational state (named G^*). G^* is in equilibrium with G and the $G-G^*$ transition depends upon the neighboring bases (Daune et al. 1985). The steric hindrance encountered in the interaction between N-AcO-AAF and the G state is also encountered by N-AcO-AGlu-P-3. We assume that N-AcO-AGlu-P-3 reacts only with G^* since in the binding spectrum of AGlu-P-3 modified relaxed DNA, all the guanines are not equally reactive. Thus, N-AcO-AGlu-P-3 as N-AcO-AAF can probe the dynamic structure of DNA. The binding spectra of AAF and AGlu-P-3 are not identical. This probably reflects the differences in chemical structures of both carcinogens which do not stabilize the same intermediate transient states.

Both carcinogens react with $poly(dG-br^5dC).poly(dG-br^5dC)$ in the Z conformation (Marrot et al.). They do not react with $(dC\text{-}dG)_{16}$ or $(dG\text{-}dT)_{15}$ inserts when the inserts are in the Z conformation. Assuming no preferential interaction between the carcinogens and br^5 cytosines (or m^5 cytosines), an explanation is that the dynamic properties of $(dC\text{-}dG)_n$ fragments depend how these fragments are stabilized in the Z conformation. Several experiments suggest that in crystals, left-handed Z-DNA exists in a mixture of conformations (Rich et al., Holbrook et al., Drew et al., Chevrier et al.). In solution, at temperatures well below Tm, methylated and brominated oligo (dC-dG) can adopt two forms in equilibrium, forms which belong to the Z family (Laigle et al., Hartmann et al.).

Both carcinogens behave similarly with respect to their reactivity with the guanines on the 3' side of the $(dC\text{-}dG)_{16}$ insert as a function of the degree of superhelicity of pLP32. Two points deserve to be underlined : i) these guanines are much more reactive in the native supercoiled plasmid than in the relaxed plasmid, ii) the hyperreactivity decreases in the highly negatively supercoiled plasmid. The structure of the junction in the native plasmid is still unknown but experiments with chemical reagents are in favor of a tight and well-defined structure (Herr, Rio et al. 1986, Johnston et al., Kang et al., Galazka et al.). The hyperreactivity of the carcinogens do not reflect a local denaturation. N-AcO-AAF reacts slightly better with dDNA than with nDNA and N-AcO-AGlu-P-3 reacts equally well with dDNA and nDNA (Daune et al. 1985, Marrot et al.). On the other hand, at higher superhelical density, the guanines at the 3'-junction are much less reactive. This supports the finding that the location and/or the structure of the junction depends upon the degree of superhelical density of the plasmid (Johnston et al.).

The reactivity of the carcinogens with the $(dC\text{-}dG)_{16}$ insert has been studied in conditions in which the torsional stress is sufficient to induce the B-Z transition of the insert. The experiment is made possible because the rate of the B-Z transition is slow at $20°C$, the half-time of the transition being about 1 hour (Peck et al. 1986, Pohl). The stability of the active metabolites is not known accurately but it is also of the order of 1 hour. The guanines in the insert are modified as the guanines outside the insert. The carcinogens are not able to detect any hyperreactive transient states. On the other hand, the guanines on the 3' side of the insert are hyperreactive. In first approximation, the intensity of the bands is similar to that found when all the inserts are in the Z form (figure 2). After 40 minutes of incubation, less than 50 % of the inserts are in the Z form (it is expected that this percentage is even smaller at $16°C$ than at $23°C$). These experiments suggest that the two guanines on the 3' side of the $(dC\text{-}dG)_{16}$ insert adopt rapidly under the topological stress, the structure which is hyperreactive with both carcinogens. This hyperreactive structure might be important in chemical carcinogenesis since some specificity in the reaction of carcinogens and DNA can arise from the polymorphism of DNA. Moreover, both carcinogens have been shown to stabilize the Z form of poly(dG-dC).poly(dG-dC) and poly(dA-dC).poly(dG-dT) (Sage et al., Santella et al. 1981, Wells et al. 1982, Hebert et al.) and it is known that some defects well-repaired in B-DNA are not repaired in Z-DNA (Lagravere et al.).

In conclusion, the two isosteric chemical carcinogens N-AcO-AAF and N-AcO-AGlu-P-3 behave similarly with respect to their reactivity with DNA and the distortions induced by the adducts. They can be useful to analyse the polymorphism of DNA at the nucleotide level. The non-reactivity of Z-DNA and the hyperreactivity of some sequences under topological stress might have some importance in chemical carcinogenesis.

Acknowledgements

We are indebted to Dr B. Malfoy and to Dr J.C. Wang for the gift of the plasmids. This work was supported in part by Ministère de la Recherche et de l'Enseignement (contract n° 86T0629) and by La Ligue Nationale Contre le Cancer.

Abbreviations

2-N,N-acetoxyacetylaminofluorene, N-AcO-AAF ; 3-N,N-acetoxyacetylamino-4,6-dimethyldipyrido[1,2-a:3',2'-d] imidazole, N-AcO-AGlu-P-3.

References

Bichara, M., and Fuchs, R.P.P. (1985) J. Mol. Biol 183, 341-351.

Camerini-Otero, R., and Felsenfeld, G. (1977) Nucl. Acids Res. 4, 1159-1181.

Chevrier, B., Dock, A.C., Hartmann, B., Leng, M., Moras, D., Thuong, N.T., and Westhof, E. (1986) J. Mol. Biol. 188, 707-719.

Commoner, B., Vithayathil, A.J., Dolora, P., Nair, S., Madyastha, P., and Cuca, G.C. (1978) Science (Wash.) 201, 913-916.

Daune, M., Fuchs, R.P.P., and Leng, M. (1981) Natl. Cancer Inst. Monogr. 58, 201-210.

Drew, H., Takano, T., Tanaka, S., Itakura, K., and Dickerson, R.E. (1980) Nature (London) 286, 567-573.

Fuchs, R., and Daune, M. (1972) Biochemistry 11, 2659-2666.

Fuchs, R.P.P. (1984) J. Mol. Biol. 177, 173-180.

Galazka, G., Palecek, E., Wells, R.D., and Klysik, J. (1986) J. Biol. Chem. 261, 7093-7098.

Grunberger, D., and Weinstein, I.B. (1979) in Chemical Carcinogenesis and DNA, ed Grover, P.L., CRC, West Palm Beach, Vol 2, pp 59-93.

Hartmann, B., Genest, D., Thuong, N.T., Ptak, M., and Leng, M. (1986) Biochimie 68, 739-744.

Hebert, E., Loukakou, B., Saint-Ruf, G., and Leng, M. (1984) Nucl. Acids Res. 12, 8553-8566.

Herr, W. (1985) Proc. Natl. Acad. Sci., USA 82, 8009-8013.

Holbrook, S.R., Wang, A.H.J., Rich, A., and Kim, S.H. (1980) J. Mol. Biol. 187, 429-440.

Johnston, B.H., and Rich, A. (1985) Cell 42, 713-724.

Kang, D.S., and Wells, R.D. (1985) J. Biol. Chem. 260, 7783-7790.

Laigle, A., Chinsky, L., Turpin, P.Y., Hartmann, B., Thuong, N.T., and Leng, M. (1986) Nucl. Acids Res. 14, 3425-3434.

Lagravère, C., Malfoy, B., Leng, M., and Laval, J. (1984) Nature 310, 798-800.

Loukakou, B., Hebert, E., Saint-Ruf, G., and Leng, M. (1985) Carcinogenesis 6, 377-383.

Malfoy, B., Rousseau, N., Vogt, N., Viegas-Pequignot, E., Dutrillaux, B., and Leng, M. (1986) Nucl. Acids Res. 14, 3197-3214.

Marrot, L., Hebert, E., Saint-Ruf, G., and Leng, M. submitted.

Maxam, A.M., and Gilbert, W. (1980) Methods Enzymol. 65, 499-560.

Nordheim, A., Lafer, E.M., Peck, L.J., Wang, J.C., Stollar, B.D., and Rich, A. (1982) Cell 31, 309-318.

Peck, L.J., Nordheim, A., Rich, A., and Wang, J.C. (1982) Proc. Nat. Acad. Sci., USA 79, 4560-4564.

Peck, L.J., Wang, J.C., Nordheim, A., and Rich, A. (1986) J. Mol. Biol. 190, 125-127.

Pohl, F.M. (1986) Proc. Natl. Acad. Sci. USA 83, 4983-4987.

Rich, A., Nordheim, A., and Wang, A.H. (1984) Annu. Rev. Biochem. 53, 791-846.

Rio, P., and Leng, M. (1983) Nucl. Acids Res. 11, 4947-4955.

Rio, P., and Leng, M. (1986) J. Mol. Biol. 191, 569-572.

Saint-Ruf, G., Keravis, G., Loukakou, B., and Hebert, E. (1986) J. Hetero-cyclic Chem. 23, 425-431.

Santella, R.M., Grunberger, D., Weinstein, I.B., and Rich, A. (1982) Biochem. Biophys. Res. Comm. 106, 1226-1232.

Sage, E., and Leng, M. (1980) Proc. Natl. Acad. Sci., USA 77, 4597-4601.

Singer, B., and Grunberger, D. (1983) Molecular Biology of Mutagens and Carcinogens, Plenum Press, New-York.

Spodheim-Maurizot, M., Malfoy, B., and Saint-Ruf, G. (1982) Nucl. Acids Res. 10, 4423-4430.

Sugimura, T., and Nagao, M. (1979) CRC Crit. Rev. Toxicol. 8, 189-209.

Wells, R.D., Miglietta, J.J., Klysik, J., Larson, J.E., Stirdivant, S., and Zacharias, W. (1982) J. Biol. Chem. 257, 10166-10171.

INHERENTLY CURVED DNA AND ITS STRUCTURAL ELEMENTS

E. N. Trifonov and L. E. Ulanovsky

Department of Polymer Research
The Weizmann Institute of Science
Rehovot 76100, Israel

Existence of DNA with inherently curved molecular axis has been
predicted on the basis of periodicity of chromatin DNA sequences (Trifonov
& Sussman, 1980; Trifonov, 1980). First physical observations (Marini et
al, 1982) have supported the idea while recent circularization experiments
(Ulanovsky et al, 1986) and electron microscopy (Griffith et al., 1986)
provided direct demonstration that such curved DNA, indeed, exists. Numerous
studies on the structure and function of the curved DNA have been published
recently and continue to proliferate. Several full size and mini-reviews
on this subject are already available (Trifonov, 1985; Lilley, 1986; Eisenberg, 1986; Diekmann, 1987). This paper outlines most recent developments
(Sections 1 to 4) including our own data on the quantitative characterization
of the AA·TT wedge elements and their influence on electrophoretic mobility
of DNA (Sections 5 to 11). Here and below the term "curved DNA" relates to
DNA molecules with their axes curved without any deformation forces applied
(also called "inherently curved DNA", "intrinsically curved DNA", "DNA with
sequence-dependent curvature"). The term "bent DNA" would relate to forcibly
deformed molecules, either by proteins bound to it, or by thermal motion.
Intrinsic curvature is, thus, a property of a relaxed state of the curved
DNA.

1. ELECTROPHORETIC ANOMALY.

Strongly AA-periodical kinetoplast DNA fragment from Leischmania
tarentolae shows also anomalous retardation on polyacrylamide gel (Marini
et al, 1982; Wu & Crothers, 1984; Diekmann & Wang, 1985). Its curvature,
apparently, causes some additional friction for the reptating DNA molecule.
A review of many other published cases of the anomalously slow migrating
DNA fragments (Trifonov, 1985) revealed that all these fragments have a

common feature – a periodical distribution of AA(TT) dinucleotides with the period about 10.5 bases. Since then many more cases have been studied (Bossi & Smith, 1984; Thompson et al, 1984; Zahn & Blattner, 1985; Leong et al, 1985; Ray et al, 1986; Anderson, 1986; Snyder et al, 1986; Koepsel & Khan, 1986; Kitchin et al, 1986; Ryder et al, 1986; Levinger & Nass, 1986; Silver et al, 1986; Poljak & Gralla, 1987; Christiansen et al, 1987), all showing the same periodical property.

One important observation is made by J. Anderson (1986) who found that some DNA fragments migrate rather somewhat faster than expected on the basis of their known lengths. Such reverse anomaly has been observed before with some synthetic DNA fragments (e.g. Koo et al, 1986), which do not possess the 10.5 base periodicity. These fragments, thus, should not have any overall curvature, being straight in a relaxed state. Natural DNA molecules with no 10.5 base periodicity in the distribution of AA and TT dinucleotides are also expected to be straight. Indeed, autocorrelation sequence analysis of these anomalously fast fragments (yeast centromere DNA) revealed that AA dinucleotides are found at all distances one from another, except multiples of 10.5 bases, contrary to periodically distributed AA dinucleotides in the curved fragments, migrating slowly (Anderson, 1986). This result also means that the "normal" fragments which migrate slower than the real straight ones are actually slightly curved. We return to this aspect in the concluding section.

Many synthetic fragments of DNA have been recently studied (Hagerman, 1985;1986; Ulanovsky et al, 1986; Diekmann, 1986; Koo et al, 1986). Major conclusions of these studies are: 1) High electrophoretic anomaly effect (retardation) is achieved when adenines are clustered in runs of 4 to 6 bases, and 2) The rotational separation of the runs of A along the molecule is crucial; their distance along the sequences has to be multiple of DNA helical repeat.

2. BIOLOGICAL SIGNIFICANCE OF CURVED DNA.

The spectrum of biological structures of which the curved DNA appears to be an important component is growing fast. The curved DNA seems to be typical for kinetoplast DNA in general (Ray et al, 1986; Kitchin et al, 1986; Silver et al, 1986) and not only for L. tarentolae (Marini et al, 1982). The curved DNA is found upstream of prokaryotic promoters (Bossi & Smith, 1984; Plaskon & Wartell, 1987) and in the promoters themselves (Trifonov, 1983, 1985). Anomalously slowly migrating fragments of DNA are located in close vicinity from enhancers and appear to form nuclear matrix attachement sites (Anderson, 1986). Many of the curved fragments are involved in the origins of replication (Thompson et al, 1984; Trifonov, 1985; Zahn & Blattner, 1985; Anderson, 1986; Snyder et al, 1986; Koepsel & Khan, 1986). The curved DNA is found to be a binding site for replication initiator protein (Koepsel & Khan, 1986). It also forms the O-somes in the ORI lambda region (Dodson et al, 1986). Binding site of SV40 large T-antigen is found to be curved (Ryder et al, 1986). Attachement sites for bacteriophages φ80 and P22 are also curved (Leong et al, 1985). They are involved in specific nucleosome-like particles, intasomes (Echols, 1986). Sequence periodical and, thus, curved DNA is located in the nucleosomes (Trifonov & Sussman, 1980; Trifonov, 1980; see also Sections 6 and 11).

As it follows from the above the curved DNA is likely to be frequently involved in all protein-DNA complexes in which DNA is wrapped around the protein or in which the protein binds specifically to only one side of the DNA molecule. The sequence-dependent local curvature of DNA, thus, appears to be an important structural feature for sequence-specific recognition of DNA by proteins.

3. MODELS OF DNA CURVATURE.

Two principal models of the formation of the curved DNA have survived till this day: wedge model and junction bend model. The wedge model (Trifonov & Sussman, 1980;'Trifonov, 1980) is based on assumption that different stacks of neighboring base-pairs have different dihedral angles between the base-pairs, as if some small sequence-dependent wedges were inserted between them. The spatial sum of the wedges would result in a curvilinear sequence-dependent path of the DNA axis. Repetition of strong wedges of the same kind (say, AA·TT) at distances multiple of the DNA helical repeat would result in the DNA unidirectional curving. The junction model (Levene & Crothers, 1983; Koo et al, 1986) assumes that the net curvature is caused by deflection of the axis at the junctions between locally somewhat different sequence-dependent forms of DNA, very much like inclination of DNA axis at the junction between B-form and A-form (Selsing et al, 1979) the latter having base-pairs appreciably tilted towards the axis.

The junction bend model actually can be described in terms of wedge model since the wedge presentation provides rather general description of any succession of stacked elements irrespective of whether any local unusual DNA structure is formed or whether the wedge elements are constant or variable. This equivalence has been elegantly demonstrated for the case of constant wedges (Prunell et al, 1984). The junction model, nevertheless,

175

is appealing due to the fact that the electrophoretic anomaly and (by implication) curvature are more pronounced when runs of adenines are present. On the other hand, the helical structure of polydA·polydT molecules, indeed, appears to be different from canonical B-form of DNA (Arnott et al, 1983; Alexeev et al, 1987). The dependence of the electrophoretic anomaly on the DNA curvature, however, could well be non-linear, and small curvatures caused by isolated AA-dinucleotide wedges might not be easily detected. As it is demonstrated in Section 9, this seems to be the case.

Experiments with synthetic fragments, as interpreted in the original papers (Hagerman, 1986; Koo et al, 1986), indicated that the wedge model is not applicable to the data. We have found, however, that these very data, on the contrary, can be consistently described by and provide very strong support to the wedge model (Ulanovsky & Trifonov, 1987; see also Sections 5 to 10).

4. ENERGY CALCULATIONS.

Both wedge model and junction model are only simplified descriptions of the curved DNA. The path of the DNA axis is, after all, a result of complex interplay of intramolecular forces and the exact solution can be, in principle, obtained by energy calculations. Such calculations indicate that, indeed, as soon as the angle between the base planes is allowed to change during the minimization, the DNA axis appreciably deviates from straight line (Zhurkin, 1985; Jernigan et al, 1986, 1987; Tung & Burks, 1987; von Kitzing & Diekmann, 1987). Results of such calculations for inter-spersed runs of adenines (von Kitzing & Diekmann, 1987) are in favor of the junction model, suggesting also some further detalization of the model. Hydrophobic interactions between methyl groups of thymines seem to contribute to the DNA curvature (Jernigan et al, 1986,1987). The direction of the roll angle for AA·TT element predicted on this basis, however, turns out to be wrong (see Section 7). Energy calculations by Zhurkin (1985) favor RY·RY and YR·YR elements as major structural components of the curved DNA. This is in conflict with experiments on synthetic DNA fragments in which various $(R_5Y_5)_n$ sequences are compared, clearly demonstrating that AA and TT di-nucleotides are the major elements responsible for DNA curvature (Hagerman, 1985; see, however, Section 10). Recent calculations, supported by experiments, indicate that amino group at 6th position of adenine heterocycle and its absence at the second position are determining factors in formation of roll angle between base-pairs of the AA·TT minihelix, opening towards DNA major groove (Diekmann et al, 1987).

176

5. MAGNITUDES OF THE AA·TT WEDGE COMPONENTS.

Generally speaking, each of 10 possible combinations of adjacent base-pairs is characterized by certain wedge angle with its own direction of opening. Such wedge angle can be decomposed in two orthogonal components: roll angle opening towards one of the grooves, and tilt angle which opens towards one of two backbones of the DNA molecule. This is illustrated schema-tically in Fig. 1 where the twist angle between the base-pairs is set equal to zero, for clarity. The axes drawn in the Figure are the dyad axes of chemical symmetry of the stacked base-pairs. Rotation by 180° around this axis, i.e. replacement of the dinucleotide (5'-AA in this case) by its complementary (5'-TT) results in reversal of the tilt component, while direction of the roll wedge remains unchanged. The values and directions of the tilt and roll angles can be estimated by designing various 10 - 11 base periodical DNA fragments containing AA and TT dinucleotides at different distances (and, thus, orientations) and measuring the curvature of the fragments. Together with vectorial sums of the unknown roll and tilt wedges for different fragments this would provide a simple system of equations to calculate the unknowns. Once the angles are small and only AA·TT elements are assumed to be responsible for DNA curvature, the problem can be in principle reduced to only two experiments. One such experiment (Ulanovsky et al, 1986) allowed to measure the absolute value of the AA·TT wedge, 8.7°, which equals to square root of ($r^2 + t^2$) where r and t are values for roll and tilt wedges, respectively. This is the only absolute measurement

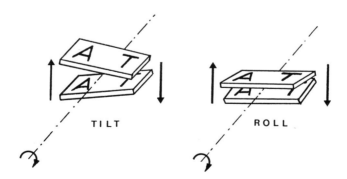

Figure 1. Tilt and roll wedge components of the stack of DNA base-pairs (AA·TT in this case). Helical twist angle is set equal to zero, for simpli-city. 5' - 3' backbone directions are indicated by arrows.

of DNA curvature available so far. The data on electrophoretic
mobility of various fragments are not calibrated yet to be used
for the curvature measurements. The absolute measurements can
be excluded, however, if one could find at least one pair of
different AA and TT containing fragments which show the same
electrophoretic mobility (equal curvatures), or a single fragment
with AA and TT dinucleotides which does not show any anomaly
(zero curvature). These rather special demands turn out to be
met by recently published experiments (Hagerman, 1985,1986).

In one of these experiments (Hagerman, 1985) two fragments,
$(nAAAATTTTn)_{10}$ and $(nnAAATTTnn)_{10}$ show identical electrophoretic
anomaly. They possess the same AAATTT core and differ only by
extra AA dinucleotide in the first fragment separated by six
bases from extra TT dinucleotide. The wedge angles of these two
AA·TT elements, thus, cancel one another. Orientation-wise the
first element can be transformed into the second one by two
rotations around two different axes (Fig. 2): $6 \cdot 36^{\circ} = 208^{\circ}$
around DNA helix axis, equivalent to helical translation of the
first AA dinucleotide into position of the last TT dinucleotide;
and a 180° rotation around the groove-to-groove dyad passing
through the AA·TT stack, equivalent to replacement of AA by TT.
The last operation leaves the roll wedge vector unchanged, while

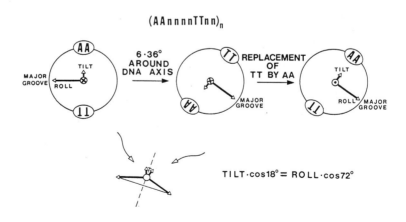

Figure 2. Vectorial summation of AA·TT and TT·AA wedges six bases apart
(center-to-center). Each circle corresponds to the minihelix viewed along
DNA axis. Directions (opening) and magnitudes of the roll and tilt wedge
components are represented by orthogonal vector pairs.

the tilt component reverses its sign (see also Fig. 1). Two roll-tilt vector pairs corresponding to the two cancelling elements are shown together at the bottom of Fig. 2. From the condition that total wedge angle in this case equals zero, one arrives to a unique ratio $r/t = 3.1$. More accurate calculation which takes into account also the DNA curvature of the 6 base-pairs separating the AA·TT elements yields the value $r/t = 3.5 \pm 0.5$. Combining this with the earlier result, $r^2 + t^2 = (8.7^o)^2$, we obtain, finally, $r_{AA} = 8.4^o$; $t_{AA} = 2.4^o$. The calculations described give, actually, two solutions for the orientations of the wedge components: either roll AA opens towards major groove and tilt - towards the AA dinucleotide (like in Fig. 2), or roll AA opens towards minor groove, while tilt - towards TT dinucleotide. On the choice of the directions - see Section 7.

Second experiment, with the fragment $(nTTTTAAAAn)_n$ which displays only a very small electrophoretic anomaly (Hagerman, 1986), provides another possibility for independent calculation of the r/t value which turns to be identical to the value above (Ulanovsky & Trifonov, 1987). This fragment can also be presented in equivalent form $(AAAAnnTTTT)_n$ which makes obvious the fact that it is actually built of three pairs of AA and TT elements separated by 6 bases. According to above calculations each pair should have zero net contribution to the DNA curvature which, indeed, is the case.

6. tyrT FRAGMENT IS CURVED.

Recent study by Drew and Travers (1985) has shown that in circularized tyrT DNA fragment certain bases are oriented preferentially outwards while others, half period away, occupy inner side of the circle. This has been described by Drew and Travers in terms of sequence-dependent deformational anisotropy of DNA. Such anisotropy, as has been discussed earlier (Trifonov, 1980) is a result of contributions of non-deformed structure (curved DNA) and of the deformability term. Since the tyrT DNA fragment "does not appear to be bent very much in the absence of protein" (Drew & Travers, 1985) the authors concluded that the observed anisotropy of circularization is due to DNA deform-ability. Since now the values for the tilt and roll components of the AA wedge are determined (see previous Section) we can

179

calculate how the DNA path of the tyrT fragment looks like. The corresponding
plot, calculated as described earlier (Ulanovsky & Trifonov, 1987), is shown
in Fig. 3. The roll$_{AA}$ angle is chosen to open towards the major groove, in
which case the tilt$_{AA}$ opens towards AA dinucleotide (see above). The plot
reveals that according to the wedge model the tyrT DNA fragment is well
curved. To find out whether the direction of curvature is in any relation
with the direction of circularization we mapped on the same plot the sites
of DNaseI attack (Drew & Travers, 1985) in the circularized molecule (black
dots, Fig. 3). Practically all these sites are located on the outer side
of the curved fragment. In other words, the fragment circularizes exactly
in the direction of its curvature which is, thus, a major if not sole
component of the sequence-dependent anisotropy of circularization in case
of the tyrT fragment.

7. ORIENTATIONS OF AA·TT WEDGE COMPONENTS.

As we have found in Section 5, there are only two possible orientations for the roll and tilt components of the AA·TT wedge - either roll
and tilt open towards major groove and towards AA dinucleotide, respectively,
or both in opposite directions - towards minor groove and TT dinucleotide.
The analysis of the circularization experiment of Drew and Travers described
above definitely points to the first choice. Changing the signs of the
roll and tilt results in the curving of the plot in opposite direction (not
shown), so that the nuclease attack sites appear at the inner side of the
arc.

PolydA·polydT fiber x-ray data (Arnott et al, 1983; Alexeev et al,
1987) indicate that the major groove of the helical structure of polydA·
polydT is broader as compared to normal B-DNA. This is also consistent with

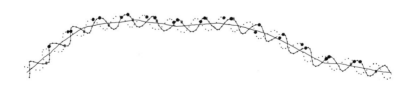

Figure 3. DNA path of the tyrT fragment. For this calculation the roll angle
of the AA·TT elements is taken equal to 8.4°, opening towards DNA major
groove, tilt (2.4°) opens towards AA dinucleotide. The left end of the
continuous sinusoidal curve (path of phosphates) corresponds to the 5'-end
of the original nucleotide sequence (Drew & Travers, 1985). The bonds
attacked by DNaseI in the circularized tyrT fragment are shown by black dots.

the above choice, where the roll wedge of the AA·TT element opens towards the major groove.

Yet another independent estimate is provided by energy calculations and experiments with various substitutions for adenine in the curved synthetic fragments (Diekmann et al, 1987). Here, too, the roll wedge of AA·TT is found to open towards major groove.

The final magnitudes and orientations, thus, are as follows:

Roll$_{AA}$ = 8.4^{O}, opening towards major groove.

Tilt$_{AA}$ = 2.4^{O}, opening towards AA dinucleotide.

Noteworthy that these are wedge parameters in free non-constrained DNA in solution. AA·TT wedge in complexes with proteins and in the nucleosome, in particular, does not have to be the same and, apparently, is not (see Section 11).

8. ELECTRON MICROSCOPY OF CURVED DNA.

A particularly curved 219 bp long kinetoplast DNA fragment from C. fasciculata (Kitchin et al, 1986) has been chosen by J. Griffith et al (1986)

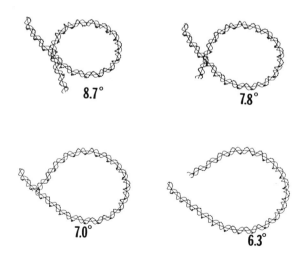

Figure 4. DNA paths of 219 bp fragment of kinetoplast DNA from Crithidia fasciculata calculated for different values of total AA·TT wedge angle. For all four cases r/t = 3.5. In each case the upper left end of the dotted sinusoidal path of phosphates corresponds to the 5'-end of the nucleotide sequence of the fragment (Kitchin et al, 1986).

to directly visualize its curvature by electron microscopy. These fragments, indeed, appear in the micrographs either as C-shape molecules or as circles which are, probably, a mixture of α-shape and nearly closed C-shape molecules.

These experiments provide a critical test for every theoretical or experimental estimation of the DNA curvature. The calculated shape of the kinetoplast DNA fragment has to be at least semiquantitatively consistent with the EM data. We have calculated the DNA path for the fragment from its sequence using AA·TT wedge parameters given in the previous Section, as well as smaller total wedge angles, within 2^O error bar of the original estimate of 8.7^O (Ulanovsky et al, 1986), with the same r/t ratio of 3.5. The corresponding plots are presented in Fig. 4. Given low experimental accuracies of both techniques (circularization and EM) we consider the test to be successful.

The plots as above (Figs. 3 and 4) calculated for DNA fragments with any given sequence can be considered, therefore, as realistic picture of the path of DNA and rotational positions of its structural elements.

9. CALIBRATION OF ELECTROPHORETIC ANOMALY DATA.

Assuming that the DNA curvature is primarily caused by the AA·TT wedge elements we calculated the curvatures of a series of synthetic fragments for which their electrophoretic mobility is estimated in identical conditions (Koo et al, 1986). The corresponding plot anomaly - curvature is shown in Fig. 5. Here the curvature is expressed in DNA curvature units. This natural unit of DNA curvature is defined as the average curvature of DNA in the nucleosome core particle, $1/42.8\overset{o}{A}$ (Bentley et al, 1984; Richmond et al, 1984; Uberbacher & Bunick, 1985).

Despite an appreciable scatter of the points around the best fit curve, several conclusions can be made about the dependence of the retardation effect on the DNA curvature. The dependence is, obviously, non-linear, small curvatures being of appreciably less influence on the electrophoretic mobility of corresponding fragments. The curve does not start from the origin (zero) since some synthetic fragments manifest reverse effect - faster than expected mobility. This property is known as well for some natural DNA fragments (Anderson, 1986; see Section 1). Apart from purely experimental causes for the scattering of the points, there are two possible physical reasons for the scattering. First, many of the curved fragments are non-planar due to small difference between sequence periodicity of the fragments (10 or 10.5 bases) and their actual helical twist. The effect of non-planari-

182

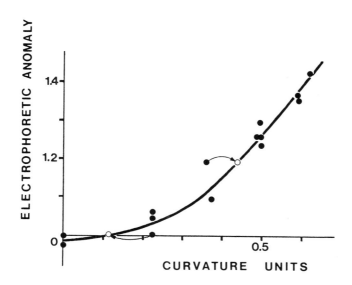

Figure 5. Calibration of DNA electrophoretic anomaly. The anomaly is expressed as apparent to actual molecular length ratio. The corresponding values are read from Fig. 3 of the Koo et al paper (1986), at the fragment length 90 bp. Corrected points (open circles) correspond to the sequences (CA9)9 (upper point) and (GGCAACAACG)9 (bottom). See Section 10 for explanation.

ty on the DNA mobility has been described earlier (Koo et al, 1986; Ulanovsky & Trifonov, 1987). Taking this effect into account, indeed, results in some improvement of the scatter around the curve (data not shown). The second reason is a possible influence of wedge elements other than the AA·TT wedge which could also make some contribution to the DNA curvature. It is clear, however, that already at this stage the curve shown in the Fig. 5 can be used for an approximate (± 50% in worst cases) estimation of absolute DNA curvature by measuring its electrophoretic mobility in certain standard conditions. This would be a technique by far simpler and less time-consuming than DNA circularization experiments.

10. AC AND/OR CA DINUCLEOTIDES MIGHT CONTRIBUTE TO DNA CURVATURE.

Two points, probably, most deviating from the calibration curve in Fig. 5 (see arrows) correspond to two sequences shown in Fig. 6 one of which is expected to be more curved while the other is expected to be almost straight. Both fragments apart from the curvature generating AA·TT elements have common ACA·TGT element which is in one case half-period away from the

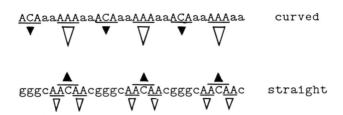

ACA wedge

ACAaaAAAaaACAaaAAAaaACAaaAAAaa   curved

gggcAACAAcgggcAACAAcgggcAACAAc   straight

Figure 6. In-phase and out-of-phase contributions of the ACA·TGT element to DNA curvature (see text).

centers of the A-runs, while in the other one it is in the same phase with AA dinucleotides. The behaviour of these two fragments can be understood if it is speculated that the common ATA·TGT element provides some additional curvature, of opposite sign to what is contributed by AA·TT in the same phase. The appropriate calculations give for the components of the (combined) ACA wedge the values about twice smaller than for AA·TT wedge. Thus, at least the roll wedge for either AC·GT or CA·TG elements could be as large as 3 - 5°, opening towards minor groove. As indicated by energy calculations (Zhurkin, 1985) the CA element is expected to have such geometry.

11. DIRECTIONS OF DNA BENDING IN THE NUCLEOSOME AND DNA CURVATURE IN
    SOLUTION ARE DIFFERENT.

In free DNA the roll wedge of the AA·TT element is 3.5 times larger than the tilt wedge, as estimated above. In other words, the DNA curvature in solution is, primarily, due to the roll component of the AA.TT wedge. From the analysis of nucleotide sequences of chromatin DNA (Trifonov, 1980) it is known also that AA and TT dinucleotides in the nucleosomes appear to be separated, typically, by about 4.5 bases. The roll angles at such separation almost cancel one another. Therefore, if caused by the AA(TT) periodicity, the DNA bending in the nucleosome is due to, primarily, contribution of the tilt components. An immediate important conclusion follows from this difference: direction of the DNA bending in the nucleosome is not the same as direction of free DNA curvature. The angular difference should be somewhat less than 90°, due to small contributions of rolls in one case and tilts in another one. A direct verification of this effect is provided by

the experiments of Drew and Travers (1985) on the 169 bp tyrT DNA fragment
in which the direction of its circularization is accurately determined by
DNaseI digestion, as well as its rotational setting in the nucleosome. As
we discussed above (Section 6) the circularization direction and direction
of curvature in this case are the same. Nucleotide positions oriented to
outer and inner surfaces of the tyrT DNA in the circle and in the nucleosome,
presented in Fig. 7, are taken from pp. 781 and 784 of the Drew and Travers
paper without any change. These data clearly show that the nucleotide positions which are oriented outwards in the nucleosome are systematically
shifted 1 to 4 bases upstream relative to corresponding positions in the
circle. The positions oriented inwards show similar shifts. The average
difference in the orientation of the DNA in the nucleosome relative to the
DNA in the circle is about 73°, in accordance with what one would expect.
This rotation is smaller at the left end of the molecule ($\approx 35^O$) and larger
at the right end ($\approx 115^O$) due to small difference in DNA helical repeat
between free DNA and DNA in the nucleosome.

Although the rotation of the DNA around its axis in the nucleosome
by 73° relative to the free DNA circle indicates a switch from roll to tilt
induced curvature this angular difference was not considered significant
by the authors of the paper (Drew & Travers, 1985).

One reason for the possible larger role of tilts in the nucleosome
might be the electrostatic neutralization of phosphates facing the histones,
introducing additional tilt components of the wedges.

14	24	35	45	56	66	77	87	98	108	119	129	140		– OUTSIDE, CIRCLE
13	23	33	44	55	65	74	84	94	106	116	126			– OUTSIDE, CORE-PARTICLE
1	1	2	1	1	1	3	3	4	2	3	3			– DIFFERENCE

AVERAGE ANGULAR DIFFERENCE $= 2.2$ bp (75°)

9	19	30	40	51	61	72	82	93	103	114	124	135	145	– INSIDE, CIRCLE
	18	28	39	49	60	70	80	90	100	112	121	131		– INSIDE, CORE-PARTICLE
	1	2	1	2	1	2	2	3	3	2	3	4		– DIFFERENCE

AVERAGE ANGULAR DIFFERENCE $= 2.1$ bp (71°)

Figure 7. Angular separation between DNA in solution and in the nucleosome.
In the first two lines of each subset the sites are listed (Drew & Travers,
1985) which are oriented outwards and inwards in the circularized tyrT DNA
and in the linear fragment reconstituted with histone octamers. Calculated
average angular difference is indicated.

12. CONCLUDING REMARKS.

How wide spread is the curved DNA? Partial answer to this question is provided by studies of J. Anderson (1986) whose experiments have indicated that, actually, straight pieces of DNA are rather exceptional, the bulk of natural DNA apparently being always at least slightly curved, as one could judge from electrophoretic mobility experiments. With the DNA plotting program at our disposal we were able to screen large ensemble of the DNA sequences, searching for particularly curved pieces. We, too, came to the same conclusion: the distribution of curvatures in natural DNA shows that straight molecules are rare and the rest of DNA is always somewhat curved (unpublished). Similar distribution is obtained by Tung and Burks (1987). Thus, the curved DNA is a common structure, the straight one being unusual.

This work has been supported in part by Minerva Foundation. Discussions with S. Diekmann and P. Hagerman are highly appreciated.

REFERENCES

Alexeev, D.G., A.A. Lipanov and I.Y. Skuratovskii (1987), Nature 325: 821-823.
Anderson, J.N., (1986), Nucl. Acids Res. 14: 8513-8533.
Arnott, S., R. Chandrasekaran, I.H. Hall and L.C. Puigjaner (1983), Nucl. Acids Res. 11: 4141-4155.
Bentley, G.A., A. Lewit-Bentley, J.T. Finch, A.O. Podjarny and M. Roth (1984) J. Molec. Biol. 176: 55-75.
Bossi, L., and D.M. Smith (1984), Cell 39: 643-652.
Christiansen, K., B.J. Bonven and O. Westergaard (1987), J. Molec. Biol. 193: 517-525.
Diekmann, S., (1986), FEBS Letters 195: 53-56.
Diekmann, S., (1987), In F. Eckstein & D.M.J. Lilley (Eds.), Nucleic Acids and Molecular Biology, Springer Verlag, Heidelberg, in press.
Diekmann, S., E. von Kitzing, L. McLaughlin, J. Ott and F. Eckstein (1987), submitted for publication.
Diekmann, S., and J.C. Wang (1985), J. Molec. Biol. 186: 1-11.
Dodson, M., H. Echols, S. Wickner, C. Alfano, K. Mensa-Wilmot, B. Gomes, J. LeBowitz, J.D. Roberts and R. McMacken (1986), Proc. Natl. Acad. Sci. USA 83: 7638-7642.
Drew, H.R., and A.A. Travers (1985), J. Molec. Biol. 186: 773-790.
Echols, H., (1986), Science 233: 1050-1056.
Eisenberg, H., (1986), Trends Biochem. Sc. 11: 350-351.
Griffith, J., M. Bleyman, C.A. Rauch, P.A. Kitchin and P.T. Englund (1986), Cell 46: 717-724.
Hagerman, P.J., (1985), Biochemistry 24: 7033-7037.
Hagerman, P.J., (1986), Nature 321: 449-450.
Jernigan, R.L., A. Sarai, K.-L. Ting and R. Nussinov (1986), J. Biomolec. Str. Dyn. 4: 41-48.
Jernigan. R.L., A. Sarai, B. Shapiro and R. Nussinov (1987), J. Biomolec. Str. Dyn. 4: 561-567.
Kitchin, P.A., V.A. Klein, K.A. Ryan, K.L. Gann, C.A. Rauch, D.S. Kang,

R.D. Wells and P.T. Englund (1986), J. Biol. Chem. 261: 11302-11309.
Koepsel, R.R., and S.A. Khan (1986), Science 233: 1316-1318.
Koo, H.-S., H.-M. Wu and D.M. Crothers (1986), Nature 320: 501-506.
Leong, J.M., S. Nunes-Duby, C.F. Lesser, P. Youderian, M.M. Susskind and A. Landy (1985), J. Biol. Chem. 260: 4468-4477.
Levene, S.D., and D.M. Crothers (1983), J. Biomolec. Str. Dyn. 1: 429-435.
Levinger, L.F., and G.S. Nass (1986), FEBS Letters 209: 340-346.
Lilley, D.M.J., (1986), Nature 320: 487-488.
Marini, J.C., S.D. Levene, D.M. Crothers and P.T. Englund (1982), Proc. Natl. Acad. Sci. USA 79: 7664-7668.
Plaskon, R.R., and R.W. Wartell (1987), Nucl. Acids Res. 15: 785-796.
Poljak, L.G., and J.D. Gralla (1987), Biochemistry 26: 295-303.
Prunell, A., I. Goulet, V. Jacob and F. Goutorbe (1984), Eur. J. Biochem. 138: 253-257.
Ray, D.S., J.C. Hines, H. Sugisaki and C. Sheline (1986), Nucl. Acids Res. 14: 7953-7965.
Richmond, T.J., J.T. Finch, B. Rushton, D. Rhodes and A. Klug (1984), Nature 311: 532-537.
Ryder, K., S. Silver, A.L. DeLucia, E. Fanning and P. Tegtmeyer (1986), Cell 44: 719-725.
Selsing, E., R.D. Wells, C.J. Alden and S. Arnott (1979), J. Biol. Chem. 254: 5417-5422.
Silver, L.E., A.F. Torri and S.L. Hajduk (1986), Cell 47: 537-543.
Snyder, M., A.R. Buchman and R.W. Davis (1986), Nature 324: 87-89.
Thompson, R., L. Taylor, K. Kelly, R. Everett and N. Willets (1984), EMBO J. 3: 1175-1180.
Trifonov, E.N., (1980), Nucl. Acids Res. 8: 4041-4053.
Trifonov, E.N., (1983), Cold Spring Harb. Symp. Quant. Biol. 47: 271-278.
Trifonov, E.N., (1985), CRC Crit. Rev. Biochem. 19: 89-106.
Trifonov, E.N., and J.L. Sussman (1980), Proc. Natl. Acad. Sci. USA 77: 3816-3820.
Tung, C.-S., and C. Burks (1987), J. Biomolec. Str. Dyn. 4: 553-559.
Uberbacher, E.C., and G.J. Bunick (1985), J. Biomolec. Str. Dyn. 2: 1033-1055.
Ulanovsky, L.E., M. Bodner, E.N. Trifonov and M. Choder (1986), Proc. Natl. Acad. Sci. USA 83: 862-866.
Ulanovsky, L.E., and E.N. Trifonov (1987), Nature 326: 720-722.
von Kitzing, E., and S. Diekmann (1987), Eur. Biophys. J., in press.
Wu, H.-M., and D.M. Crothers (1984), Nature 308: 509-513.
Zahn, K., and F.R. Blattner (1985), EMBO J. 4: 3605-3616.
Zhurkin, V.B., (1985), J. Biomolec. Str. Dyn. 2: 785-804.

DNA Flexibility Under Control:
The Jumna Algorithm and its Application to BZ Junctions

Richard Lavery
Institut de Biologie Physico-Chimique
13 rue Pierre et Marie Curie
Paris 75005, France

When our laboratory received a high-resolution color graphics terminal about two years ago I spent several happy months in front of the screen admiring the beautiful images of nucleic acid structure that we were able to create. These images had several clear advantages over molecular models: they could be created quickly, they didn't need supports to keep them in place, they were accurate, they could be colored to reveal many interesting properties and nobody tried to steal them in order to build something else. However it soon became clear that one thing that was lacking in these images, namely that one cannot get hold of them and pull them into new conformations. This is a property of molecular models that is very useful and often exploited, especially in dealing with the subject matter of this meeting – "Unusual DNA Structures".

This disadvantage of graphic images is also shared by the mathematical algorithms which are most commonly used in the computer simulation of the properties of DNA. In the molecular mechanics approach each atom of the molecule under study is considered to be free in space and moves with 3 degrees of freedom under the influence of the force field generated by the other atoms. The force field itself contains, apart from electrostatic and Lennard-Jones interactions, terms representing the distorsion energy associated with chemical bonds, valence angles and torsion angles, but these are not direct variables of the algorithm and consequently they cannot be constrained individually or collectively. In fact, the only way the molecular conformation can be controlled in this approach, apart from choosing the starting point, is through energy penalty functions (for example on the distance separating pairs of atoms) which may subsequently have unpleasant effects on the course of the optimisation since these penalties have to be very strong in order to ensure that they will be satisfied.

I would now like to describe our efforts to overcome this problem, which can be more formally described as the control of DNA flexibility during computer simulation, and discuss what a new approach can bring to the theoretical study of nucleic acid conformation.

The starting point of our work was the requirement that the variables describing a fragment of DNA during energy optimisation should be appropriate for making large and defined changes in conformation. This necessitates two types of variables, firstly, the direct use of "chemical" or "internal" coordinates, that is to say bond lengths, valence angles and torsion angles to describe the conformation of the nucleotides forming the fragment and, secondly, helicoidal variables to describe the position of these nucleotides in space. Note that the use of helicoidal variables does not imply that we will only be able to treat regular helical DNA's. Although this was the case with our initial methodology, I will show that this restriction can easily be removed to open the way for studies of even the most irregular and distorted nucleic acid conformations.

The development of algorithms based on these requirements occurred in several stages in our laboratory and I would like to briefly summarize our first two models before passing to the details of the most advanced technique.

(I) SIR

The SIR (Successive Infinitesimal Rotations) model was developed in collaboration with Heinz Sklenar from the Central Institute of Molecular Biology in Berlin GDR and was aimed at energy optimising the conformations of regular helical DNA (Sklenar 1983, Sklenar et al 1986, Lavery et al 1986a,b). In the case of a DNA which has a symmetry repeat per nucleotide, the structure can be fully described by a single dinucleotide monophosphate unit. The relative position of the bases in this unit must be described by six variables, three translations and three rotations, which can be conveniently chosen as the helicoidal parameters detailed in the next section. The backbone conformation is described by the valence angles and torsion angles defining the sugar puckering and the phospho-diester conformation. Note that in this approach and those that follow bond length variations are excluded and valence angle changes are limited to the endo-cyclic and exo-cyclic angles of flexible rings. Although this choice limits to some extent the full flexibility of the macromolecules studied the loss is not very significant since such changes are usually associated with high distorsion energies. By making this approximation, on the other hand, we can achieve a very important reduction in the number of variables to be treated and a considerable simplification of the mathematical algorithm.

The fundamental problem in mathematically modelling nucleic acid conformational change is illustrated by the basic dinucleotide monophosphate unit. Any change of a single helicoidal variable or a backbone torsion or valence angle will necessarily result in breaking the backbone linkage between the two bases. Moreover, if the torsion or valence angle falls within one of the sugar rings, the ring closure will also be destroyed. It thus becomes clear that such changes must be coupled together in a specific way in order to maintain the integrity of the nucleic acid structure. In principal to solve this problem one should develope the analytical equations representing the backbone torsions and the sugar puckering as a function of the helicoidal parameters which position the bases. Attempts have been made in this direction (see, for example, Zhurkin et al 1981) based on the early work of Go and Scheraga, however, these equations are very complicated, non-linear and coupled and do not lend themselves easily to computer solution.

The SIR approach avoids this problem by making any changes in conformation through a series of very small steps. Since the starting conformation of the nucleic acid represents one solution of the set of equations described, changes to this solution induced by very small changes in backbone conformation or in the helicoidal variables can be determined by a set of linear differential equations which ignore all higher order terms. The solution of linear differential equations is clearly much simpler and faster. The programs we have developed achieve this solution by what amounts to a stepwise numerical integration and are capable of performing even large changes in DNA conformation with only modest amounts of computer time.

The details of this approach are given in full in the articles referenced. Those readers interested will see that the procedure is somewhat complicated by the introduction of a set of curvilinear variables, obtained by a transformation of the basic helicoidal parameters. This operation however allows the energy optimisation to follow a smooth path through allowed DNA conformations without encountering boundary points. The SIR methodology has recently been employed to study both isolated DNA flexibility and the role of this flexibility in the interaction of DNA with a variety of ligands (Lavery et al 1986a,b,c, Zakrzewska et al 1986).

(II) Cinflex

The Cinflex (Constrained Internal Coordinate Flexibility) model was developed in collaboration with Ian Parker and John Kendrick of the ICI New Science Group in England (Lavery et al 1986d). This model was designed to study any molecule or molecular complex containing flexible rings, including nucleic acids, by profiting from developments in the field of numerical analysis which have led to very efficient techniques for the optimisation of constrained systems (Powell 1982,1983).

The problem encountered in describing ring systems using internal coordinates is that the full set of these coordinates represents an over complete description of the system in question. Consider the sugar ring shown in figure 1. If we take the bond lengths to be constant then it is easy to see that fixing any two torsion angles (such as T1 and T2) and any three valence angles (such as A1, A2 and A3) fixes the position of all five ring atoms. The remaining three torsions (U1,U2,U3) and two valence angles (B1,B2) must therefore be dependent and cannot be used as variables during optimisation. A simple solution to this state of affairs is to consider one bond of the sugar ring to be "broken" (C4'-O1' in this case). If the length of this bond (d) can be included as a constraint during optimisation then the ring can effectively be thought of as a linear system whose conformation is described by five variables or, more generally, by 2N-5 variables for an N-membered ring. One should note however that the bond constraint mentioned is a mathematical constraint dealt with by the optimisation algorithm and not an energy penalty function. This is the basis of the Cinflex approach. The only complexity which arises with this technique is that minimizer requires the derivatives of the energy and of any constraints with respect to all independent variables. In the case of the energy derivatives this implies that it is also necessary to know the differential change of the dependent variables with respect to the independent variables because the dependent angles and torsions naturally contribute to the total conformational energy of the ring. We will return to this point in the following section.

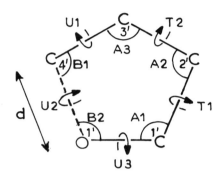

Figure 1. Treatment of sugar ring closure as a constraint.

The Cinflex algorithm thus overcomes one of the principal drawbacks of the SIR approach, the necessity of symmetry within the nucleic acid structure, but it has the disadvantage that control over the helicoidal variables has been lost. It was with the aim of correcting this restriction that the latest model for studying nucleic acid flexibility was developed. This more sophisticated technique, termed Jumna (Junction Minimisation of Nucleic Acids), combines the best elements of both of the above models and, I hope, justifies the optimism of the title of the present article. In the following section I will give the details of this model before discussing its application to a problem which has been troubling both theoreticians and experimentalists since 1979 - how do the B and Z conformations of DNA co-exist at a junction site and how does the transition between these two strikingly different conformations occur?

Methodology

In order to make this section easy to follow it has been divided into three sections: firstly, a careful description

192

of the helicoidal variables employed, secondly, an explanation of the constraint approach applied to DNA and, thirdly, details of the energy formulation that we have developed.

(a) Helicoidal variables

The first requirement for a rigorous definition of a set of helicoidal variables to be used for the construction and optimisation of DNA fragments is that they commute. That is to say that the same conformation will be obtained from a given set of such variables whatever the order in which they are applied. This is not the case for the common definitions based on axes between base atoms nor for simple rotations around cartesian axes and thus we must introduce some changes. It should also be noted that there is no escaping from the necessity of having 3 translational and three rotational parameters to fully define the position of the bases within DNA, thus certain definitions, such as that of Arnott (Arnott 1970) which only have two translational variables (separation of the bases along the helical axis and their displacement in a plane perpendicular to this axis) cannot be complete and must be discarded for our present purposes.

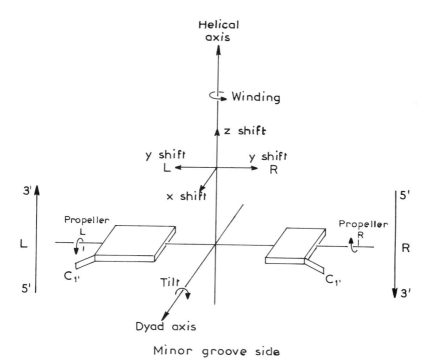

Figure 2. Definition of a set of commuting helicoidal variables.

Our proposition for a complete set of helicoidal variables is
given in figure 2. The translational position of any base is
fixed by 3 variables: Zsh along the helical axis, Xsh along
the local dyad axis and Ysh along the axis mutually
perpendicular to the helical and dyad axes. The values of
these shifts could be measured with respect to any base atom,
but in order to be closer to existing definitions we use a
reference point which corresponds to the position at which
the helical axis passes through the base pair plane in
Arnott's latest fibre coordinates for the B-DNA conformation
(Arnott et al 1980). Consequently, Xsh and Ysh are zero for
this reference structure. The rotational position of each
base is also given by 3 variables: Wdg (winding or twist)
around the helical axis, Tlt (tilt) around an axis parallel
to the local dyad axis, but passing through the base
reference point described above and Pro (propeller) around an
axis initially parallel to the Ysh axis, but passing through
the base reference point and rotated by the current tilt
angle around the Tlt axis.

Although these definitions may seem unnecessarily complicated
they are essential to obtain a complete commuting set of
helicoidal variables. However, it should be remarked that
they have been chosen in order to stay as close as possible
to the conventional meaning of tilt, propeller and
displacement. This aim explains the changing sense of Ysh
and propeller in the two strands, the direction of the dyad
vector toward the minor groove and the left handed sense of
the tilt rotation.

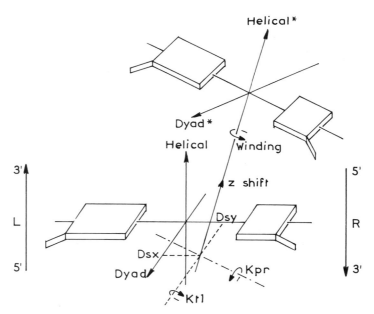

Figure 3.   Definition of a set of commuting junction
variables.

While these definitions are sufficient to describe the
relative position of any number of bases with respect to a
single helical axis, imagine what happens if this axis is
broken or bent at some point. The relative position of two
bases on either side of this interruption are given by only
two parameters Zsh and Wdg, whereas in the most general case
six parameters are again required. This problem can be
solved by clearly defining the relative orientation of the
two helical axis segments as shown in figure 3. Two further
translations Dsx and Dsy (dislocations along the Xsh (dyad)
axis and along the Ysh axis respectively) as well as two
rotations Ktl (kink tilt, around the shifted dyad axis) and
Kpr (kink propeller, around the tilted and shifted Ysh axis)
are defined in keeping with our previous conventions. (Note
that the dyad and Ysh axes referred to are the local axes of
the base preceding the kink in the 5'-3' strand). These new
variables also commute with the previous set. This full set
of variables now allows any DNA conformation to be described
easily including structures that are very far from regular
helices such as junctions between different conformations,
open base pairs, loops and so on.

It should finally be remarked that the definition of kink
variables for the helical axis could in principal be avoided
since the helical axis segment associated with each base or
group of bases is simply a geometrical construct and has no
reality in terms of the conformational energy of the
corresponding DNA fragment. In fact even a very irregular
structures could be defined uniquely by the six helicoidal
parameters described above with respect to an arbitrary
linear helical axis. However, as will be seen in the
application described later, it is often very convenient to
allow for changes of direction or dislocations in the helical
axis.

(b) The "Jumna" Constraint Approach

The aim of the Jumna approach is to combine the control over
the helicoidal parameters of DNA, as in SIR method, with the
rapidity and ease of treating ring closure derived from the
Cinflex procedure. In order to achieve this we first divide
the nucleic acid fragment (single or double stranded) to be
studied into 3' monophosphate nucleotides as shown in figure
4. These nucleotides are then positioned with respect to the
helical axis using the helicoidal parameters described above,
which consequently become direct variables of the
minimisation. In addition there are a total of 8 independent
backbone variables per nucleotide (1 glycosidic angle, 5
sugar variables and two backbone torsions, C3'-O3' and O3'-
P). It is then the task of the constraint minimizer to use
these variables to satisfy a total of 4 constraints per
nucleotide: the sugar ring closure distance between C4' and
O1', the backbone closure distance between O5' and C5', and
two valence angles P-O5'-C5' and O5'-C5'-C4'.

The minimisation algorithm that we use subsequently requires
derivatives of the conformational energy and of the
constraints with respect to all independent variables. The
constraint derivatives can be calculated as in the Cinflex
procedure by obtaining the differential motion of the atoms

involved in the constrained bond or angle with respect to each of the helical and backbone variables. (The necessary vector algebra is described in Lavery et al 1986d). The energy derivatives are obtained by first calculating the force on each atom of the DNA fragment generated by the energy functional discussed below. This is an easy task, but one should note that a slight complication is introduced by the angular dependence of the hydrogen bonding terms which results in the creation of torques on the atoms involved in such interactions. The total derivative can then be obtained by calculating the vector products of the atomic forces on the moving atoms with vectors linking these atoms to the appropriate center of rotation and summing the components of these products in the direction of the rotation axis. In the case of helical translation variables the derivatives are simpler, being just the sum of the components of the atomic forces on the moving atoms in the direction of the translation. One should note however that this procedure is made somewhat more difficult by the fact that changing any independent variable results in changing the dependent backbone torsions (C4'-C5', C5'-P and P-O5') for at least one junction. Since these torsions also contribute to the conformational energy of the DNA fragment their derivatives must also be calculated and added to those generated by the independent variable moves (see again Lavery et al 1986d).

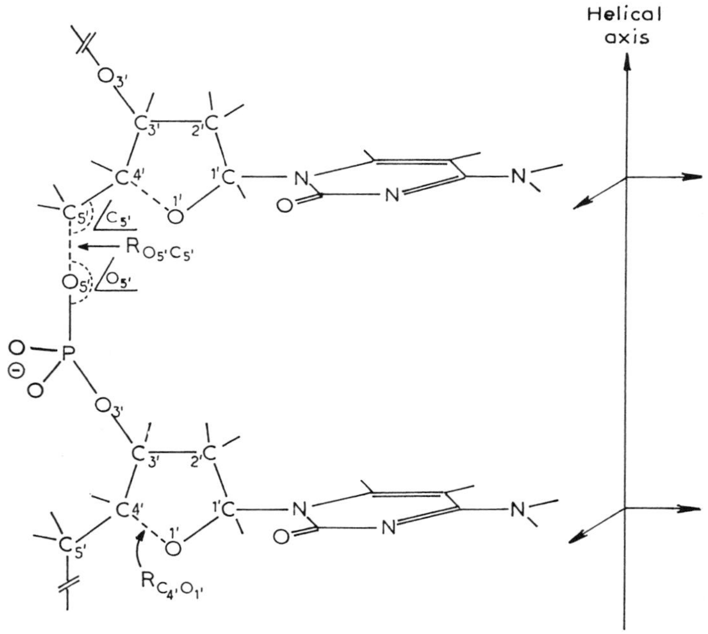

Figure 4. The constraints used in the Jumna procedure for a dinucleotide monophosphate unit.

The calculation of these derivatives, although mathematically simple, represents quite a lot of vector algebra and we have found that coding such procedures is greatly helped by checking each analytic derivative or derivative component numerically by making a very small move in the appropriate independent variable.

From the user's point of view the Jumna program is very easy to exploit. DNA fragments are build from a library of nucleotides which have standard geometries (based on B-DNA fibre coordinates) and standard helical orientations (Xsh=Ysh=Tlt=Pro=0). It is possible to build the desired fragment with any base sequence, any backbone and sugar geometry and any helical parameters, with, if desired, kinks or dislocations at any point along the helical axis. If, as is usually the case, this set of choices does not correspond to a closed sugar-phosphate backbone, the constraint minimizer will ensure correct closure during energy optimisation. The true power of this approach comes from the fact that it is not only possible to easily generate any starting point, but that it is also possible to fix any desired variables by simply removing them from the optimisation process. In this way one can guarantee that the final minimum energy conformation found will have any structural properties that are desired – a certain backbone kink, given sugar puckers, a given twist angle, an intercalation site and so on. Moreover, helical symmetry can easily be introduced by simply equating helicoidal or backbone variables along the fragment. The present implementation of Jumna is equipped in this way to impose mono or dinucleotide symmetry at will, with, in addition, the choice of identical of different strands for duplex fragments. It is these possibilities which set Jumna apart from normal molecular mechanics procedures. Their usefulness will hopefully be apparent from the example concerning junctions between different conformations of DNA described below.

From a numerical point of view the Jumna approach is rather fast. A symmetric six base pair fragment of DNA, which represents a problem of roughly 100 variables and 40 constraints can be energy optimised in about 4hr on a Vax750 or about 20min on a NAS 9080. In this respect, the breaking down of the DNA fragment into nucleotides has the considerable advantage of decoupling the internal variables. When this is not the case, as in the Cinflex approach, the large atom motions produced far from a small bond rotation can considerably slow down convergence.

(c) Energy Formulation

The conformational or interaction energy of the molecules we study is formulated below as a series of terms which are pairwise additive with the exception of the polarization energy between molecules. The first term of this formula is the electrostatic energy, calculated as the sum of interactions between atomic monopoles Q_i damped by a dielectric function $\epsilon(R)$ which is described below. The monopoles we employ are calculated by the Huckel-Del Re method which was specially reparameterised to obtain the best possible fit with the electrostatic distributions around the nucleic acid subunits calculated by quantum chemical methods (Lavery et al 1984). A second parameterisation has more recently been carried out for polypeptides and peptide-like molecules (Zakrzewska et al 1985). The next three terms represent the dispersion-repulsion energy calculated with a 6-12 dependence and using in part the parameter set developed by the group of Poltiev (Zhurkin et al 1980). Hydrogen bonds are dealt by the latter two of these terms, which take into account angular dependence by mixing together two sets of A,B parameters using a cosine function of the angle formed by the vectors X→H and H→Y for a bond X-H...Y. All of these terms are summed only over pairs of atoms separated by at least three chemical bonds in order to avoid calculating contributions which cannot change within our model. The next two terms take into account the distorsion energy associated with torsion angles τs (including anomeric effects) and valence angles σa and the final term represents the polarization energy based on average atomic polarisabilities αi parameterised by Kang and Jhon (Kang et al 1982). The polarization energy requires the calculation of the total electrostatic field $F i$ generated at atom i and is only calculated as part of DNA-ligand interaction energies, the internal polarization of DNA itself having been found to be very small. The reader is referred to (Lavery et al 1984,1986a and 1986d) for more details of the parameter set used. Note that since we consider bond lengths to be constant there is no term representing bond length distorsion as in classical molecular mechanics formulations.

$$E = \Sigma \; QiQj/\epsilon(R)Rij + \Sigma \; (\; -Aij/Rij^6 \; + \; Bij/Rij^{12} \;)$$

$$+ \; \Sigma \; [\; Cos\Theta \; (\; -Aij/Rij^6 \; + \; Bij/Rij^{12})$$

$$+ \; (1-Cos\Theta) \; (\; -Aij/Rij^6 \; + \; Bij/Rij^{12}) \;]$$

$$+ \; \Sigma \; Vs/2 \; (\; 1 \pm Cos \; Ns\tau s \;) + \Sigma \; Va \; (\; \sigma a - \sigma a^\circ)^2$$

$$- \; 1/2 \; \Sigma \; \alpha i \; Fi^2$$

The dielectric function $\epsilon(R)$ we use is based on a model of the dielectric damping of the electrostatic interaction between two charges in a polar solvent developed by Hingerty (Hingerty et al 1985). We have reformulated this function as shown in figure 5 so that it is possible to vary both the plateau value of the dielectric reached at long distance (D) and the slope of the sigmoidal segment of the function (S). Note that S may also be written, following our previous publications, as H/2.674, where H is the distance corresponding to the half height of the curve. Figure 5 shows two forms of this function, the upper curve corresponding to S=0.356 being a close fit to that proposed by Hingerty and representing rather strong damping and a lower curve corresponding to S=0.16. We will return to the effect of these changes on DNA conformational energy in the following section.

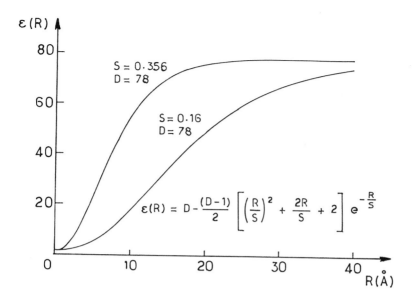

$$\varepsilon(R) = D - \frac{(D-1)}{2}\left[\left(\frac{R}{S}\right)^2 + \frac{2R}{S} + 2\right] e^{-\frac{R}{S}}$$

Figure 5. The dielectric function $\varepsilon(R)$ for two different slopes.

Application of Jumna to the creation of BZ junctions

The first step performed to study how the B and Z allomorphs of DNA could be put together consisted of optimising the structures of these two conformations separately with starting geometries from the fibre results of Arnott (Arnott et al 1980) and the single crystal results from the group of Rich (Wang et al 1980) respectively. In both cases we studied fragments of duplex DNA five nucleotides long with the sequence CGCGC. Since our aim was to study a junction between these conformations, we attempted to find conditions under which B- and Z-DNA would have similar stabilities. Two parameters were varied in order to achieve this, the slope of our dielectric function and the net charge on the phosphate group. Table 1 shows the results for just two conditions, firstly, with a slope of 0.16 and phosphate charges of -1.0 corresponding to weak electrostatic screening and, secondly, with a slope of 0.356 (close to Hingerty's curve) and a phosphate charge of -0.5 corresponding to rather strong screening. Note that in this table the pairs of values correspond respectively to the cytidine and guanosine nucleotides and also that the glycosidic angle and sugar puckers are average values since the dinucleotide symmetry restrictions imposed on these fragments do not apply to the backbone variables.

As might be expected the latter set of conditions succeed in considerably stabilizing the Z conformation which predominates in high ionic concentrations and these were adopted for the junction studies which follow. (See also similar theoretical results obtained by Kollman et al 1982).

Table 1. Optimised conformations of B- and Z-DNA as a function of electrostatic screening.

Conformation	B	Z	B	Z
$\epsilon(R)$ slope	0.16	0.16	0.356	0.356
Phos Charge	-1.0	-1.0	-0.5	-0.5
Xsh (Å)	2.2 / 2.3	4.5 / 4.3	3.6 / 3.7	2.7 / 2.5
Ysh (Å)	-0.4 / 0.5	-2.6 / 2.8	-0.3 / 0.4	-2.2 / 2.3
Zsh (Å)	3.6 / 3.2	2.9 / 4.6	3.5 / 3.0	2.8 / 4.3
Tlt (°)	-3.0 / 3.1	5.6 / -8.3	-1.1 / 3.6	9.1 / 1.8
Pro (°)	1.2 / 1.3	-177.7 /172.8	2.4 /-0.2	176.4 /175.1
Wdg (°)	30.2 /38.3	-51.0 / -4.4	28.9/32.8	-49.8 / -9.5
Gly (°)	37 / 70	30 /-125	36 / 62	28 /-122
Sugar	O1'endo/	C2'endo/	O1'endo/	C2'endo/
	C2'endo	C3'endo	C2'endo	C3'endo
Energy	-303.4	-283.4	-49.2	-46.3
δEnergy Z-B		20.0		2.9
(kcal/mole)				

It is also interesting to note that the best correlation between the optimised conformations and the standard B and Z geometries are obtained under the conditions where they are experimentally most stable. As table 1 shows, increasing the screening for B-DNA increases the displacement of the base pairs from the helical axis (Xsh) and decreases the average winding from 34° to 31°, conversely, decreasing the screening for the Z-DNA leads to the same effects.

One important point should be mentioned concerning the helicoidal variables given in table 1. The care which was taken in defining these variables so that they would form a commuting set leads to what may, at first, seem a surprising result. While the parameters for the B conformations appear normal, the parameters for the Z conformations exhibit propeller angles close to 180° and tilts opposite in sign to those usually given. These values simply indicate that the base pairs in Z-DNA have "flipped-over" with respect to the B conformation. This fact is by no means clear from the usual conventions employed for defining helicoidal parameters and should be considered as an advantage of the present approach.

The second step of actually forming a BZ junction can now be carried out by putting together 3 nucleotide pairs from each of the optimised DNA conformations. This can be done in two different ways. The junction of the 5'-3' strand can occur between a cytidine in the B conformation and a guanosine in the Z conformation giving an overall sequence CGC-GCG (termed type I from now on) or between a guanosine in the B conformation and a cytidine in the Z conformation giving the sequence GCG-CGC (termed type II). If one forms either of these junctions using a common linear helical axis there is clearly no way of linking together the B and Z backbones whatever the twist angle between the two conformations without considerable distorsion. This is clear from figure 6 which shows (for a type 1 junction) that the error of closure

in the O5'-C5' bond is approximately 10Å for the optimal Wdg
(or twist) angle of roughly 0°. This however causes no
problem for the Jumna program since we may freely set kink
parameters at the junction site. A little experimentation
shows that a much better approximation to closure can be
achieved by simply sliding the helical axis of the Z fragment
roughly 10Å towards the minor groove of the B fragment (i.e.
Dsx=+10 following the conventions defined above).

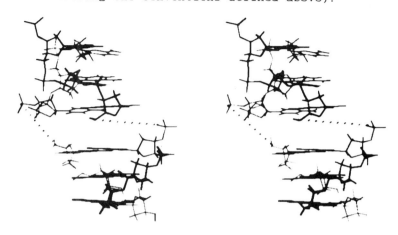

Figure 6. The error in backbone closure (indicated by the
dotted lines) for coaxial fragments of B-DNA (below) and Z-
DNA (above).

This manoeuver yields starting conformations which are
sufficiently good to be energy minimized. These
minimizations were carried out at first without allowing any
change in the helicoidal variables within either of the
fragments. The results in the first two columns of table 2
show that closure can easily be achieved under these
circumstances. If we take as reference energy the mean
energy of the B and Z conformations obtained above, (but
extended to six nucleotide pairs for consistency) 58.5
kcal/mole, then the junction formation costs 29.7 kcal/mole
and 19.3 kcal/mole respectively for type I and type II.

Table 2. Energy optimised BZ junction conformations.

Junction Helical var.	I Locked	I Locked	II Free	II Free
Dsx (Å)	10.5	9.4	13.3	8.7
Dsy (Å)	-1.4	-0.8	-1.2	-1.0
Ktl (°)	15.6	-4.7	13.8	1.8
Kpr (°)	-15.8	-10.6	-42.1	27.6
Zsh (Å)	5.2	5.5	7.1	1.3
Wdg (°)	-1.0	-16.2	4.3	-17.2
Energy (kcal/mole)	-28.8	-37.0	-39.2	-48.2

It is now possible to go further and release the helicoidal
variables. It was decided however to keep the variables of
the terminal pairs fixed in order to confine the junction
perturbations to the four inner base pairs, which seems a
reasonable assumption following experimental indications
(Wells et al 1983). Under these circumstances the junctions
further stabilize to yield formation energies of 19.3
kcal/mole (type I) and 10.3 kcal/mole (type II) and there are
noticeable changes in conformation around the junction site.
Although these values are larger than the current
experimental estimates of junction free energies (roughly 5
kcal/mole, Frank-Kamenetskii et al 1984 and references
therein), they are quite small on the scale of theoretical
conformational energy changes. Some changes in these values
can certainly be expected from solvent or counterion effects
not included in the present calculation. The kink parameters
obtained are given in the latter two columns of table 2 and
the helicoidal parameters for the base pairs on either side
of the junction are given in table 3 (again the pairs of
values refer respectively to the cytidine and guanosine
nucleotides). The junctions are also illustrated by the
stereo molecular graphics in figures 7 and 8.

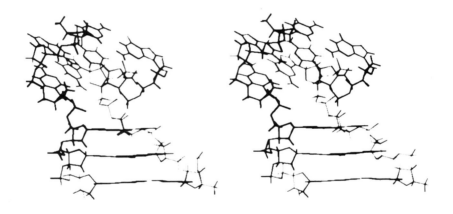

Figure 7. Stereo graphic of the optimised Type I junction
looking into the major groove side of the B fragment (below).

Comparing tables 1 and 3 it can be seen that there is
surprisingly little change in the helicoidal parameters of
the base pairs either above or below the junction with
respect to the isolated B and Z conformations. The only
notable exceptions to this are the reduced propeller angles
of the C on the Z side of the type I junction and of the G on
the Z side of the type II junction. The majority of the
structural adaption necessary occurs in the parameters of the
junction itself (table 2 and table 3, Wdg and Zsh of the GC
pair on the Z side of the junction) and in the sugar puckers
of the bases on the Z side (table 2).

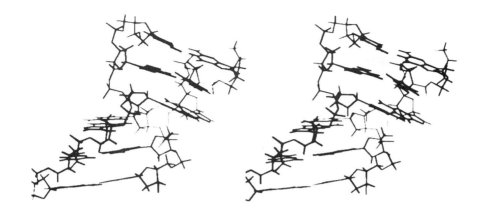

Figure 8. Stereo graphic of the optimised Type II junction looking into the <u>minor groove side</u> of the B fragment (below).

The most remarkable feature of the junction parameters is the difference observed between the two types of junction. While both have large displacements towards the minor groove (Dsx), their kinking directions are quite different. The type I (C-5'3'-G) junction exhibits a strong kink towards the major groove of the B fragment (Kpr= -42°), while the type II (G-5'3'-C) junction bends the opposite way, towards the minor groove of the B fragment (Kpr= 28°).

Table 3. Optimised conformations of base pairs on either side of the BZ junctions.

Junction	I	I	II	II
Side	B	Z	B	Z
Xsh (Å)	3.5 / 3.6	2.3 / 2.1	3.3 / 3.4	2.4 / 2.2
Ysh (Å)	-0.2 / 0.4	-1.6 / 1.7	-0.6 / 0.7	-1.9 / 2.1
Zsh (Å)	3.4 / 3.5	7.5 / 7.1	3.6 / 3.4	1.3 / 1.3
Tlt (°)	0.4 / 0.2	9.7 / -2.0	-2.3 /-4.6	10.2 / 4.9
Pro (°)	4.0 / 1.2	166.3 /179.5	-7.8 / 9.1	179.2 /169.9
Wdg (°)	29.2 /28.6	6.0 / 4.3	35.6/35.6	-17.2 /-18.1
Gly (°)	33 / 26	30 /-125	23 / 55	18 /-113
Sugar	O1'endo/	C3'exo /	O1'endo/	C1'exo /
	O1'endo	C4'endo	C2'endo	C1'exo

These differences can be explained by the half twist which the cytidine sugar undergoes in passing from the B to the Z conformation. This twist, which is right handed when viewed along a GC pair from G to C, greatly shortens the linkage distance between the cytosine and the preceding guanine in the B fragment and obliges the Z fragment to tilt in a left handed sense to re-establish the phosphodiester bond. When the Z side cytosine occurs in the 3'-5' strand, as for type I junctions, left handed rotation of the Z fragment brings it naturally towards the major groove side of the B fragment,

while the reverse happens for type II junctions. This is
represented schematically in figure 9. Such kinking has
another consequence which further distinguishes the two types
of junction. As can be seen from figures 7 and 8, for the
type I junction, the major groove kink leads to the 3'-5'
strand of the Z fragment approaching the base pairs of the B
fragment, which in turn forces the rise of the junction to
become very large (Zsh= 7.1Å). In contrast the opposing kink
of the type II junction brings the Z fragment into a space on
the minor groove side of the B fragment and leads to a much
more compact (Zsh= 1.3Å) and more stable conformation.

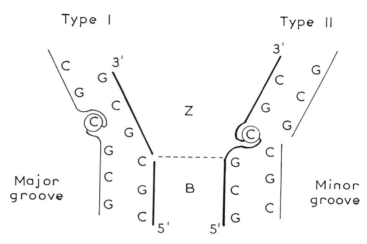

Figure 9. Schematic representation of the BZ junction kinks.

Since available experimental evidence seems to indicate that
the BZ transition occurs stepwise along the double helix
(Wells et al 1983, Callahan et al 1986 and references
therein) by rotation of successive base pairs (probably
without hydrogen bond breakage, although very recent results
suggest that this may in fact occur, Manzini et al 1987),
calculations are now underway to go one step further and
complete the inversion of one nucleotide pair from the B to
the Z conformation. Again this task is made relatively easy
by the Jumna algorithm since the change of the base propeller
angles from roughly 0° to 180° which exemplify this
transition can be made in easy stages, locking an
intermediate value at each calculation, and thus overcoming
the difficulty of passing what is clearly a rather large
energy barrier. In this respect it is interesting to note
that one overall feature of this transition already seems
apparent. As Harvey has clearly pointed out (Harvey 1983)
the backbone twist associated with the cytidine nucleotides
of Z-DNA implies a sense of rotation for the bases in passing
from B to Z. Using the terminology developed above, this
means that, at a type I junction, the terminal B base pair
must rotate under towards the lower B pair in order to
finally stack with the first Z pair (right handed rotation
around the G→C vector), while at the same time the junction
kink must rotate in the opposite sense to finally lean

204

towards the minor groove. The final stage of this motion creates a type II junction. Continuing the process, the bases of this junction must rotate over towards the Z pairs before finally stacking under the first such pair (again right handed rotation around the G→C vector), while the junction kink rocks back towards the major groove side. The travelling transition zone should thus involve a "flip-flop" motion where the kink and base rotations nicely counterbalance one another. One can also remark that the lack of stacking between the B and Z fragments at the junctions described can offer some explanation of the cooperativity of the transition since the base pairs at the junction can certainly start to rotate more easily than if they we closely held by stacking interactions on both sides.

Conclusions

The Jumna constraint algorithm which has been described offers very extensive possibilities for building easily even very "unusual" DNA conformations. The fact that it has control over both the helicoidal parameters of the bases and the backbone geometry means that it is very well adapted to simulating large changes in DNA conformation. The additional possibility of kinking or dislocating the helical axis of the fragments studied opens the route for the investigation of junctions of all types, loops, open pairs or other strong perturbations of regular helical structure.

The results presented demonstrate that Jumna has no difficulty in obtaining a model for the junction between B- and Z-DNA and moreover brings to light a number of interesting structural features including notably different stabilities and geometries depending on whether the junction is formed at a C-5'3'-G pair or a G-5'3'-C pair.

References

S. Arnott 1970 Progress in Biophys. and Mol. Biology 21,265.

S. Arnott, R. Chandrasekharan, D. L. Birdsall, A. G. W. Leslie and R. L. Ratliff 1980 Nature 283,743 and coordinates communicated to our laboratory by S. Arnott.

L. Callahan, F-S. Han, W. Watt, D. Duchamp, F. J. Kézdy and K. Agarwal 1986 Proc. Natl. Acad. Sci. USA 83,1617.

M. D. Frank-Kamenetskii and A. V. Vologodskii 1984 Nature 307,481.

N. Go and H. A. Scheraga 1970 Macromolecules 3,178.

S. C. Harvey 1983 Nucl. Acids Res. 11,4867.

B. Hingerty, R. H. Richie, T. L. Ferrel and J. E. Turner 1985 Biopolymers 24,427.

Y. K. Kang and M. S. John 1982 Theoret. Chim. Acta 61,41.

P.Kollman, P.Weiner, G.Quigley and A.Wang 1982 Biopolymers 21,1945.

R.Lavery, K.Zakrzewska and A.Pullman 1984 J.Comp.Chem. 5,363.

R.Lavery, H.Sklenar, K.Zakrzewska and B.Pullman 1986a J.Biomol. Struct. and Dynam. 3,989.

R.Lavery, H.Sklenar and B.Pullman 1986b J.Biomol.Struct. and Dynam. 3,1015.

R.Lavery, K.Zakrzewska and B.Pullman 1986c J.Biomol.Struct. and Dynam. 3,1155.

R.Lavery, I.Parker and J.Kendrick 1986d J.Biomol.Struct. and Dynam. 4,443.

G.Manzini, L.E.Xodo, F.Quadrifoglio, J.H.van Boom and G.A.van der Marel 1986 J.Biomol.Struct. and Dynam. 4,651.

M.D.J.Powell 1982 Report DAMPT NA4 Dept. of Applied Mathematics and Theoretical Physics, University of Cambridge, England.

M.D.J.Powell 1983 Report DAMPT NA17 Dept. of Applied Mathematics and Theoretical Physics, University of Cambridge, England.

H.Sklenar 1983 Studia Biophysica 93,175.

H.Sklenar, R.Lavery and B.Pullman 1986 J.Biomol.Struct. and Dynam. 3,967.

A.H-J.Wang, G.J.Quigley, F.J.Kolpak, G.van der Marel, J.H.van Boom and A.Rich 1980 Science 211,171.

R.D.Wells, R.Brennan, K.A.Chapman, T.C.Goodman, PA.Hart, W.Hillen, D.R.Kellogg, M.W.Kilpatrick, R.D.Klein, J.Klysik, P.F.Lambert, J.E.Larsen, J.J.Miglietta, S.K.Neuendorf, T.R.O'Connor, C.K.Singleton, S.M.Stridivant, C.M.Veneziale, R.M.Wartell and W.Zacharias 1983 Cold Spring Harbour Symposia 47,77.

K.Zakrzewska and A.Pullman 1985 J.Comp.Chem. 6,265

K.Zakrzewska and B.Pullman 1986 J.Biomol.Struct. and Dynam. 4,127.

V.B.Zhurkin, V.I.Poltiev and V.L.Florent'ev 1980 Molekulyarnaya Biologiya 14,116.

V.B.Zhurkin, Yu.P.Lysov and V.I.Ivanov 1981 Biopolymers 17,377.

BASE SEQUENCE EFFECTS IN CURVED AND RODLIKE DNA

Wilma K.Olson, A. R. Srinivasan, Rachid C. Maroun, Ramon Torres & William Clark
Department of Chemistry
Rutgers, the State University of New Jersey
New Brunswick, New Jersey 08903

INTRODUCTION

A number of intriguing structural studies have prompted efforts to understand how the sequence of heterocyclic bases governs the conformation and properties of double helical DNA. The side groups introduce subtle irregularities in the local geometries of crystalline oligomers (Conner *et al.*; Dickerson & Drew 1981a; Fratini *et al.*; Kneale *et al.*; McCall *et al.* 1985, 1986; Shakked *et al.*; Viswamitra *et al.*; Wang *et al.* 1982a, 1982b) and affect the observed twisting of adjacent residues in solution (Kabsch *et al.*; Peck & Wang; Rhodes & Klug; Strauss *et al.*). The chain sequence also influences the overall dimensions of the DNA, the alternating poly d(GC) duplex being more extended (Thomas & Bloomfield) and the alternating poly d(AT) duplex more compact (Chen *et al.*) than random sequence B-DNA (Borochov *et al.*; Hagerman 1981; Kam *et al.*; Kovacic & van Holde; Rizzo & Schellman). Moreover, certain sequences curve naturally in solution, moving more slowly than expected on nondenaturing polyacrylamide gels (Anderson; Bossi & Smith; Challberg & Englund; Diekmann 1985, 1986; Diekmann &Wang; Hagerman 1985, 1986; Kitchin *et al.*; Koepsel & Khan; Koo *et al.*; Ryder *et al.*; Simpson; Snyder *et al.*; Wu & Crothers; Zahn & Blattner 1985a, 1985b), exhibiting unusual rotational relaxation properties (Hagermann 1984; Levene *et al.*; Marini *et al.*), and adopting arc-like shapes under the electron microscope (Griffith *et al.*; Ray *et al.*). Finally, the chemical composition of the bases determines the extent to which intercalated fluorescent dyes wobble between adjacent residues, those within the poly dA · poly dT duplex fluctuating to a much greater extent than those within the poly dG · poly dC complex (Hogan *et al.*).

While the rules governing the spatial patterns observed in crystallographic examples are not yet well understood, simple steric arguments have already proven quite useful in rationalizing some features of chain structure (Calladine; Calladine & Drew 1984, 1986; Dickerson; Dickerson & Drew 1981b; Dickerson *et al.* 1982, 1983). The solid state data, however, are still too limited (Shakked & Rabinovich) to offer a reliable description of sequence dependent helical structure or to extract a set of rules of sequence governed geometric variations. Moreover, the popular static models of double helical structure (Arnott *et al*. 1983; Hagerman 1984; Koo *et al*.; Trifonov; Trifonov & Sussman) are unable to account for the intrinsic curvature of certain DNA sequences in solution and the observed sequence dependent variations of average chain dimensions. With one recent exception (Calladine & Drew 1986), the X-ray derived models are also generally not concerned with the sequence dependent flexibility of nucleotide residues. The known differences in dye wobbling observed with triplet anisotropy decay measurements in different DNA's have yet to be rationalized by a molecular model.

As part of a program to understand how local structural irregularities are translated at the macromolecular level in DNA, we have recently completed a series of calculations (Maroun & Olson; Srinivasan *et al*.) probing the structure and properties of the double helix. We have carried out potential energy calculations on the simplest dimeric fragments that can be used to build long DNA's, computing the energy as a function of the orientation and displacement of various base pair combinations (Srinivasan *et al*.). As a first approximation, we have ignored the long range interactions between more distant neighbors (Jernigan *et al*.) which may affect the local chain configuration. The energies, while approximate, reflect local structure more accurately than many of the intuitive models used to date. We have subsequently incorporated the energies in statistical mechanical calculations (Maroun & Olson) to treat the average extension and orientation of the chain and in rotational isomeric state modeling studies (Srinivasan *et al*.) to predict the intrinsic curvature of different polymeric sequences. In this paper we illustrate how the local energies depend upon base sequence and how these differences influence overall properties of the double helix. We also present a new interpretation of DNA curvature based upon the potential energy analysis.

METHODS

Local Base Pair Energies

The local flexibility of the double helix is estimated in terms of the energies of interaction of free base pairs. Rough approximations of this sort are not too different from more realistic models of DNA flexibility that include the interactions of the chain backbone as well as those of adjacent residues (Srinivasan *et al*.). The energies are measured as a function of the roll, tilt, and twist between adjacent base pairs using standard semiempirical potential functions. The roll angle ρ is the rotation about the long, in-plane axis of a base pair, the tilt angle λ that about the dyad or short axis relating complementary residues of the complex, and the twist angle τ that about the normal to the base pair plane. The angle convention used here is opposite in the sign of ρ from that used in X-ray analyses (Arnott *et al*. 1969; Fratini *et al*.) but similar to that used in earlier theoretical studies (Ulyanov & Zhurkin 1984a, 1984b). The twist angle τ is identical to the helical rotation of an unbent helix, positive values corresponding to right-handed structures and negative values to left-handed ones. A positive value of ρ is chosen to increase the separation between atoms along the major groove edges of adjacent base pairs, while a positive value of λ is seen to rotate the normal of base i+1 in a clockwise sense relative to that of base i. Coordinates are chosen so that the standard B-DNA geometry is reproduced when $\rho = 0°$, $\lambda = 0°$, and $\tau = 36°$. Bond lengths and valence angles are fixed at accepted values (Taylor and Kennard) so that only nonbonded contributions are considered.

The potential energy function is a sum of pairwise van der Waals' repulsions, London attractions, and electrostatic interactions between the jth atoms of residue i and the kth atoms of residue i+1.

$$V_{i,i+1}(\rho, \lambda, \tau) = \sum_j \sum_k \left[\frac{A_{jk}}{r_{jk}^{12}} - \frac{B_{jk}}{r_{jk}^6} + \frac{332\ \delta_j \delta_k}{\varepsilon r_{jk}} \right] \tag{1}$$

Nonbonded parameters -- including van der Waals' constants A_{jk}, London constants B_{jk}, and partial atomic charges δ_j -- are values detailed elsewhere (Olson 1978, 1982; Srinivasan & Olson 1980; Srinivasan et al.; Taylor & Olson) that reproduce the hydrogen bonding energies and base stacking distances of model compounds. The numerical constant appearing in equation 1 is chosen to yield energies in kcal/mole when the distances of separation are expressed in Ångstroms and the partial atomic charges in fractions of one electronic charge. The dielectric constant ε is assigned a value of 4. As a first approximation, individual base pairs are kept in an ideal planar Watson-Crick arrangement, ignoring the possible effects of propeller twisting, buckling, or other small perturbations of the hydrogen bonded complex.

Virtual Bond Model

Chain extension and flexibility can be related to the local structure and motions of the DNA duplex using a simple virtual bond scheme that describes the orientation and displacement of adjacent nucleotide residues. The end-to-end vector $\mathbf{r}$ connecting chain termini, for example, is a sum of successive virtual bond vectors, one such vector $\mathbf{v}_i$ taken per nucleotide repeating unit. Specifically,

$$\mathbf{r} = \sum_{i=1}^{x} \mathbf{v}_i \tag{2}$$

where the virtual bonds are expressed in a common coordinate system and x is the number of chain repeating units. The coordinate transformation is achieved with the transformation matrices $\mathbf{T}_{i,i+1}(\rho, \lambda, \tau)$ that relate the orientation of adjacent residues. The $\mathbf{T}_{i,i+1}(\rho, \lambda, \tau)$ are arbitrarily defined by the products of right-handed rotations $\mathbf{Z}(\tau)\ \mathbf{X}(\rho)\mathbf{Y}(\lambda)$ about the twist (z), roll (x), and tilt (y) axes, respectively, of residue i. Virtual bond vectors are drawn along the local z-axis of the base pairs from a central point on one residue to a corresponding point on the next. The backbone is not explicitly treated, although it is possible to identify numerous sugar-phosphate conformations that fit a particular base-base arrangement (Srinivasan & Olson 1987). The orientational and translational fluctuations of adjacent residues that are permitted are small enough so that no major chain backbone changes are expected.

In practice, a different set of fixed orientational and translational parameters can be entered for each residue of the DNA chain, generating specific spatial arrangements of the double helix. By introducing the energy weighted transformation matrices $< T_{i,i+1} >$ associated with each residue, the average unperturbed extension of the double helix can be obtained. The i,i+1 dimeric chain sequence is presumed to be independent of all dimer fragments in this approach. In other words, the average orientation and displacement of successive base pairs is assumed independent of their position and sequence in the polynucleotide chain. This is generally thought to be a good approximation in double helical DNA although there is little structural evidence (Shakked & Rabinovich) in its support.

The resulting average end-to-end vector $< r >$ is the so-called persistence vector of the chain (Flory, 1973), the angle brackets denoting chain averages. The projection of this vector on the first virtual bond in the chain is the persistence length a, a measure of the distance over which the direction of the DNA chain is maintained. This parameter is normally associated with a chain molecule of infinite length (Kratky & Porod) but can be expressed for finite chains as follows:

$$a = < r \cdot (v_1/v) > = v \left[1 + \sum_{k=1}^{x-1} \left[\prod_{i=1}^{k} < T_{i,i+1} > \right]_{33} \right] \qquad (3)$$

The subscript in this expression denotes the 3,3 element of the specified matrix products, while v is the virtual bond length (3.4 Å). Execution of the summation is readily accomplished with standard matrix generator techniques (Flory, 1969). The averaged matrices are Boltzmann averages evaluated on the basis of the $V_{i,i+1}(\rho, \lambda, \tau)$ potential energy function.

The double helix can also be characterized in terms of the average spatial orientation of terminal base pairs (Flory *et al.*). These angles are obtained from the serial product of sequential average transformation matrices. The mean overall bending of the chain ends $< \gamma >$, as measured by the scalar product of terminal base normals (e.g., z-axes), is given by the 3,3 element of the serial product of the $< T_{i,i+1} >$'s.

$$< \gamma > = < (v_1/v) \cdot (v_x/v) > = \left[\prod_{i=1}^{x-1} < T_{i,i+1} > \right]_{33} \qquad (4)$$

RESULTS

Sequence Dependent Potential Energies

The dimeric energy surfaces used to describe the local flexibility of the poly dA · poly dT and the poly dG · poly dC helices are reproduced in figure 1. The energies are calculated at

5°-increments in roll (ρ) and tilt (λ) with twist (τ) fixed at 36°. Contours are drawn by interpolation at 1 kcal/mole intervals from 1 to 5 kcal/mole relative to the lowest energy states marked by ×. The presumed twist angle is close to that found to characterize poly dA · poly dT in solution (Kabsch *et al.*; Peck & Wang; Rhodes & Klug; Strauss *et al.*) but is somewhat greater than that (33-34°) reported to describe poly dG · poly dC under the same conditions (Peck & Wang). As noted elsewhere (Srinivasan *et al.*), both the $(G \cdot C)_2$ and $(A \cdot T)_2$ energy maps are only slightly altered from those in figure 1 when τ is varied by ±3°.

As evident from figure 1, the $(A \cdot T)_2$ map is characterized by a broader range of low energy states than the $(G \cdot C)_2$ surface. The areas lying within the energetically accessible 5 kcal/mole contour of each map, however, are roughly comparable in size. In both instances residues are found to roll more easily about their long axes than to tilt about their short (dyad) axes. Similar bending motions are indicated by the anisotropic temperature factors of the crystalline B-DNA dodecamer (Holbrook & Kim). The range of rolling motions within the 5 kcal/mole contours is roughly three times that of the tilting motions. The range of base pair rolling at low energies (e.g., less than 1 kcal/mole), however, is considerably greater for $(A \cdot T)_2$ than for $(G \cdot C)_2$. The tendencies toward chain bending are also more symmetrically distributed for $(A \cdot T)_2$ than for $(G \cdot C)_2$. The relative contributions of positive versus negative roll to the total chain partition function are 0.59:0.41 in the former system and 0.92:0.08 in the latter. The ratios of positive versus negative tilting contributions, however, are comparable in the two systems at 0.34:0.66 and 0.37:0.63, respectively.

The greater flexibility of $(A \cdot T)_2$ compared to $(G \cdot C)_2$ in figure 1 is also consistent with the observed triplet anisotropy (Hogan *et al.*) of poly dA · poly dT and poly dG · poly dC. The computed dimeric partition functions at 298 °K are 7.1 and 2.7, respectively. According to more elaborate conformational energy estimates (Ulyanov & Zhurkin 1984a, 1984b), however, $(G \cdot C)_2$ is somewhat more flexible than $(A \cdot T)_2$, at least when bent up to 4°. A comparable potential energy study of the base sequence dependent anisotropy of chain bending at larger angles, however, has never been reported.

The optimum orientation of adjacent base pairs is clearly skewed on the two surfaces, the minimum energy configurations occurring in figure 1 at $\rho,\lambda = (5°, 0°)$ for $(A \cdot T)_2$ and at $\rho,\lambda = (10°, 0°)$ for $(G \cdot C)_2$. The more skewed minimum on the latter surface can be traced to the 2-amino groups of G which encounter less favorable interactions with atoms of the neighboring base pair when $\rho < 0°$. The energies are dramatically altered if G is replaced by I, its 2-deoxyamino analog, in the calculations (Srinivasan *et al.*). In contrast to earlier suppositions (Prunell *et al.*; Trifonov & Sussman), the preferred $(A \cdot T)_2$ dimeric configuration is little affected by the 5-methyl groups of successive T's. With the assumed twist of 36° and the standard B-DNA translation, the methyls are essentially free of restrictive nonbonded contacts with adjacent residues. Indeed, an almost identical energy surface is obtained if T is replaced by U in these calculations (Srinivasan *et al.*). The hydrophobic contacts of methyl groups of T, however, are an important factor in the stabilization of positive over negative roll on the $(A \cdot T)_2$ energy surface as recently hypothesized (Jernigan

211

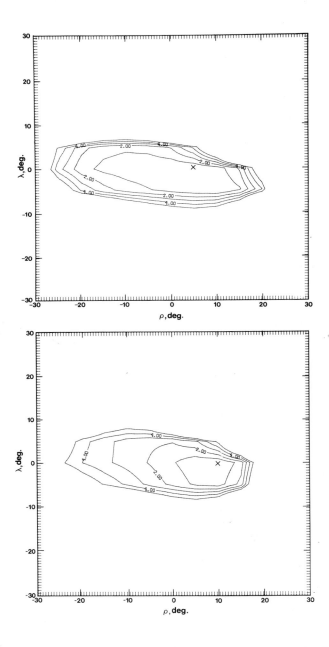

Figure 1. Contour diagrams of the total potential energy V associated with the rolling (ρ) and tilting (λ) motions of adjacent (A · T) and (G · C) base pairs in the B-DNA duplex. The angular twist τ is fixed at 36° and the distance between base pair centers at 3.4 Å. Contours are drawn at 1 kcal/mole intervals from 1 to 5 kcal/mole relative to the lowest energy conformations located at ×. The (A · T)$_2$ surface is above the (G · C)$_2$ map in the figure.

et al.). Hydrophobic interactions of thymine methyl groups are also thought (Sundaralingam & Rao) to allow the polynucleotide chain backbone to flex through close interactions with neighboring sugar rings.

The symmetric conformational energy map describing the preferred orientation of successive (A · T) base pairs, however, is not consistent with current static models (Arnott *et al.* 1983; Hagerman 1984; Koo *et al.*; Trifonov; Trifonov & Sussman) of DNA curvature. With successive residues twisted at an angle of 36°, there is limited bending asymmetry in the energetic analysis of the dimer complex. The (A · T) base pairs are presumed to be perturbed from the classical B-DNA structure in the various static models. The (A · T) residues described by the energy function, however, are similar to those of an isotropically bent rod with no preference for positive over negative bending, although there is a computed preference for bending (e.g., rolling) into the major and minor grooves of the duplex over other directions. The theoretically predicted orientation of successive (G · C) base pairs, on the other hand, is noticeably skewed with a marked preference for residues to roll in a positive direction about their long axes. Regular bending of this sort is known to distort the B-DNA double helix reducing the number of residues per helical turn and tilting base planes with respect to the helical axis (Prunell *et al.*; Olson *et al.*). A structure so generated can be used to construct curved DNA's (Selsing *et al.*; Jernigan *et al.*).

If inserted between rodlike segments, such as those of flexible poly dA · poly dT, partial turns of the poly dG · poly dC helix are expected to bend the DNA. Since the bending is maximized at half turns of the duplex, poly $d(A_5G_5)$ · poly $d(T_5C_5)$ should be a strongly curved DNA. While the curvature of this sequence has never been characterized experimentally, it has been suggested that half turn periodicities in (A · T) and (G · C) sequence content are used to facilitate the tight bending of nucleosomal DNA around the histone octamer (Drew & Travers). The (A · T)/(G ·C) and (G · C)/(A ·T) dimers found at the junctions of the (A_5G_5) · (T_5C_5) repeat are energetically similar to the $(G · C)_2$ and (A · T)$_2$ dimers, respectively (Srinivasan *et al.*). The chain is accordingly constructed from five "straight" units and five bent ones.

Chain Extension and Flexibility

The relative extension of the poly dG · poly dC, poly dA · poly dT, and poly $d(A_5G_5)$ · poly $d(T_5C_5)$ double helices subject to the local dimeric bending motions outlined above is described in figure 2. The persistence length *a* is reported as a function of the chain length for duplexes of up to 500 repeating residues. The data points are plotted at 10-residue increments and connected by straight lines. As evident from the figure, chain extension of the homopolymers is inversely proportional to the local flexibility of the double helices. The broader (A · T)$_2$ energy surface is seen to generate considerably less extended polymers than the more limited (G · C)$_2$ map at all chain lengths. In addition, the poly dA · poly dT chain is found to approach its limiting persistence length of 288 Å more rapidly than the poly dG · poly dC chain is seen to attain its limit of 805 Å, the former occurring at roughly 650

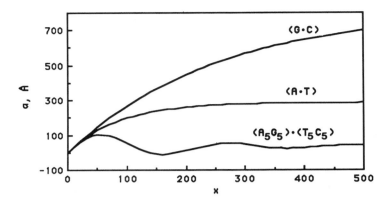

Figure 2. Persistence length a of poly dA · poly dT, poly dG · poly dC, and poly d(A$_5$G$_5$) · poly d(T$_5$C$_5$) as a function of chain length x.

residues and the latter at approximately 2,500 residues (Maroun & Olson). Interestingly, the limiting persistence lengths of the two duplexes are values which evenly bracket the accepted values of a (500-600Å) for DNA in aqueous salt solution (Borochov et al.; Hagerman 1981; Kam et al.; Kovacic & van Holde; Rizzo & Schellman). The standard persistence length of DNA can be reproduced by a random copolymeric duplex containing approximately 15% poly dA · poly dT and 85% poly dG · poly dC. Such a chain is represented by a mean repeating unit (Olson & Flory) computed from the average transformation matrices describing the $(A \cdot T)_2$ and $(G \cdot C)_2$ energy surfaces.

The computed limiting persistence length of the poly dG · poly dC duplex is remarkably close to the value of 837 Å reported in light scattering studies at 0.1M NaCl (Thomas & Bloomfield) of the related alternating duplex, poly d(GC). That predicted for the poly dA · poly dT duplex is also very similar to the persistence lengths of 200-250 Å derived from transient electric dichroism measurements on the alternating poly d(AT) sequence (Chen et al.). Similar measurements, however, have yet to be performed on either DNA homopolymer in aqueous salt solution to provide a firm test of the predictions in figure 2. The poly dA · poly dT helix is apparently less extended than poly dG · poly dC according to the published interpretation (Hogan et al.) of triplet anisotropy measurements on the two systems. The persistence lengths reported in the latter study are 425-510 Å and 884-1700 Å for poly dA · poly dT and poly dG · poly dC, respectively, and 884-1360 Å for chicken DNA, the last values somewhat larger than normally accepted for random sequence DNA.

The persistence length a of the curved poly d(A$_5$G$_5$) · poly d(T$_5$C$_5$) duplex is quite different from those of the rodlike homopolymer duplexes. As expected for a chain that changes direction, the computed values of a are seen to fluctuate in a periodic fashion with x. Moreover, over the range x = 140-180, the projection is nearly antiparallel to the initial chain direction (e.g., a < 0). The rodlike DNA's, in contrast, are characterized by continuous increases in a with chain length. The persistence length of poly d(A$_5$G$_5$) · poly d(T$_5$C$_5$) is comparable to those of poly dA · poly dT and poly dG · poly dC only at very short chain lengths. Differences become apparent at roughly chain length 50 when σ of the curved polymer is first seen to drop with increasing x. The limiting magnitude of the persistence length of the poly ($A_5G_5 \cdot T_5C_5$) sequence (38 Å) is roughly an order of magnitude smaller than those of the homopolymers. As might be expected from its copolymeric composition, this asymptotic limit is approached more slowly than in poly dA · poly dT but more rapidly than in poly dG · poly dC.

The direction of the persistence vector is of special interest in that it serves to characterize the bending, if any, in the double helix and to locate the region of highest probability of chain termination. According to these calculations, both poly dA · poly dT and poly dG · poly dC are typical rodlike duplexes at short chain lengths. There is little deviation between $< \mathbf{r} >$ and the maximum end-to-end extension $(0, 0, r_{max})$ of ideal B-DNA in chains up to about 80 residues. The principal component of the persistence vector is also directed along the initial B-DNA helical axis. Furthermore, there are limited fluctuations of chain termini about $< \mathbf{r} >$ in the homopolymers as measured by the average second-order tensor $< \rho^{x2} >$ formed from the displacement vector $\rho = \mathbf{r} - < \mathbf{r} >$ (Maroun & Olson). Such stiffness is consistent with the inability of poly dA · poly dT and poly dG · poly dC to wrap around the histone octamer (Rhodes; Simpson & Kunzler).

The large variations in direction of the persistence vector of poly $d(A_5G_5) \cdot$ poly $d(T_5C_5)$ and the relatively slow attainment of the limiting value $< \mathbf{r} >_\infty$ with chain length are expected of curved DNA. The limit of $< \mathbf{r} >_\infty$ is obtained within 99% in chains of 1280 residues, almost midway between the corresponding asymptotes of the persistence vectors of poly dA · poly dT and poly dG · poly dC. Furthermore, the major component of $< \mathbf{r} >_\infty$ in poly $d(A_5G_5) \cdot$ poly $d(T_5C_5)$ is found to be nearly antiparallel to the roll (x) axis of the initial base pair of the chain. The direction cosine described by the projection of $< \mathbf{r} >_\infty$ along this reference axis (Γ_x) is -0.86. The projection of the vector along the initial B-DNA helical axis (Γ_z) is only 0.28, while that along the original tilt axis (Γ_y) is -0.43. The major components of the limiting persistence vector of the homopolymeric chains, in contrast, are directed almost exclusively along the initial B-DNA helical axis $(\Gamma_z = 0.98$ in poly dG · poly dC and 0.999 in poly dA · poly dT).

Orientational Correlations

The character of the local base sequence is also evident in the orientational correlations between vectors in the terminal residues of the DNA duplex. The bending correlations for poly dA · poly dT, poly dG · poly dC, and poly $d(A_5G_5) \cdot$ poly $d(T_5C_5)$ computed at one-residue increments up to 500 base pairs are reported in figures 3. In the absence of any chain bending the terminal residues will be aligned at an angle of 0° with $< \gamma > = 1$. Values of $< \gamma >$ in this range in the figure are therefore indicative of rigid rod behavior. If the orientation of chain ends is totally uncorrelated, however, the average of the function is zero. As evident from the figure, this random limit is attained within 0.1 in poly dA · poly dT

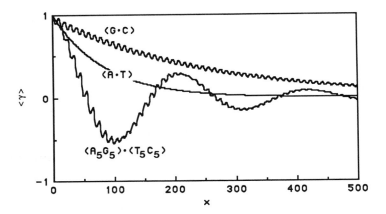

Figure 3. Orientational correlation $<\gamma>$ of poly dA · poly dT, poly dG · poly dC, and poly d(A_5G_5) · poly d(T_5C_5) as a function of chain length x.

chains of roughly 200 residues and in poly dG · poly dC chains greater than 500 residues. These lengths are well below the ideal limits described by $< r >$, illustrating how bond orientation and spatial orientation may be independent of one another in DNA. As noted below, values of an orientational correlation function near zero may sometimes be a sign of curvature in short duplexes rather than an indication of random terminal bond orientations. The gradual decrease of $< \gamma >$ with chain length in the homopolymers, however, is typical of rodlike rather than curved DNA. The small sinusoidal fluctuations of $< \gamma >$ with x are tied to the twisting of the double helix. The variations are more noticeable and are slower to die out in poly dG · poly dC where the local flexibility is more restricted than in poly dA · poly dT.

The curved nature of poly d(A_5G_5) · poly d(T_5C_5) sequence is immediately evident from the large sinusoidal fluctuations in $< \gamma >$ with x. Sign changes are found to recur with the same 200-fold periodicity found in the variations of the persistence length. Maxima are found between residues separated by complete superhelical turns and minima between residues separated by half turns. Large positive correlations are found when the residues are aligned on average in parallel, while negative correlations are seen when the mean angular orientation is more nearly antiparallel. Values of $< \gamma >$ near zero are found both at long chain lengths where terminal residues are completely uncorrelated and at shorter chain lengths where the chain is curved through roughly 90°. The 200-residue superhelical oscillations, while considerably damped out, are still appreciable ($\approx$ 0.1) at chain length 500. Superimposed upon the large oscillations are smaller fluctuations in $< \gamma >$ which follow the tenfold helical repeat of the chain. Only the latter perturbations are present in the bending correlations of the rodlike homopolymers. The superhelical periodicity of the curved duplex may be estimated by quadrupling the shortest chain length at which $< \gamma >$ is equal to zero. In the case of poly d(A_5G_5) · poly d(T_5C_5), the value is 4 · 56 = 224 residues.

Table 1.

Rotational Isomeric State Models of Selected Dimers

Dimer Sequence	ρ_0, deg	λ_0, deg	$T_{i,i+1}(\rho_0, \lambda_0, \tau_0)$			$< T_{i,i+1}(\rho, \lambda, \tau) >$		
$(A \cdot T)_2$	1.0	-2.0	0.809	-0.588	-0.011	0.800	-0.581	-0.011
			0.587	0.809	-0.032	0.583	0.810	-0.031
			0.028	0.019	0.999	0.028	0.019	0.988
$(A \cdot T)/(G \cdot C)$	6.9	-2.6	0.811	-0.584	0.034	0.812	-0.581	0.034
			0.583	0.803	-0.123	0.581	0.799	-0.122
			0.044	0.120	0.992	0.044	0.119	0.986
$(G \cdot C)/(A \cdot T)$	2.4	-0.9	0.809	-0.587	0.011	0.810	-0.583	0.011
			0.587	0.808	-0.044	0.584	0.802	-0.043
			0.016	0.042	0.999	0.016	0.042	0.990
$(G \cdot C)_2$	7.1	-1.3	0.810	-0.583	0.054	0.810	-0.581	0.054
			0.585	0.803	-0.113	0.584	0.800	-0.112
			0.022	0.123	0.992	0.022	0.123	0.988

Rotational Isomeric State Modeling

The sequence dependent conformational preferences can also be described in terms of the average spatial orientations of adjacent base pairs. These angles are readily obtained from the orthogonal matrices $T_{i,i+1}(\rho_0, \lambda_0, \tau_0)$ which best approximate the average transformation matrices $< T_{i,i+1}(\rho, \lambda, \tau) >$ used to relate the coordinate frames of residues i and i+1. The rotational isomeric states are chosen so that the T's associated with the specific states differ negligibly from the averaged $< T >$'s computed on the basis of the energy surfaces (Srinivasan *et al.*).

The rotational isomeric states which best match the ρ, λ behavior of the $(A \cdot T)_2$, $(A \cdot T)/(G \cdot C)$, $(G \cdot C)/(A \cdot T)$, and $(G \cdot C)_2$ dimer sequences at $\tau = 36°$ are detailed in table 1. The orthogonal matrices generated from the rotational isomeric states and the average transformation matrices associated with the sequences are also included in the table. As evident from the data, the rotational isomeric states provide better approximations of the asymmetric energy surfaces than the symmetric ones do. The states describing the symmetric $(A \cdot T)_2$ and $(G \cdot C)/(A \cdot T)$ energy surfaces are essentially unperturbed, perfectly stacked states near $\rho, \lambda = (0°, 0°)$. These dimers, however, are capable of bending up to $\pm 15°$ without increasing the potential energy more than 1 kcal/mole. Terms in the transformation matrices which involve sin ρ are therefore not well reproduced. The minimum energy conformations are even worse approximations than the rotational isomeric states of the symmetric energy surfaces since they are slightly bent arrangements. The states detailing the asymmetric $(A \cdot T)/(G \cdot C)$ and $(G \cdot C)_2$ maps, on the other hand, are noticeably skewed with roll angles of approximately 7°. Low energy fluctuations about these states are smaller than those on the symmetric surfaces and are confined to the +,- quadrant of (ρ, λ) configuration space. Individual terms of $< T_{i,i+1}(\rho, \lambda, \tau) >$ are accordingly better matched by $T_{i,i+1}(\rho_0, \lambda_0, \tau_0)$.

The bending described by the rotational isomeric state models of the $(A \cdot T)/(G \cdot C)$ and $(G \cdot C)_2$ dimers is very limited and difficult to discern by eye. The fixed spatial geometries associated with the four dimers detailed in table 1 are illustrated in figure 4. The very small perturbations of local structure, however, are found to have pronounced effects on the overall conformation of the DNA duplex, if correctly phased with the tenfold helical repeat. Oligomeric structures associated with the eleven possible combinations of the $(A_i G_{10-i}) \cdot (T_i C_{10-i})$ repeating sequence are illustrated in figure 5. Chains of 50 residues are drawn relative to the xz plane of the initial base pair as reference. The homopolymeric sequences where i = 0 (poly dG · poly dC) and i = 10 (poly dA · poly dT) are rodlike structures. Base pairs in the latter duplex are essentially perpendicular to the helical axis (the base pair normal and helix axis forming an angle of 3°), while those in the former structure are oriented at an angle of 11° with respect to the long axis of the chain. Because of the tilting of base pair planes, $(G \cdot C)$ residues separated by half turns of the helix are found to be oriented at angles of 23° with respect to one another. Adjacent base pair normals of the poly dG · poly dC duplex, however, are oriented no more than 7.4° with respect to one another and residues separated by a complete helical turn are parallel to one another. The poly dA · poly dT model is a tenfold helix of pitch 34 Å like standard B-DNA, while the poly dG · poly dC structure is slightly compacted with 9.8 residues per helical turn and 32.8 Å pitch.

The double helix is seen to curve with the insertion of a single $(G \cdot C)$ base pair between stretches of the $(A \cdot T)$ duplex. Terminal residues of the 50 base pair $(A_9 G) \cdot (T_9 C)$ repeating structure in figure 5 are aligned at an angle of 32°. The curvature is even more pronounced as $(A \cdot T)$ base pairs are replaced by additional $(G \cdot C)$'s in the tenfold sequential repeat, the maximum effect occurring in the $(A_5 G_5) \cdot (T_5 C_5)$ repeating chain. Terminal residues of the latter structure are oriented at an angle of 84° with respect to one another. The bending of this chain is also confined to the plane of the paper (e.g., the xz plane of the initial base pair). As evident from figure 5, DNA curvature is rapidly reduced as $(G \cdot C)$ base pairs begin to outnumber $(A \cdot T)$ base pairs in the sequential repeat. The $(AG_9) \cdot (TC_9)$ repeating structure is an extended superhelix only slightly altered from the poly dG · poly dC complex.

The sequences detailed in figure 5 are similar, although not identical, to a series of curved DNA's recently screened on the basis of gel electrophoretic mobilities (Koo *et al.*). The experimental systems are constructed with various combinations of G's and C's between the oligo $(A \cdot T)$ stretches of the duplexes. Duplexes of repeating $(A_i X_{10-i}) \cdot (T_i Y_{10-i})$ sequences have been examined for cases i = 3-6, 8, and 9 with X given as G or C and Y as the complement of X. All combinations are more curved than the corresponding oligo $(A \cdot T)$ duplex of the same chain length up to 200 base pairs. The degree of curvature in chains of 40-60 residues is found to increase, in qualitative agreement with the models in figure 5, with values of i in order $3 < 9 \cong 4 < 8 < 5 < 6$.

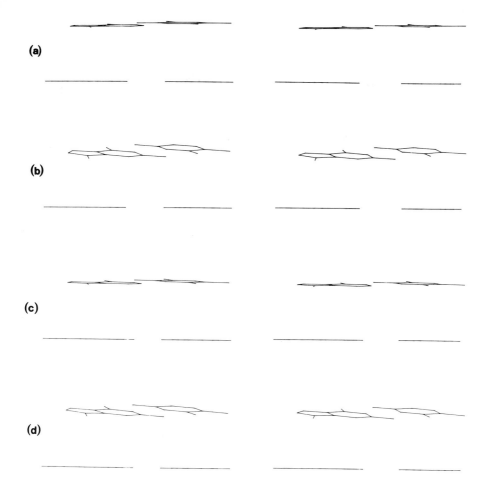

Figure 4. Stereo views of the rotational isomeric states which best match the ρ, λ energy surfaces of (a) $(A \cdot T)_2$, (b) $(A \cdot T)/(G \cdot C)$, (c) $(G \cdot C)/(A \cdot T)$, and (d) $(G \cdot C)_2$ dimers. Structures are drawn in the xz plane of the initial base pair to emphasize the orientational differences among sequences.

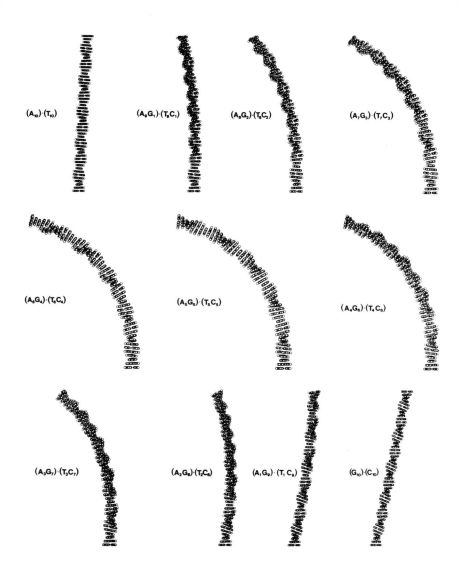

Figure 5. Computer generated representations of 50 base pairs fragments of DNA of repeating sequence $(A_iG_{10-i}) \cdot (T_iC_{10-i})$ where i = 0-10. Structures are drawn in a common coordinate frame (e.g., the xz plane of the lowest base pair) to emphasize differences in overall chain morphology.

DISCUSSION

The sequence dependent differences in overall chain properties of various DNA's are easily understood in terms of the local energy surfaces that describe the mutual orientation of successive Watson Crick base pairs. Some maps, like that of $(A \cdot T)_2$, are symmetric while others, like that of $(G \cdot C)_2$, are skewed with base pairs more likely to roll in a positive rather than a negative sense. The symmetric surfaces are analogous to the residues of a perfectly elastic DNA, while the asymmetric surfaces are more like locally kinked models of double helical bending. These differences, together with the degree of local flexibility, are able to account for the observed base sequence dependence of chain extension and flexibility. The combination of local structural symmetry and asymmetry is also found to describe the known curvature of certain DNA's.

According to these calculations, the tendency of a DNA sequence to curve is found in both the components of the persistence vector $< r >$ and the average orientational correlation $< \gamma >$ of terminal bond vectors. The projection of the persistence vector in the initial direction of the double helix is a direct measure of chain stiffness. If this projection, the persistence length, is comparable to the magnitude of $< r >$, the DNA is rodlike at short chain lengths. This is what is seen with poly dA · poly dT and poly dG · poly dC. If the components of $< r >$ perpendicular to the helix axis are greater than a, however, the chain is found to be curved. A curved chain like poly $d(A_5G_5) \cdot$ poly $d(T_5C_5)$ is also seen to exhibit large sinusoidal fluctuations of $< \gamma >$ as a function of chain length, the parameter adopting both positive and negative values. The average orientation of terminal bond vectors in the rodlike homopolymeric duplexes is found, in contrast, to decrease gradually from unity to zero with only small sinusoidal variations as chain length increases.

The computational studies emphasize the fact that curved DNA is more compact than random sequence DNA and illustrate how the simple replacement of one residue $(A \cdot T)$ by another $(G \cdot C)$ can alter the extension and curvature of the double helix. The limiting persistence lengths of the rodlike duplexes exceed those of curved chains by an order of magnitude. Indeed, at some chain lengths, the computed persistence lengths of curved DNA equal zero. The published interpretation of experimental data in terms of a significantly increased persistence length in curved DNA (Levene *et al.*), however, conflicts with the present calculations. The calculations also appear to account satisfactorily for the sequence dependent dimensions of GC and AT rich duplexes. The persistence length of poly dA · poly dT duplex is compacted compared to standard B-DNA in solution while that of poly dG· poly dC chain is extended.

The principal difference between the present energetic description of curved DNA and earlier static models (Arnott *et al.* 1983; Hagerman 1984; Koo *et al.*; Trifonov; Trifonov & Sussman) is the predicted bending of GC rather than AT helical stretches. The more skewed minima of the former sequences are traced to steric and electrostatic contacts involving the

2-amino group of guanine and the 2-carbonyl of cytidine. The 5-methyl group of thymine, on the other hand, is not found to conflict with adjacent residues when the bases are twisted at 36° as in B-DNA. Similar maps are generated for GG, GC, CG, and CC dimer combinations so that the computed persistence lengths are independent of bending sequence (Maroun & Olson). The more pronounced bending of GC versus AT sequences is also evident in the X-ray crystallographic structures of oligonucleotide fragments (Conner *et al.*; Dickerson & Drew 1981a; Fratini *et al.*; Kneale *et al.*; McCall *et al.* 1985, 1986; Shakked *et al.*; Viswamitra *et al.*; Wang *et al.* 1982a, 1982b). It should be noted, however, that there are many more examples of bends involving G and C than A and T in the X-ray literature (Shakked & Rabinovich).

The interpretation of structure in terms of skewed and straight helical stretches is analogous to the macroscopic bending introduced at junctions of A- and B-DNA (Selsing *et al.*) or by heteronomous models of DNA (Arnott *et al.* 1983). The energetic picture, however, is more consistent than A/B or heteronomous models with the observed nmr and spectroscopic properties (Alexeev *et al.*; Behling & Kearns 1985, 1986; Jolles *et al.*;Sarma *et al.* 1985a, 1985b; Steely *et al.*; Thomas & Peticolas; Wartell *et al.*) of poly dA · poly dT, which do not support the occurrence of a duplex composed of different backbone conformations. Finally, the bent structural fragments of curved DNA are predicted to have fewer residues per helical turn than the rodlike segments. Recently observed differences in the hydroxyl radical cutting patterns of curved DNA (Burkhoff & Tullius) may be related to this variation in twist.

ACKNOWLEDGMENT

Sponsorship of this research by the U. S. Public Health Service under grant GM-20861 is gratefully acknowledged. Computational resources were provided by the Chemistry Computational Center at Rutgers.

REFERENCES

Alexeev, D. G., Lipanov, A. A. & Skuratovskii, I. Ya. (1987) *Nature* **325**, 821-823.
Anderson, J. N. (1986) *Nucleic Acids Res.* **14**, 8513-8533.
Arnott, S., Dover, S. D. & Wonacott, A. J. (1969) *Acta Cryst.* **B25**, 2192-2206.
Arnott, S., Chandrasekaran, R., Hall, I. H. & Puigjaner, L. C. (1983) *Nucleic Acids Res.* **11**, 4141-4155.
Behling, R. W. & Kearns, D. R. (1985) *Biopolymers* **24**,1157-1167.
Behling, R. W. & Kearns, D. R. (1986) *Biochem.* **25**, 3335-3346.
Borochov, N., Eisenberg, H. & Kam, Z. (1981) *Biopolymers* **20**, 231-235.
Bossi, L. & Smith, D. (1984) *Cell* **39**, 643-652.
Burkhoff, A. M. & Tullius, T. D. (1987) *Biophys. J.* **51**, 510a.
Calladine, C. R. (1982) *J. Mol. Biol.* **161**, 343-352.
Calladine, C. R. & Drew, H. R. (1984) *J. Mol. Biol.* **178**, 773-782.
Calladine, C. R. & Drew, H. R. (1986) *J. Mol. Biol.* **192**, 907-918.
Challberg, S. S. & Englund, P. T. (1980) *J. Mol. Biol.* **138**, 447-472.
Chen, H. H., Rau, D. C. & Charney, E. (1985) *J. Biomol. Str. & Dynamics* **2**, 709-719.

Conner, B. N., Yoon, C., Dickerson, J. L. & Dickerson, R. E. (1984) *J. Mol. Biol.* **174**, 663-695.
Diekmann, S. (1986) *FEBS Letters* **195**, 53-56.
Diekmann, S. (1987) *Nucleic Acids Res.* **15**, 247-265.
Diekmann, S. & Wang, J. C. (1985) *J. Mol. Biol.* **186**, 1-11.
Dickerson, R. E. (1983) *J. Mol. Biol.* **166**, 419-441.
Dickerson, R. E. & Drew, H. R. (1981a) *J. Mol. Biol.* **149**, 761-786.
Dickerson, R. E. & Drew, H. R. (1981b) *Proc. Natl. Acad. Sci. USA* **78**, 7318-7322.
Dickerson, R. E., Kopka, M. L. & Drew, H. R. (1982) In *Conformation in Biology*, Srinivasan, R. & Sarma, R. H., Eds., Adenine Press, Guilderland, New York, pp.227-257.
Dickerson, R. E., Kopka, M. L. & Pjura, P. (1983) *Proc. Natl. Acad. Sci. USA* **80**, 7099-7103.
Drew, H. R. & Travers, A. A. (1985) *J. Mol. Biol.* **186**, 773-790.
Flory, P. J. (1969) *Statistical Mechanics of Chain Molecules*, Interscience, New York, Chapter 4.
Flory, P. J. (1973) *Proc. Natl. Acad. Sci. USA* **70**, 1819-1823.
Flory, P. J., Suter, U. W. & Mutter, M. (1976) *J. Am. Chem. Soc.* **98**, 5733-5739.
Fratini, A. V., Kopka. M. L., Drew, H. R. & Dickerson, R. E. (1982) *J. Biol. Chem.* **257**, 14686-14707.
Griffith, J., Bleyman, M., Rausch, C. A., Kitchen, P. A. & Englund, P. T. (1986) *Cell* **46** 717-724.
Hagerman, P. J. (1981) *Biopolymers* **20**, 1503-1535.
Hagerman, P. J. (1984) *Proc. Natl. Acad. Sci. USA* **81**, 4632-4636.
Hagerman, P. J. (1985) *Biochem.* **24**, 7033-7037.
Hagerman, P. J. (1986) *Nature* **321**, 449-450.
Hogan, M., LeGrange, J. & Austin, B. (1983) *Nature* **304**, 752-754.
Holbrook, S. R. & Kim, S.-H. (1984) *J. Mol. Biol.* **173**, 361-388.
Jernigan, R. L., Sarai, A., Ting, K.-L. & Nussinov, R. (1986) *J. Biomol. Str. & Dynamics* **4**, 41-48.
Jolles, B., Laigle, A., Chinsky, L. & Turpin, P. Y. (1985) *Nucleic Acids Res.* **13**, 2075-2085.
Kabsch, W., Sander, C. & Trifonov, E. N. (1982) *Nucleic Acids Res.* **10**, 1097-1104.
Kam, Z., Borochov, N. & Eisenberg, H. (1981) *Biopolymers* **20**, 2671-2690.
Kitchin, P. A., Klein, V. A., Ryan, K. A., Gann, K. L., Rauch, C. A., Kang, D. S., Wells, R. D. & Englund, P. T. (1986) *J. Biol. Chem.* **261**, 11302-11309.
Kneale, G., Brown, T., Kennard, O. & Rabinovich, D. (1985) *J Mol. Biol.* **186**, 805-814.
Koepsel, R. R. & Khan, S. A. (1986) *Science* **233**, 1316-1318.
Koo, H.-S., Wu, H.-M. & Crothers, D. M. (1986) *Nature* **320**, 501-505.
Kovacic, R. T. & van Holde, K. E. (1977) *Biochem.* **16**, 1490-1498.
Kratky, O. & Porod, G. (1949) *Rec. Trav. Chim. Pays-Bas* **68**, 1106-1122.
Levene, S. D., Wu, H.-M. & Crothers, D. M. (1986) *Biochem.* **25**, 3988-3995.
Marini, J. C., Levene, S. D., Crothers, D. M. & Englund, P. T. (1982) *Proc. Natl. Acad. Sci. USA* **79**, 7664-7668.
Maroun, R. C. & Olson, W. K. (1987) *Biopolymers*, submitted.
McCall, M. J., Brown, T. & Kennard, O. (1985) *J. Mol. Biol.* **183**, 385-396.
McCall, M., Brown, T., Hunter, W. N. & Kennard, O. (1986) *Nature* **322**, 661-664.
Olson, W. K. (1978) *Biopolymers* **17**, 1015-1040.
Olson, W. K. (1982) *J. Am. Chem. Soc.* **104**, 278-284.
Olson, W. K. & Flory, P. J. (1972) *Biopolymers* **11**, 1-66.
Olson, W. K., Srinivasan, A. R., Cueto, M. A., Torres, R., Maroun, R. C., Cicariello, J. & Nauss, J. L. (1985) In *Biomolecular Stereodynamics IV*, Sarma, R. H. & Sarma, M. H., Eds.,Adenine Press, Guilderland, New York, pp. 75-100.
Peck, L. J. & Wang, J. C. (1981) *Nature* **292**, 375-378.
Prunell, A., Goulet, I. Jacob, Y. & Goutorbe, F. (1984) *Eur. J. Biochem.* **138**, 253-257.
Ray, D. S., Hines, J. C., Sugisaki, H. & Sheline, C. (1986) *Nucleic Acids Res.* **14**, 7953-7965.
Rhodes, D. (1979) *Nucleic Acids Res.* **6**, 1804-1816.
Rhodes, D. & Klug, A. (1981) *Nature* **292**, 378-380.
Rizzo, V. & Schellman, J. A. (1981) *Biopolymers* **20**, 2143-2163.

Ryder, K., Silver, S., DeLucia, A. L., Fanning, E. & Tegtmeyer, P. (1986) *Cell* **44**, 719-725.

Sarma, M. H., Gupta, G. & Sarma, R. H. (1985a) in *Proceedings Fourth Conversation in Biomolecular Stereodynamics* , Sarma, R. H., Ed., Institute of Biomolecular Stereodynamics, Albany, New York, pp. 77-78.

Sarma, M. H., Gupta, G. & Sarma, R. H. (1985b) *J. Biomol. Str. & Dynamics* **2**, 1057-1084.

Selsing, E., Wells, R. D., Alden, C. J. & Arnott, S. (1979) *J. Biol. Chem.* **254**, 5417-5422.

Shakked, Z. & Rabinovich, D. (1986) *Prog. Biophys. & Molec. Biol.* **41**, 159-195.

Shakked, Z., Rabinovich, D., Kennard, O., Cruse, W. B. T., Salisbury, S. A. & Viswamitra, M. A. (1983) *J. Mol. Biol.* **166**, 183-201.

Simpson, L. (1979) *Proc. Natl. Acad. Sci. USA* **76**, 1585-1588.

Simpson, R. T. & Kunzler, P. (1979) *Nucleic Acids Res.* **6**, 1387-1415.

Snyder, M., Buchman, A. R. & Davis, R. W. (1986) *Nature* **324**, 87-89.

Srinivasan, A. R. & Olson, W. K. (1980) *Fed. Proc.* **39**, 2199.

Srinivasan, A. R. & Olson, W. K. (1987) *J. Biomol. Str. & Dynamics*, in press.

Srinivasan, A. R., Torres, R., Clark, W. & Olson, W. K. (1987) *J. Biomol. Str. & Dynamics*, submitted.

Steely, H. T., Jr., Gray, D. M. & Ratliff, R. L. (1986) *Nucleic Acids Res.* **14**, 10071-10090.

Strauss, F., Gaillard, C. & Prunell, A. (1981) *Eur. J. Biochem.* **118**, 215-222.

Sundaralingam, M. & Rao, S. T. (1983) *Int. J. Quantum Chem., Quantum Biol. Symp.* **10**, 301-308.

Taylor, E. R. & Olson, W. K. (1983) *Biopolymers* **22**, 2667-2702.

Taylor, R. & Kennard, O. (1982) *J. Am. Chem. Soc.* **104**, 3209-3212.

Thomas, G. A. & Peticolas, W. L. (1983) *J. Amer. Chem. Soc.* **105**, 993-996.

Thomas, T. J. & Bloomfield, V. A. (1983) *Nucleic Acids Res.* **11**, 1919-1930.

Trifonov, E. N. & Sussman, J. L. (1980) *Proc. Natl. Acad. Sci. USA* **77**, 3816-3820.

Trifonov, E. N. (1985) *CRC Crit. Rev. Biochem.* **19**, 89-106.

Ul'yanov, N. B. & Zhurkin, V. B. (1984a) *Mol. Biol. USSR* (Eng. Ed.) **18**, 1366-1384.

Ulyanov, N. B. & Zhurkin, V. B. (1984b) *J. Biomol. Str. & Dynamics* **2**, 361-385.

Viswamitra, M. A., Kennard, O., Jones, P. G., Sheldrick, G. M., Salisbury, S., Falvello, L. & Shakked, Z. (1978) *Nature* **273**, 687-688.

Wang, A. H.-J., Fujii, S., van Boom, J. H. & Rich, A. (1982) *Proc. Natl. Acad. Sci. USA* **79**, 3968-3972.

Wang, A. H.-J., Fujii, S., van Boom, J. H., van der Marel, G. A., van Boeckel, S. A. A. & Rich, A. (1982) *Nature* **299**, 601-604.

Wartell, R. M. & Harrell, J. T. (1986) *Biochem.* **25**, 2664-2671.

Wu, H.-M. & Crothers, D. M. (1984) *Nature* **308**, 509-513.

Zahn, K. & Blattner, F. R. (1985a) *Nature* **317**, 451-453.

Zahn, K. & Blattner, F. R. (1985b) *EMBO J.* **4**, 3605-3616.

SEQUENCE-DEPENDENT CURVATURE OF DNA

PAUL J. HAGERMAN

Department of Biochemistry and Biophysics
University of Colorado Medical Center
4200 East 9th Avenue
Denver, Colorado 80262

A large variety of DNA molecules, derived as restriction fragments from both prokaryotic (Ross et al., 1982; Stellwagen, 1983; Bossi and Smith, 1984; Zahn and Blattner, 1985) and eukaryotic (Simpson, 1979; Challberg and Englund, 1980; Israelewski, 1983; Schmidt, 1984; Kidane et al., 1984; Garrett and Carroll, 1986; Ray et al., 1986; Ryder et al., 1986) sources, display abnormal electrophoretic behavior. In particular, such molecules run more slowly in acrylamide gels than would be expected on the basis of their sizes (by sequence). An additional example of such electrophoretic behavior is displayed in figure 1. This example is of historical interest in that the HindII+III digest of SV40 was the first published restriction digest. The retardation of the F fragment was originally believed to be due to a chemical interaction of the fragment with the acrylamide gel. The example is also noteworthy in that the sequences giving rise to the anomalous behavior lie entirely within the coding sequences for the major capsid protein (VP1) gene.

The most notable features of axially-curved DNA are: (1) a reduced electrophoretic mobility (relative to normal DNA controls), with the degree of reduction increasing with increasing acrylamide gel concentration (figure 1; Marini et al., 1982) (the mobilities of curved fragments are virtually normal on agarose gels, except at very high gel concentrations), and (2) an enhancement of the degree of retardation of curved fragments by either lowering the temperature of the gel or by adding magnesium to the gel buffer (Diekmann and Wang, 1985; Diekmann, 1987). Other physical findings pertaining to curved DNA molecules are unremarkable at this point, primarily since only a few molecules have been investigated.

THE EXISTENCE OF STABLE AXIAL CURVATURE

The first issue to be dealt with regarding the electrophoretically-abnormal

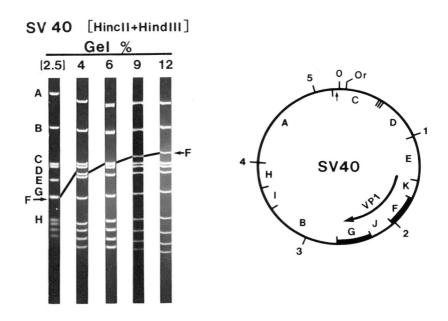

Figure 1. Gel electrophoresis patterns of simian virus 40 (SV40) DNA that has been digested with the restriction endonucleases HincII (an isoschizomer of HindII) and HindIII. The percent acrylamide comprising each gel is indicated above each pattern, with the exception of the left-most gel lane, "[2.5]", which displays the pattern obtained on 2.5% agarose. The gel patterns were approximately aligned with respect to the C,D doublet. Bands were labeled according to the convention of Danna and Nathans (1971). Note, however, that the 4% gel pattern differs from the pattern of Danna and Nathans with respect to the relative positions of the E and F fragments. This slight difference could be due to slightly elevated temperatures in the previous gels (Danna and Nathans, 1971; Mertz and Berg, 1974) or to the presence of sodium dodecyl sulfate in those gels. In the present study, the identity of the F fragment in the various gels was verified by secondary digestion of the HincII+HindIII digest with the restriction endonuclease EcoRI, which cleaves the SV40 DNA once, at a location within the F fragment. The right portion of the figure displays a map of the locations of the restriction fragments referred to in the gel patterns. The numbers indicate sequence position (in kilobase pairs) and follows the convention of Fiers et al (1978); Or refers to the origin of SV40 replication (Buchman et al, 1981; Fiers et al, 1978). The arrow labeled "VP1" indicates the coding portion of the major capsid protein mRNA. The arrow located within fragment C indicates the position of the 17 bp T antigen binding site in the SV40 origin region I, previously reported to be associated with curvature (Ryder et al, 1986). The positions of restriction fragments F and G have been highlighted by heavier lines.

DNA molecules is whether, in fact, their behavior is due to axial curvature, rather than some other phenomenon. Studies using permuted DNA restriction fragments derived from trypanosome kinetoplast minicircles demonstrated that base-composition effects (eg, through chemical interaction of the DNA with the acrylamide gel matrix) were not responsible for the reduced mobility of

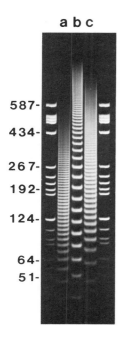
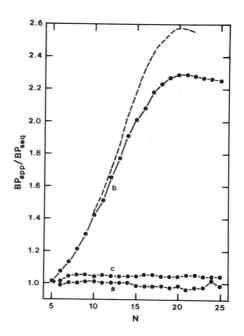

Figure 2. Gel electrophoretic behavior of synthetic duplex DNA polymers of the form (a) $5'-[GA_3T_3C]_N$ (8-mer repeat), (b) $5'-[G_2A_3T_3C_2]_N$ (10-mer repeat), and (c) $5'-[G_3A_3T_3C_3]_N$ (12-mer repeat). For each of the three polymer series, electrophoretic mobilities are normal for N < 6. Polymers of varying N were constructed by partial ligation of the basic oligomer duplexes specified within the square brackets. Lanes a–c are flanked by a HaeIII digest of pBR322, with representative fragment sizes specified to the left of the gel pattern. Synthetic oligomers were produced manually by using the phosphite triester approach (For further experimental details, see: Hagerman, 1985). The gel was 12% polyacrylamide (12% acrylamide, 37:1 monomer to bis ratio), and was run at approximately 22°C. To the right of the figure are plotted the relative electrophoretic patterns of the oligomer ladders displayed in the gel. In each case, the ratio of the apparent number of base pairs (BP_{app}) to the actual number of base pairs (BP_{seq}) is plotted as a function of the degree of oligomer polymerization, N. Letter designations correspond to those of the gel lanes. The dashed line represents the pattern observed for $[GA_4T_4C]_N$ at 22°C. (From Hagerman, 1985. Reprinted with permission. © Am. Chem. Soc.)

the abnormal fragments (Hagerman, 1984; Wu and Crothers, 1984). Using synthetic DNA fragments of defined sequence arrangement, Hagerman (1985) confirmed experimentally the earlier suggestion (Wu and Crothers, 1984) that the appropriate sequence periodicity was essential for gel retardation (figure 2). In particular, it was shown that phasing of blocks of A residues with respect to the helix repeat is required for gel retardation (Hagerman, 1985; figure 2). This observation also rules out the possibility of simple increased flexibility of runs of A residues, since such an effect would not be expected to manifest a pronounced phase-dependence. An asymmetric flexure (eg, into

227

the major groove of the helix) cannot formally be ruled out on the basis of those experiments; however, the observation that the gel abnormality increases with decreasing temperature would tend to militate against the flexibility model. Moreover, the rotational diffusion studies of Hagerman (1984a) at low ionic strength, where stiffness is dominated by electrostatic effects, argues in favor of a stably curved helix axis. In their analysis of a larger series of synthetic molecules, Koo et al. (1986) confirmed the phasing observation of Hagerman (1985). Those authors, along with Diekmann (1986), also demonstrated the requirement for several contiguous A residues as a minimal unit required for the generation of curvature.

In addition to the sequence studies mentioned above, there are three lines of experimental evidence which support the notion of stable axial curvature, namely, (1) rotational diffusion measurements, (2) ligase-catalyzed DNA circle formation, and (3) electrom micrographic studies. Using the technique of differential decay of birefringence (DDB; Hagerman, 1984a; 1984b), which compares the rotational diffusion times of electrophoretically normal and abnormal fragments of the same number of base pairs, Hagerman (1984a) demonstrated that the abnormal fragment was, in fact, less extended in solution than its electrophoretically-normal counterpart. A similar report by Marini et al (1982) was apparently the result of an experimental error in their dichroism measurement, leading them to conclude erroneously that the consequences of curvature would have profound effects on the rotational diffusion times. However, those authors later repeated their measurements (Levene et al., 1986), and found decay behavior in qualitative agreement with the conclusions of Hagerman (1984a). The problem inherent with the studies of Hagerman (1984a) as well as Levene et al (1986) is that a model is required for precise estimates of axial curvature from the diffusion data, and since such a model has not yet been fully realized, diffusion studies with DNA fragments derived from natural sources (with undefined bend centers) must remain qualitative.

In an interesting study of the T4 DNA ligase-catalyzed formation of small circles from synthetic DNA fragments, Ulanovsky et al. (1986) demonstrated that DNA molecules, designed to be curved on the basis of placement of A residues, formed small DNA circles at a rate orders of magnitude greater than rates observed for electrophoretically-normal DNA molecules (Shore et al., 1981). Ulanovsky et al (1986) quite reasonably proposed that their increased closure rates were a consequence of axial curvature of their DNA molecules.

The third line of evidence in support of curvature comes from the electron micrographic studies of natural kinetoplast sequences by Ray et al.

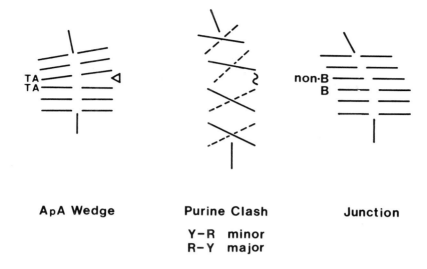

ApA Wedge **Purine Clash** **Junction**

Y–R minor
R–Y major

Figure 3. Models for sequence-directed curvature (details in the text). ApA wedge: This model invokes non-coplanarity between adjacent AT base pairs (A´s on the same strand) as the primary determinant of axial curvature. Purine clash: This model states that, as a consequence of positive propeller twist between bases in a base pair, cross-chain purines at 5´-pur-pyr (R–Y) or 5´-pyr-pur (Y–R) transitions will experience unfavorable steric interactions ("clash") in either the major grove (for R–Y transitions) or minor groove (for Y–R transitions). The figure displays an untwisted helix, viewed from a direction perpendicular to the dyad axis of the helix (view through the long axis of the base pairs. The wavy line indicates cross-chain clash with attendent roll to relieve the clash. Junction: junction models propose that as a consequence of the joining of two helical segments having differing helix parameters (eg, tilts and rolls), base stacking at the transition point gives rise to a change in helix axis direction.

(1986) and Griffith et al. (1986). Those groups have provided graphic confirmation of the inherent curvature of the kinetoplast fragments, while taking pains to rule out spreading artifacts.

In aggregate, therefore, the above lines of evidence provide a persuasive case for stable axial curvature. It should be noted that "stable" does not imply "static", only that the conformation of minimum free energy is not linear.

MODELS FOR CURVATURE OF THE HELIX AXIS

Purine–Clash Model On the basis of an extensive x-ray diffraction analysis of crystals of the DNA dodecamer, 5´-CGCGAATTCGCG (duplex), by Wing et al. (1980), Calladine (1982) and Dickerson (1983) proposed a set of rules for alterations in helix twist, roll, and tilt, such alterations being driven by unfavorable steric interactions between cross-chain purines in adjacent base pairs (figure 3). In its simplest form, the model does not distinguish between A and G purines; both should give rise to equal helix distortions in reducing unfavorable interactions. However, one might expect that cross-chain interactions at $5'$-CG transitions would be somewhat more severe, due to minor groove clash between the exocyclic amino groups of the two G residues in the adjacent base pairs. In any case, the sequence pattern expected to generate maximal curvature for this model is that of a purine–pyrimidine block copolymer of the form $[R_5Y_5]_N$. For this copolymer, positions of major groove clash would occur five base pairs (approximately one–half helical turn) away from positions of major groove clash.

Using the approach depicted in figure 2, we have examined a series of block copolymers of the above form, having various A-to-G ratios (figure 4). Three features of the figure should be noted: (1) G_5C_5 and G_4ATC_4 polymers are electrophoretically normal, thus arguing against the simple notion of cross-chain purine clash (various forms of intra-chain clash are not excluded, however), and in favor of some other AT-specific phenomenon. This AT-dependence has also been demonstrated elegantly by Koo et al. (1986), who observed that the curvature of A_5 blocks was substantially eliminated by the replacement of the central AT base pair by a GC base pair. Diekmann has demonstrated a similar AT-dependence (Diekmann, 1986). (2) The $G_3A_2T_2C_3$ polymers are clearly abnormal, thus indicating that runs of A residues (on the same strand) as short as two can give rise to curvature. Furthermore, it should be noted that the central hexamer, GAATTC, is the recognition sequence for the restriction endonuclease EcoRI. This observation bears on the conclusions of Frederick et al. (1984), who observed in their x-ray diffraction study of the EcoRI restriction endonuclease, complexed with a DNA oligomer containing the EcoRI recognition sequence, that the DNA in the complex was bent. From figure 4, it can be concluded that the recognition sequence is, to some extent, curved before it interacts with the endonuclease. (3) The A_4T_4-containing polymers are distinctly more abnormal than their A_3T_3 counterparts. Furthermore, both are more abnormal than the A_5T_5 polymers. Although differences in axial twist might account for some of the differences in those series, we believe, on the basis of experiments of the type discussed below (figure 6), that the gel patterns represent real differences in axial curvature. These last observations are pertinent to the subsequent discussion of the ApA wedge model.

ApA Wedge Model This model (figure 3), historically the first one proposed in the context of DNA curvature, was set forth by Trifonov and

230

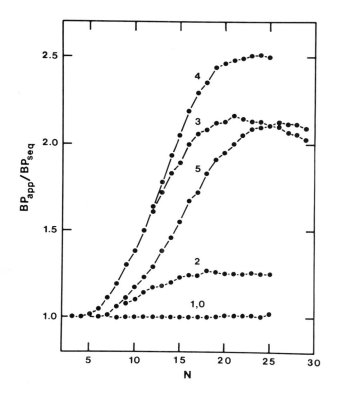

Figure 4. Plot of the relative electrophoretic patterns of a series of 10-mer repeats of the form: $5'-[G_{5-i}A_iT_iC_{5-i}]_N$ (all R_5-Y_5 block copolymers) as a function of N. The values for i ($\equiv$ 1 to 5) are given next to each curve.

Sussman (1980). As originally formulated, the model envisioned a base tilt between adjacent AT base pairs (AA/TT). The notion of a pure tilt wedge was ruled out on the basis of the experimental non-equivalence of $5'-A_4T_4$ and $5'-T_4A_4$-containing polymers (Hagerman, 1986). Ulanovsky and Trifonov (1987; this monograph) then extended their wedge model to include a substantial component of roll, approximately 3.5:1 roll-to-tilt ratio between adjacent AT base pairs. Their elegant treatment of this wedge model was able to interpret, at least qualitatively, nearly all of the experimental work of Hagerman (1985; 1986) and of Koo et al (1986).

Although this model will be discussed in more detail elsewhere in this monograph, several points pertaining to the model should be noted. In addition to explaining the difference between the A_4T_4 and T_4A_4-containing polymers (Hagerman, 1986), the model makes the interesting prediction that polymers containing the decamer repeat, $C_2T_3A_3G_2$ should be more curved than their

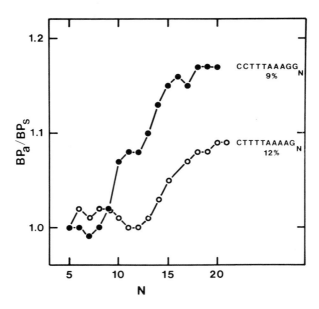

Figure 5. Plot of the relative electrophoretic patterns of the series $5'-[C_2T_3A_3G_2]_N$ and $5'-[CT_4A_4G]_N$ as a function of N. The "T_3A_3" series was run on a 9% polyacrylamide gel, whereas the "T_4A_4" series was run on a 12% polyacrylamide gel in order to increase the apparent degree of abnormality of the latter series.

CT_4A_4G counterparts, even though the former series has fewer AT base pairs. In fact, this prediction is borne out (figure 5). It should also be noted that we have observed (Hagerman, unpublished results) that runs of two A residues, when spaced appropriately, do give rise to a small, but detectable, curvature, a prediction also made by the ApA wedge model. Finally, the modified ApA wedge model correctly predicts that the A_5T_5-containing polymer (figure 4) is less curved than its "4" and "3" counterparts. These predictions arise as a simple consequence of the vector sums of tilts and rolls in the helix (Ulanovsky and Trifonov, 1987; this monograph).

Despite the dramatic success of the ApA wedge model in being able to interpret, at least qualitatively, the experimental results, it will have to be modified further: (1) the value of 3.5:1 for the roll-to-tilt ratio was based, in part, on the equivalence of the degree of curvature of the A_4T_4 and A_3T_3-containing polymers (figure 2); however, as was noted above, the former is significantly more curved than the latter. Final determination of such a ratio will have to await quantitative information on the relative degrees of curvature for those species. (2) The magnitude of the angles, based on the analysis of the formation of small DNA circles by Ulanovsky et al. (1986), is

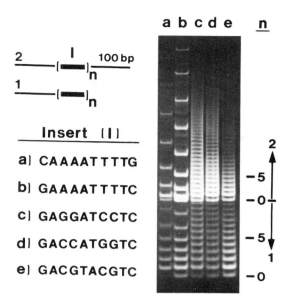

Insert (I)

a) CAAAATTTTG

b) GAAAATTTTC

c) GAGGATCCTC

d) GACCATGGTC

e) GACGTACGTC

Figure 6. Gel patterns for several 10-mer sequences inserted (in copy number n) between electrophoretically-normal 100 base pair DNA molecules. As shown in the figure, "2" refers to central (2-arm) insertions and "1" refers to end (1-arm) additions. Molecules of each of the above classes are located by arrows to the right of the gel pattern. The inset numbers (0,5) indicate the positions expected for electrophoretically-normal 1 and 2-arm molecules having 0 and 5 10-mer inserts. Lanes c-e display normal electrophoretic patterns for all values of n.

subject to some uncertainty, due to the assumption of perfect torsional phasing of the oligomers, and the assumption of a close proximity of the ends of the molecules prior to circle formation. These assumptions have yet to be critically tested for the oligomers of Ulanovsky et al.. However, despite this concern, the roll-to-tilt ratio would not be affected by the circle-formation studies. (3) The ApA wedge model does not take into consideration effects of the base pairs flanking the AT runs. In this regard, it should be noted (figure 6, lanes a and b) that flanking bases do have a significant effect on curvature. Replacement of a C residue by a G residue at the 5´ end of a run of A residues clearly reduces the degree of curvature. The experiments depicted in figure 6 were carried out in a manner that minimizes the contribution of differences in twist: a small number of decamers were placed in the center of an otherwise normal DNA molecule (Hagerman, manuscript in preparation).

In sum, while the ApA model will need to be modified to account for the above observations, it provides a surprisingly simple and self-consistent explanation for most of the experimental observations reported thusfar. Two things are needed at this point: quantitative information on the degree of curvature for several well-defined molecules of the type depicted in figure 6, and a reasonable physical basis for the formation of the wedge itself.

Junction Models Taken as a class, these models propose that a change in the direction of the helix axis is a consequence of base stacking across a transition between two helix segments, each having different base pair inclinations with respect to their individual segmental axes. This type of effect was first discussed by Selsing et al. (1979) in their analysis of an A-B DNA junction. In the context of the curvature problem, the junction model was most carefully considered by Koo et al. (1986). Although this type of model is plausible, and may well participate with wedge-type deformations in producing curvature, it has not been tested extensively against experimental data for the self-complementary sequences. This throws into question the quantitative analysis of curvature by Levene et al (1986), who assumed a pure junction model with equivalent junctions at the 5´ and 3´ ends of a run of A residues, particularly since the same laboratory had, concurrently, provided evidence that the two junctions were not equivalent (Koo et al., 1986).

PERSPECTIVE

A great deal of experimental data, derived from the analysis of gels, has been applied to the question of the origin of sequence-dependent curvature of DNA. Unfortunately, these data are qualitative, since the relationship between the degree of curvature and the magnitude of the reduction in mobility is unknown. This lack of information precludes a rigorous evaluation of the extent to which wedge-type deformations and/or junctional effects give rise to curvature of the helix axis in solution. Therefore, the focus of future work in this area should be on the quantitative determination of axial curvature for a number of well-defined sequences.

It is also essential that experiments be performed which will identify those elements of the base-pairs which give rise to curvature. Such studies might entail selective modification of the bases themselves.

Finally, an impotant aspect of the work on DNA curvature will be the assessment of its functional significance. This task is, in some respects, much more difficult than the determination of the origins of curvature, since one must clearly distinguish between the effects of curvature per se and those of sequence.

REFERENCES

Bossi, L. and Smith, D.M. (1984). Conformational change in the DNA associated with an unusual promoter mutation in a tRNA operon of Salmonella. Cell 39:643-652.

Buchman, A.R., Burnett, L. and Berg, P. (1981). The SV40 nucleotide sequence (appendix A): in DNA Tumor Viruses, J. Tooze, ed. Cold Spring Harbor Laboratory, N.Y., 799-842.

Calladine, C.R. (1982). Mechanistics of sequence-dependent stacking of bases in B-DNA. J. Mol. Biol. 161:343-352.

Garrett, J.E. and Carroll, D. (1986). Tx1: a transposable element from Xenopus laevis with some unusual properties. Mol. Cell. Biol. 6:933-941.

Challberg, S.S. and Englund, P.T. (1980). Heteroeneity of minicircles in kinetoplast DNA of Leishmania tarentolae. J. Mol. Miol. 138:447-472.

Danna, K. and Nathans, D. (1971). Specific cleavage of simian virus 40 DNA by restriction endonuclease of Hemophilus influenzae. Proc. Natl. Acad. Sci. USA 68:2913-2917.

Diekmann, S. and Wang, J.C. (1985). On the sequence determinants and flexibility of the kinetoplast DNA fragment with abnormal gel electrophoretic mobilities. J. Mol. Biol. 186:1-11.

Diekmann, S. (1986). Sequence specificity of curved DNA. FEBS Lett. 195:53-56.

Diekmann, S. (1987). Temperature and salt dependence of the gel migration anomaly of curved DNA fragments. Nucl. Acids Res. 15:247-265.

Dickerson, R.E. (1983). Base sequence and helix structure variation in B and A DNA. J. Mol. Biol. 166:419-441.

Fiers, W., Contreras, R., Haegeman, G., Rogiers, R., Van de Voorde, A., Van Heuverswyn, H., Van Herreweghe, J., Volckaert, G. and Ysebaert, M. (1978). Complete nucleotide sequence of SV40 DNA. Nature 273:113-120.

Frederick, C.A., Grable, J., Melia, M., Samudzi, C., Jen-Jacobson, L., Wang, B.-C., Greene, P., Boyer, H.W. and Rosenberg, J.M. (1984). Kinked DNA in crystalline complex with EcoRI endonuclease. Nature 309:327-331.

Griffith, J., Bleyman, M., Rauch, C.A., Kitchin, P.A. and Englund, P.T. (1986). Visualization of the bent helix in kinetoplast DNA by electron microscopy. Cell 46:717-724.

Hagerman, P.J. (1984a). Evidence for the existence of stable curvature of DNA in solution. Proc. Natl. Acad. Sci. USA 81:4632-4636.

Hagerman, P.J. (1984b) Application of Transient Electric Birefringence to the Study of DNA Structure: in Methods in Enzymology. C.H.W. Hirs and S.N. Timasheff, eds. Academic Press, Inc., New York, 198-219.

Hagerman, P.J. (1985). Sequence dependence of the curvature of DNA: a test of the phasing hypothesis. Biochem. 24:7033-7037.

Hagerman, P.J. (1986). Sequence-directed curvature of DNA. Nature 321:449-450.

Israelewski, N. (1983). Structure and function of an AT-rich, interspersed repetative sequence from Chironomus thummi: solenoidal DNA, 142 bp palindrome-frame and homologies with the sequence for site-specific recombination of bacterial transposons. Nucl. Acids Res. 11:6985-6996.

Kidane, G.Z., Hughes, D. and Simpson, L. (1984). Sequence heterogeneity and anomalous electrophoretic mobility of kinetoplast minicircle DNA from Leishmania tarentolae. Gene 27:265-277.

Koo, H.S., Wu, H.M. and Crothers, D.M. (1986). DNA bending at adenine-thymine tracts. Nature 320:501-506.

Levene, S.D., Wu, H.M. and Crothers, D.M. (1986). Bending and flexibility of kinetoplast DNA. Biochem. 25:3988-3995.

Marini, J.C., Levene, S.D., Crothers, D.M. and Englund, P.T. (1982). Bent helical structure in kinetoplast DNA. Proc. Natl. Acad. Sci. USA 79:7664-7668; Marini, J.C. and Englund, P.T. (1983). Correction. Proc. Natl. Acad. Sci. USA 80:7678.

Mertz, J.E. and Berg, P. (1974). Viable deletion mutants of simian virus 40:

selective isolation by means of a restriction endonuclease from Hemophilus parainfluenzae. Proc. Natl. Acad. Sci. USA 71:4879–4883.

Ray, D.S., Hines, J.C., Sugisaki, H. and Sheline, C. (1986). kDNA minicircles of the major sequence class of C. fasciculata contain a single region of bent helix widely separated from the two origins of replication. Nucl. Acids Res. 14:7953–7965.

Ross, W., Shulman, M. and Landy, A. (1982). Biochemical analysis of att-defective mutants of the phage lambda site-specific recombination system. J. Mol. Biol. 156: 505–529.

Ryder, K., Silver, S., Delucia, A.L., Fanning, E. and Tegtmeyer, P. (1986). An altered DNA conformation in origin region I is a determinant for the binding of SV40 large T antigen. Cell 44:719–725.

Schmidt, E.R. (1984). Clustered and interspersed repetative DNA sequence family of Chironomus: the nucleotide sequence of the Cla-elements and of various flanking sequences. J. Mol. Biol. 178:1–15.

Selsing, E., Wells, R.D., Alden, C.J. and Arnott, S. (1979). Bent DNA: Visualization of a base-paired and stacked A–B conformational junction. J. Biol. Chem. 254:5417–5422.

Shore, D., Langowski, J. and Baldwin, R.L. (1981). DNA flexibility studied by covalent closure of short fragments into circles. Proc. Natl. Acad. Sci. USA 78:4833–4837.

Simpson, L. (1979). Isolation of Maxicircle component of kinetoplast DNA from hemoflagellate protozoa. Proc. Natl. Acad. Sci. USA 76:1585–1588.

Stellwagen, N.C. (1983). Anomalous electrophoresis of deoxyribonucleic acid restriction fragments on polyacrylamide gels. Biochem. 22:6186–6193.

Trifonov, E.N. and Sussman, J.L. (1980). The pitch of chromatin DNA is reflected in its nucleotide sequence. Proc. Natl. Acad. Sci. USA 77:3816–3820.

Ulanovsky, L., Bodner, M., Trifonov, E.N. and Choder, M. (1986). Curved DNA: design, synthesis, and circularization. Proc. Natl. Acad. Sci. USA 83:862–866.

Ulanovsky, L.E. and Trifonov, E.N. (1987). Curved DNA: Wedge components estimated. Nature (in press).

Wing, R.M., Drew, H.R., Takano, T., Broka, C., Tanaka, S., Itakura, K. and Dickerson, R.E. (1980). Crystal structure analysis of a complete turn of B DNA. Nature 287:755–758.

Wu, H.M. and Crothers, D.M. (1984). The locus of sequence-directed and protein-induced DNA bending. Nature 308:509–513.

Zahn, K. and Blattner, F.R. (1985). Sequence-induced DNA curvature at the bacteriophage lambda origin of replication. Nature 317:451–453.

The author wishes to thank Drs. Ulanovsky and Trifonov for providing their manuscript prior to publication, and Drs. Wells and Harvey for providing a stimulating forum for the discussion of the curvature problem. This work was supported by grant GM28293 from NIH.

Biophysical Principles of Specificity and Stability of Protein-DNA

Interactions: Application to RNA Polymerase ($E\sigma^{70}$)-Promoter Interactions

M. Thomas Record, Jr.
Departments of Chemistry and Biochemistry
University of Wisconsin-Madison
Madison, Wisconsin 53706

Protein-DNA interactions are competitive noncovalent interactions. The rate and extent of complex formation at a specific site is affected not only by macromolecular concentration variables but also by the concentrations of low molecular weight competitors, especially electrolyte ions (Record and Mossing, 1987). Given this situation, how can one characterize the contributions of various noncovalent interactions to the stability and specificity of these complexes and to their mechanisms of formation and/or dissociation? What is the relevance of this information to understanding the regulation of these interactions in the control of gene expression? Regulation may occur at either a kinetic or thermodynamic level. For example, the control of initiation of transcription by E. coli RNA polymerase ($E\sigma^{70}$) occurs at the level of the kinetics of formation of open complexes at these promoters (kinetic control) (cf. McClure (1985) and references therein). The control of expression of the lac operon by lac repressor in conjunction with specific effectors (inducers) apparently occurs at the thermodynamic level (thermodynamic control): the equilibrium degree of occupancy of the operator site by repressor is thought to be the key factor regulating transcription of the lac genes (cf. vonHippel, 1979; Sellitti et al., 1987). Consequently, thermodynamic and mechanistic studies provide the information needed to relate structure to function in these systems.

Generalizations from Studies of Interactions of Proteins with Small
Ligands: The Hydrophobic Effect as a Fundamental Thermodynamic Driving
Force for Complex Formation

Sturtevant (1977) has surveyed the thermodynamics of a variety of
protein–ligand interactions and other processes involving proteins. From
this and more recent work (cf. Baldwin, 1987; Record and Richey (1987)
and references therein) it is clear that the hydrophobic effect provides
a general driving force for formation of protein–ligand complexes and for
the assembly processes (folding, aggregation) of proteins. The role of
the hydrophobic effect is demonstrated most unambiguously by the large
negative standard heat capacity changes ($\Delta C_p^{\,\circ}$) accompanying association/–
assembly. At sufficiently low temperatures, association/assembly is typically accompanied by a net positive (favorable) standard entropy change,
reflecting the dominance of the hydrophobic component over conformational, vibrational, and other contributions.

The interacting surfaces in site-specific protein–ligand complexes
are complementary, both sterically and at the level of functional groups.
A general thermodynamic explanation of the stability and specificity of
these interactions therefore should include the following points: a)
interaction of complementary molecular surfaces displaces (releases) H_2O
that previously hydrated these regions (hydrogen bonding, hydrophobic
hydration); b) in the absence of coupled processes (e.g. protonation/–
deprotonation, conformational changes), the interaction of fully complementary surfaces is likely to be be an enthalpically–neutral exchange
reaction (protein + ligand → complex + nH_2O) driven by the increase in
entropy of the released H_2O; and c) specificity results from the destabi-
lizing effects that would accompany juxtaposition of noncomplementary
regions in other potential complexes (i.e. steric overlaps, site binding
of H_2O or loss of interaction with H_2O without compensation).

Unique Characteristics of DNA as a Ligand that Binds to Proteins: The
Roles of Local Site Concentration and of Ion Release as Driving Forces
for Complex Formation

Substantial attention at this conference has been focused on various
levels of DNA structure (sequence-specific curvature, unusual secondary
structures, supercoiling) and their possible roles in the control of gene

expression. In addition to these structural effects, it is necessary to consider the characteristics of DNA as a chain polymer and as a polyelectrolyte. The flexibly-coiling DNA polyanion has a high local concentration of individual chain segments in the vicinity of distant segments, and a high local concentration of cations in the vicinity of the phosphate backbone. These local concentrations of segments and of ions are of key importance for the kinetics and equilibria of interactions of proteins with macromolecular DNA (Record and Mossing, 1987).

1) DNA is a flexibly-coiling chain molecule. Though the double helix is relatively stiff and hence resistant to deformation, the residual flexibility present in each segment of a high molecular weight DNA due to thermal effects (fluctuations) is sufficient to cause the polymer chain to trace out a relatively random, coiling path in solution. Therefore two regions of a DNA molecule which are separated by a substantial distance along the polymer chain are at a relatively high local concentration. A bidentate DNA-binding protein can utilize this high local concentration of one DNA site in the vicinity of another as a driving force for binding via formation of a stable looped complex (Mossing and Record, 1986; Whitson et al., 1986). (Evidence for such looped complexes has indeed been obtained for various prokaryotic repressor proteins both in vivo and in vitro (cf. Ptashne, 1986; Kramer et al., 1987; Whitson et al., 1987; and references therein).) Looping of the DNA in solution can be quantified in terms of the so-called j-factor, which describes the local concentration of one DNA site in the vicinity of another (Jacobsen and Stockmayer, 1950; Shore and Baldwin, 1983; Shimada and Yamakawa, 1984). The j-factor exhibits a relative maximum as a function of the contour length between the two sites, and should depend sensitively on the intrinsic stiffness of the intervening DNA and on the balance between segment-segment interactions, segment-solvent interactions and excluded volume effects that determines the overall degree of chain expansion or chain collapse.

For the DNA flexible coil under standard solution conditions in vitro, the relative maximum in j occurs at a contour separation of approximately 550 bp, at which length j exceeds 100 nM. For contour separations less than ~ 550 bp, the envelope of j values decreases steeply with decreasing contour separation, as a result of the intrinsic lateral stiffness of the DNA. In this region, the requirement for a correct torsional alignment of the two sites introduces a periodic oscillation into j, where the period corresponds to that of the helical repeat and the amplitude of the oscillation increases markedly with decreasing contour separation. Above ~550 bp, torsional constraints and consequent periodicity in j are less significant, and the envelope of j decreases gradually with increasing length (Shimada and Yamakawa, 1984).

Although the physical characteristics of DNA in vivo (associated with chromatin or other binding proteins, possibly under superhelical stress, and certainly in a highly compact state) are certainly very different from those of DNA in a dilute solution environment, it is reasonable to expect that some qualitative features of the above behavior (e.g. the dependence of local concentration on contour length and on helical repeat) will be preserved. For many regulatory proteins, present in small quantities per cell or involved in nonspecific DNA binding equilibria which reduce the free protein concentration, the local concentration (j) of a bound protein in the vicinity of a distant DNA site may substantially exceed the bulk free protein concentration. Any bidentate regulatory protein, with two DNA-binding sites or one DNA-binding and one protein-binding site, can in principle utilize this thermodynamic effect to drive the formation of a stable looped complex bridging two regions of DNA (Mossing and Record, 1986). The formation of stable loops provides a general (potentially quantitative) thermodynamic model for action-at-a-distance, just as transient loop formation is part of the general kinetic model for facilitated association of a bidentate protein by transfer between two DNA sites (cf. Fried and Crothers, 1984; Berg and von Hippel, 1985). The ability of the cell to vary continuously the local concentration of a bound protein by varying the spacing between its sites of interaction or by altering the intrinsic stiffness or degree of curvature of the intervening DNA provides a powerful means of regulating site-specific protein-DNA interactions. The concept of local concentrations of bound bidentate proteins and tethered DNA sites provides a means of fine-tuning protein-DNA regulatory systems which would not be possible at the level of the bulk protein concentration (Record and Mossing, 1987).

2) DNA is a highly charged polyanion. With two (structural) negative charges per 3.4 Å, the axial charge density along the DNA cylinder is so high as to cause the local accumulation of a high concentration of cations and the exclusion of anions from the vicinity of the cylinder

(cf. Manning, 1978; Anderson and Record, 1982; Mills et al., 1985; and references therein). These ion concentration gradients are equivalent thermodynamically to the neutralization of approximately 88% of the DNA structural charge. As a consequence, helical DNA behaves like a weak electrolyte $(M^+_{0.88}DNAP^-)_n$ (Record et al., 1976, 1978, 1982, 1985). The thermodynamic degree of neutralization is found to be independent of bulk univalent salt concentration, but is strongly dependent on the polyion structural charge density.

Processes which reduce the DNA structural charge density (i.e. denaturation, association of oligocations or of proteins with positively charged residues in their binding sites) are accompanied by release of ions to the solution and consequently are driven by a reduction in the bulk salt concentration (Manning, 1978; Record et al., 1978). We have proposed that this driving force for reduction of polyelectrolyte charge density at low salt concentrations be called the polyelectrolyte effect, by analogy with the hydrophobic effect in which the association of nonpolar surfaces is driven by water release (Record and Mossing, 1987; Record and Richey, 1987).

Noncovalent Interactions of Proteins and Nucleic Acids

The major classes of noncovalent interactions thought to be important for the stability and specificity of the native structures and the interactions of proteins and nucleic acids in aqueous solution are the hydrophobic effect (cf. Tanford, 1980; Hvidt, 1983; Edsall and MacKenzie, 1978), the polyelectrolyte effect (cf. Record et al., 1976; Record and Richey, 1987), aromatic ring interactions (cf. Burley and Petsko, 1985), ionic interactions (cf. Honig et al., 1986), and hydrogen bonding interactions (cf. Fersht et al., 1985). The latter three types of interactions have definite structural roles in noncovalent complexes or assemblies, and can be visualized at atomic resolution by x-ray crystallography. However, the hydrophobic effect and the polyelectrolyte effect, which make major contributions to the stability of complexes involving nonpolar and charged residues, respectively, are not visible in the crystal structures of these complexes. Instead, these effects drive formation of a macromolecular complex by the reduction of a relatively unfavorable noncovalent interaction between the uncomplexed macromolecule and a solvent or solute component. In the hydrophobic effect, the unfa-

vorable interaction of water with exposed nonpolar functional groups is
reduced when those nonpolar groups are buried in the interior of a complex and therefore removed from the aqueous environment. In the
polyelectrolyte effect, unfavorable electrostatic repulsions among closely spaced negative charges on a polyanion in solution are reduced by
the accumulation of cations and exclusion of anions from its vicinity.
This ordering of ions into locally steep concentration gradients is
itself unfavorable, relative to the random distribution of these ions
that would exist around an uncharged polymer. Neutralization of some
polyanion charge by binding an oligocation (L^{z+}) reduces or locally eliminates the requirement for the ion concentration gradients.

There is a fundamental parallelism between these two important
effects. Self-association of nonpolar groups releases water; binding of
an oligocation to a polyelectrolyte releases electrolyte ions.
Consequently both of these processes are exchange reactions in solution.

   1) Association of nonpolar groups (driven by the hydrophobic effect)
(Hvidt, 1983):

$$n\ R(H_2O)_m \rightleftharpoons R_n + nm\ H_2O \tag{1}$$

   2) Binding of an oligocation to the DNA polyanion (driven by the
polyelectrolyte effect) (Record et al., 1976; Record and Mossing, 1987):

$$L^{z+} + (M_{0.88}^+ DNAP^-)_n \rightleftharpoons complex + 0.88z\ M^+ \tag{2}$$

If the ligand is itself a polycation or site-binds anions, then anion
exchange from the ligand must be included in the overall process, as discussed below. Formation of ionic interactions (ion pairs, salt bridges)
should be an ion exchange process only when the interacting species are
highly charged polyelectrolytes, or when isolated tight-binding sites for
ions are involved. Formation of an ion pair between isolated oppositely
charged groups generally need not involve ion exchange, nor show a significant salt dependence.

   As a result of the exchange characteristics of the polyelectrolyte
effect, the hydrophobic effect, and other noncovalent interactions, the
general thermodynamic description of a site-specific protein-DNA interaction in an aqueous univalent salt (MX) solution is (Record et al., 1978;
Record and Mossing, 1987).

$$\begin{array}{l} \text{DNA site} + \text{protein site} \rightleftharpoons complex + aM^+ + cX^- + (b + d)H_2O \\ (aM^+;bH_2O) \qquad (cX^-;dH_2O) \end{array} \tag{3}$$

In eq. (3), the stoichiometric coefficients a, b, c, and d are thermodynamic coefficients which are rigorously defined in terms of preferential interaction parameters describing the extents of accumulation or exclusion of the electrolyte component in the vicinity of the DNA and protein sites and of the complex (Record et al., 1985; Anderson and Record, 1982; Mills et al., 1986). Note that eq. (3) is written for ionic species, and does not represent merely an arbitrary choice of electroneutral components. In particular, the stoichiometric coefficients a and c cannot be specified (or related to one another) by any electroneutrality condition. The thermodynamic degree of association of electrolyte ions with a cylindrical polyelectrolyte like DNA is determined by the axial charge density and not merely by the number of polyelectrolyte charges (Manning, 1978; Anderson and Record, 1982). An apparent association equilibrium constant K_{obs} may be defined in terms of the equilibrium concentrations of the macromolecular participants:

$$\text{DNA site + protein site} \overset{K_{obs}}{\rightleftharpoons} \text{complex} \qquad (4)$$

From eq. (3), it is clear that K_{obs} is a function of salt concentration. The quantitative form of this dependence is (Record et al., 1978):

$$-\frac{d \log K_{obs}}{d \log[MX]} = a + c - \frac{2[MX]}{[H_2O]}(b + d) \qquad (5)$$

Except at salt concentrations in the molar range, where even site-specific protein-DNA interactions are generally too weak to investigate conveniently, the term in eq. (5) from H_2O release is expected to be negligible, and the salt dependence of K_{obs} provides a measure of the polyelectrolyte effect, i.e. of the net amount of electrolyte ion release upon complex formation. If anion release is also negligible (c≈0), then eq. (5) provides an estimate of the number of DNA phosphate charges neutralized in the interaction (a/0.88).

For a variety of oligocations L^{z+}, including Mg^{2+}, polyamines (z=2-4), and oligolysines (z=3-8), the DNA-binding reaction appears to be a pure cation-exchange reaction (Record et al., 1976; Braunlin et al., 1982, 1986; Lohman et al., 1980). In all cases, plots of log K_{obs} as a function of log [MX] are linear over the experimentally accessible range, with slopes which agree within experimental uncertainty with the

243

stoichiometry expected for a cation exchange process ($-d\log K_{obs}/d\log$ [MX] = 0.88z). For all these oligocations, $\log K_{obs}$ extrapolates to a value near zero (K_{obs} of order of magnitude $\sim 1M^{-1}$) at the 1M salt standard state. Therefore the process

$$L^{z+} + \text{DNA site} \rightarrow \text{complex} + 0.88Z \; M^{+} \qquad (6)$$

occurs with a standard free energy change $\Delta G^{o} \sim 0$ at 1M salt. Binding of oligocations to DNA is driven at lower salt concentrations to an extent determined by the free energy (entropy) of dilution of the displaced ions (Record et al., 1976).

The profound effects of the concentrations of electrolyte ions on the kinetics and equilibria of protein-DNA interactions, as well as on other noncovalent assembly reactions of biopolyelectrolytes, have long been appreciated. These effects were originally interpreted as ionic strength effects on electrostatic interactions, by analogy with Debye-Huckel and Bronsted-Bjerrum analyses of salt effects on the interactions of non-polymeric solutes (Riggs et al., 1970a,b). However, it is now clear that ion effects on interactions involving polyelectrolytes are not general ionic strength effects, but rather are specific effects of individual cations and anions (Record et al., 1978; Leirmo et al., 1987). For thermodynamic purposes these ion effects are best treated using a competitive ion-exchange formalism (Record et al., 1977, 1978). The stoichiometric participation of small ions in processes involving highly charged polyions is the essence of the polyelectrolyte effect, which is characteristic of processes which alter the charge density of polymeric electrolytes of high charge density (and consequently is generally not a factor in the interactions of protein polyampholytes with small ligands).

The exchange characteristics of the hydrophobic effect, the polyelectrolyte effect, and of hydrogen bonding interactions have several implications of fundamental importance:

i) The stability of a noncovalent protein-DNA complex formed in an exchange reaction in solution is critically dependent on the bulk concentrations of the exchanged species (e.g. H_2O, electrolyte ions). No unique, environment-independent stability can be assigned to such a complex. Consequently it is impossible to predict the stability of a protein-DNA complex from structural considerations alone.

ii) Because individual electrolyte ions play a direct stoichiometric role in protein-DNA interactions, with stoichiometric coefficients which are generally much greater than that of the protein, the concentration and type of electrolyte can be used to great advantage as a _probe_ both of the thermodynamics and of the kinetics and mechanism of these interactions. At a thermodynamic level, information about the relative contributions of ionic interactions and nonelectrostatic interactions to the binding free energy under specified conditions may be obtained from the salt-dependence of an apparent equilibrium constant. This is analogous to the use of temperature to dissect enthalpic and entropic contributions to the binding free energy, or to the use of pH to analyze protonation effects. At a kinetic level, the division of individual ion effects between the association and dissociation rate constants provides important and perhaps otherwise unobtainable information about the number of steps in the mechanism and the nature of the important intermediates. Over the last decade, numerous studies have appeared in which these methods of analysis have been used to probe the thermodynamics and mechanisms of site-specific and nonspecific interactions of proteins, dyes and oligocations with DNA and RNA (reviewed in Lohman, 1985; Hélène and Lancelot, 1982; Record et al., 1978, 1981, 1985). An application to the site-specific binding of RNA polymerase ($E\sigma^{70}$) is given below.

What Drives Specific Binding of RNA Polymerase and of lac Repressor to DNA?

Formation of both RNA polymerase-promoter and _lac_ repressor-operator complexes is _entropically_ driven at low temperature. For the _lac_ repressor-O^+ interaction, the entropic driving force is much larger than can be accounted for by the polyelectrolyte effect at the specified salt concentration (Record et al., 1977). It is therefore likely that the release (exchange) of water which accompanies the formation of site-specific protein-DNA hydrogen bonds (or possible nonpolar contacts) provides the additional entropic driving force (Riggs et al., 1970a,b; Record and Mossing, 1987). The optimal (i.e. tightest binding) site-specific complex is presumably one in which there is maximal physical and chemical complementarity of the interacting surfaces of the sites, so that the maximum numbers of protein-DNA hydrogen bonds, nonpolar contacts and neutralized charges can be achieved, thereby generating the maximum numbers of water molecules and ions released. For this case, as discussed above, it is reasonable that the standard free energy change for the association process should be primarily entropic, since only exchange reactions involving ions and water are involved. If the conformations or arrangements of functional groups on the isolated protein and DNA binding surfaces are not completely complementary, then to form the most stable complex, a) conformational adaptations in the recognition surfaces may occur to maximize complementarity, and/or b) noncomplementary regions may be incorporated into the complex by site-binding ions or water, or by using the free energy of binding to pay the thermodynamic price of losing a favorable hydrogen bond or other noncovalent interaction without replacing it in the complex.

Specific binding of lac repressor to O^+ and O^c operators does not involve simply the addition of specific contacts with individual functional groups on the DNA bases to the electrostatic contacts with phosphates which characterize the nonspecific binding mode of lac repressor. Indeed the twofold reduction in the extent of ion release (polyelectrolyte effect) in specific binding to O^+ suggests that different numbers of positively charged amino acid side chains are involved in contacts with the DNA in the specific and nonspecific binding modes (Mossing and Record, 1985).

Estimating the Contribution of the Polyelectrolyte Effect and the Hydrophobic Effect to the Thermodynamics of Open Complex Formation by RNA Polymerase $E\sigma^{70}$ at the λP_R Promoter

1) As a simple model for the polyelectrolyte contribution to the interaction of $E\sigma^{70}$ with a promoter, consider the interaction of an oligocation L^{z+} with DNA $(M^+_{0.88}DNAP^-)_n$ for which the apparent equilibrium binding constant is K_{obs}:

$$L^{z+} + \text{DNA site} \underset{}{\overset{K_{obs}}{\rightleftharpoons}} \text{complex} \tag{7}$$

For the simplest case, where anion and release are negligible, then (cf. eq. 5)

$$-\frac{d\ln K_{obs}}{d\ln[MX]} = 0.88z \tag{8}$$

(This corresponds to the stoichiometric picture $L^{z+} + \text{DNA site} \rightleftharpoons \text{complex} + 0.88z\ M^+$ (cf. eq. 6).

To estimate the entropic contribution of the polyelectrolyte effect to a binding free energy from L^{z+} binding studies, we note that for

$(lys)_n$, polyamines, Mg^{2+}, and $Co(NH_3)_6^{3+}$ in excess univalent salt (MX):

$$K_{obs} = K_o[MX]^{-0.88z} \qquad (9)$$

and $\quad \Delta G^o_{obs} = \Delta G^o_o + 0.88zRTln[MX] \qquad (10)$

where K_o is found to be ~ 1 M^{-1} and therefore $\Delta G^o_o \sim 0$. At $[MX] \to 1$ M, the polyelectrolyte effect therefore makes little contribution to the observed binding free energy. The electrostatic driving force from the polyelectrolyte effect (relative to $[MX] = 1$ M) may therefore be estimated from the relationship

$$\Delta(\Delta G^o_{el}) = 0.88zRTln[MX] \qquad (11)$$

Table 1 compares the polyelectrolyte contribution $\Delta(\Delta G^o_{el})$ to the $E\sigma^{70}$ - λP_R promoter interaction with the net observed free energy change (ΔG^o_{obs}) at various salt concentrations.

Table 1

Estimates of Relative Electrostatic Driving Forces for Open Complex Formation by $E\sigma^{70}$ at the λP_R Promoter (25°C, 0.01 M Hepes (pH 7.5)).[a]

$[Na^+]$(M)	$K_{obs}(M^{-1})$	$-\Delta G^o_{obs}$(kcal)	$-\Delta(\Delta G^o_{el})$(kcal)
0.19	$(3.7\pm2.6)\times10^{11}$	15.5 ± 0.5	19.0 ± 1.3
0.21	$(4.1\pm1.5)\times10^{10}$	14.2 ± 0.2	17.8 ± 1.2
0.23	$(5.0\pm1.9)\times10^{9}$	13.0 ± 0.2	16.8 ± 1.1
0.25	$(1.4\pm1.0)\times10^{9}$	12.3 ± 0.4	15.8 ± 1.0
0.27	$(3.5\pm2.4)\times10^{8}$	11.5 ± 0.4	14.9 ± 1.0

($\Delta G^o_o = 3.6\pm1.1$ kcal; $-dlnK_{obs}/dln[Na^+] = 19.6\pm1.3$)

[a]Data from Roe and Record (1985).

Although $\Delta(\Delta G^o_{el})$ provides only an estimate of the polyelectrolyte effect, clearly the contributions to K_{obs} and ΔG^o_{obs} are of comparable magnitude to the net values of these quantities themselves. Consequently the polyelectrolyte effect provides a large driving force for site-specific complex formation by $E\sigma^{70}$ under the above ionic conditions.

2) To estimate the contribution of the hydrophobic effect (ΔG^o_h) to the binding free energy of the $E\sigma^{70}$ - λP_R interaction, we have utilized thermodynamic data on the process of transfer of nonpolar solutes from

water to a nonaqueous environment. The primary characteristics of such processes are a) a large negative temperature-insensitive ΔC_p^o, leading to large temperature dependences of ΔH^o and ΔS^o, and b) enthalpy-entropy compensation (because $|\Delta C_p^o| >> |\Delta S^o|$ at most temperatures) leading to a temperature-insensitive ΔG^o (reviewed in Record and Richey, 1987). For a variety of nonpolar solutes, both ΔG^o and ΔC_p^o for the transfer process correlate well with the area of nonpolar surface removed from water: $\Delta G^o \simeq -25$ cal mole^{-1} Å^{-2} near 25°C (cf. Richards, 1977; Creighton (1983) and references therein) and $\Delta C_p^o \simeq -0.33$ cal deg^{-1} mole^{-1} Å^{-2} (Gill et al. (1985) and references therein).

In a survey of heat capacity effects in processes involving proteins, Sturtevant (1977) concluded that the hydrophobic effect made the dominant contribution to ΔC_p^o. We have therefore proposed that the large negative ΔC_p^o observed for the binding of E. coli RNA polymerase (Eσ[70]) to the λP$_R$ promoter is hydrophobic in origin and have employed the relationship $\Delta G^o_h \simeq 75 \Delta C^o_p$ (suggested by the above correlations) to estimate the hydrophobic contribution to ΔG^o. (We also find that large negative values of ΔC_p^o characterize the interaction of lac repressor with various lac operator sequences (Ha and Record, in preparation).)

For the interaction of Eσ[70] with the λP$_R$ promoter to form an open complex, Roe et al. (1985) found that $\Delta C_p^o = -2400$ cal deg^{-1}. We therefore estimate that up to 7200 Å^2 of nonpolar surface are buried (removed from exposure to water) in the process of open complex formation. (This corresponds to roughly a 10% reduction in the solvent-accessible surface area of this protein, as evaluated by the method of Richards (1977).) Roe et al. proposed that this burial of nonpolar surface occurred as part of a conformational change between the intermediates RP$_c$ (closed complex) and RP$_i$ (intermediate complex) on the pathway to open complex formation:

$$R + P \rightleftharpoons RP_c \rightleftharpoons RP_i \rightleftharpoons RP_o$$

The favorable contribution to the standard free energy change of the process of open complex formation resulting from removal of 7200 Å^2 of hydrophobic surface from water is approximately -180 kcal!

Construction of a Thermodynamic Balance Sheet for Formation of Site-Specific Protein-DNA Complexes.

The above estimates of the contributions of the polyelectrolyte effect and the hydrophobic effect to the stability of the site-specific open complex between $E\sigma^{70}$ and the λP_R promoter are highly approximate. However, they demonstrate that there is a very large driving force available from these two sources, which in fact exceeds by more than an order of magnitude the net standard free energy change ($\Delta G°_{obs}$) of complexation. To complete this thermodynamic picture of stability and to understand specificity, we need to understand the various driven processes (ranging from the consequences of local steric effects and lack of complementarity of functional groups to long-range conformational changes) that utilize this driving free energy. The specific complex should be viewed as that which maximizes the magnitude of the available driving free energy, and/or that which minimizes the expenditure of this driving free energy. Either of these views contrasts greatly with those which seek to explain the thermodynamics of site-specific binding in terms of the addition of individual contributions from specific hydrogen bonds or other noncovalent contacts discerned in the crystal structure of the complex. The hydrophobic effect and the polyelectrolyte effect, which are often dismissed as relatively nonspecific factors and which would be "invisible" in the structure of the complex, nevertheless drive these site-specific association reactions.

Acknowledgements

It is a pleasure to acknowledge the many contributions of both present and former members of this research group to the ideas presented in this article (as indicated in the literature cited), and to acknowledge the assistance of Sheila Aiello and Karen Rulland in its preparation. I also want to thank Bob Wells, Steve Harvey and their staff for their efforts which made this meeting a success. Our research is supported by grants from the NSF and the NIH.

References

Anderson, C.F. and M.T. Record, Jr. (1982), Ann. Rev. Phys. Chem. 33, 191-222,

Baldwin, R.L. (1986), Proc. Natl. Acad. Sci. U.S.A. 83, 8069-8072.

Berg, O.G. and P.H. von Hippel (1985), Ann. Rev. Biophys. Biophys. Chem. 14,131-160.

Braunlin, W.H., T.J. Strick, and M.T. Record, Jr. (1982), Biopolymers 21, 1301-1314.

Braunlin, W.R., C.F. Anderson and M.T. Record, Jr.(1986), Biopolymers 25, 205-214.

Burley, S.K. and G.A. Petsko (1985), Science 229, 23-28.

Creighton, T. (1983), Proteins, W. H. Freeman, N.Y.

Edsall, J. and H.A. MacKenzie (1978), Adv. Biophys. 10, 137-207.

Fersht, A.R., J-P. Shi, J. Knill-Jones, D.M. Lowe, A.J. Wilkerson, D.M. Blow, P. Brick, P. Carter, M.M.Y. Waye and G. Winter (1985), Nature 314, 235-238.

Fried, M.G. and D.M. Crothers (1984), J. Mol. Biol. 172, 263-282.

Gill, S.J., S. F. Dec, G. Olofsson and I. Wadso (1985), J. Phys. Chem. 89, 3758-3761.

Hélène, C. and G. Lancelot (1982), Prog. Biophys. Mol. Biol. 39, 1-68.

Honig, B.H., W. L. Hubbell and R.F. Flewelling (1986), Ann. Rev. Biophys. Biophys. Chem. 15, 163-194.

Hvidt, A. (1983), Ann. Rev. Biophys. Bioeng. 12, 1-20.

Jacobsen, H. and W.H. Stockmayer (195), J. Chem. Phys. 18, 1600-1606.

Kramer, H,M. Niemöller, M. Amouyal, B. Revet, B. vonWilcken-Bergmann and B. Müller-Hill (1987), EMBO J. 6, 1481-1491.

Leirmo, S., C. Harrison, D.S. Cayley, R.R. Burgess and M.T. Record, Jr., (1987), Biochemistry 26, 2095-2101.

Lohman, T.M., P.L. deHaseth and M.T. Record, Jr. (1980), Biochemistry 19, 3522-3530.

Lohman, T.M. (1985), CRC Crit. Rev. Biochem. 19, 191-245.

Manning, G.S. (1978), Quart. Rev. Biophys. 11, 179-246.

McClure, W.R. (1985), Ann. Rev. Biochem. 54, 171-204.

Mills, P., C.F. Anderson and M.T. Record, Jr. (1985), J. Phys. Chem. 89, 3984-3994.

Mills, P., C.F. Anderson and M.T. Record, Jr. (1986), J. Phys. Chem. 90, 6541-6548.

Mossing, M.C. and M.T. Record, Jr. (1985), J. Mol. Biol. 186, 295-305.

Mossing, M.C. and M.T. Record, Jr. (1986), Science 233, 889-892.

Ptashne, M. (1986), Nature 322, 697-701.

Record, M.T. Jr., T.M. Lohman and P.L. deHaseth (1976), J. Mol. Biol. 107, 145-158.

Record, M.T. Jr., P.L. deHaseth and T.M. Lohman (1977), Biochemistry 16, 4791-4796.

Record, M.T. Jr., C.F. Anderson and T.M. Lohman (1978), Quart. Rev. Biophys. 11, 103-178.

Record, M.T. Jr., S.J. Mazur, P. Melancon, J.-H. Roe, S.L. Shaner and L. Unger (1981), Ann. Rev. Biochem. 50, 997-1024.

Record, M.T. Jr., C.F. Anderson, P. Mills, M. Mossing and J.-H. Roe (1985), Adv. Biophys. 20, 109-135.

Record, M.T. Jr. and M.C. Mossing (1987), in RNA Polymerase and Regulation of Transcription (W. Reznikoff et al., eds.) Elsevier Science Publishing Co, New York, pp 61-83.

Record, M.T. Jr. and B. Richey (in press, 1987), ACS Sourcebook for

Physical Chemistry Instructors (T. Lippincott, ed.).

Richards, F.M. (1977), Ann. Rev. Biophys. Biophys. 6, 151–176.

Riggs, A.D., S. Bourgeois and M. Cohn (1970a), J. Mol. Biol. 53, 401–417.

Riggs, A.D., H. Suzuki and S. Bourgeois (1970b), J. Mol. Biol. 48, 67–83.

Roe, J.-H., and M.T. Record, Jr. (1985), Biochemistry 24, 4721–4726.

Roe, J.-H., R.R. Burgess and M.T. Record, Jr. (1985), J. Mol. Biol. 184, 441–453.

Sellitti, M.A., P.A. Pavco and D.A. Steege (1987), Proc. Natl. Acad. Sci. U.S.A. 84, 3199–3203.

Shimada, J. and H. Yamakawa (1984), Macromolecules 17, 689–698.

Shore, D. and R.L. Baldwin (1983), J. Mol. Biol. 170, 957–981.

Sturtevant, J.M. (1977), Proc. Natl. Acad. Sci. U.S.A. 74, 2236–2240.

Tanford, C. (1980), The Hydrophobic Effect (John Wiley and Sons, New York).

von Hippel, P.H. (1979), in: Biological Regulation and Development, R.F. Goldberger, ed. (Plenum, New York) pp. 279–347.

Whitson, P.A, J.S. Olson and K.S. Matthews (1986), Biochemistry 25, 3852–3858.

Whitson, P.A., W.-T. Hsieh, R.D. Wells and K.S. Matthews (1987), J. Biol. Chem. 262, 4943–4946.

IMMUNOCHEMICAL APPROACHES TO THE DEFINITION OF UNUSUAL NUCLEIC ACID STRUCTURES

B. David Stollar

Department of Biochemistry, Tufts University Health Science Campus,
Boston, MA 02111.

PRINCIPLES

Immunochemistry has a long history as a means for comparing conformations of proteins, polysaccharides and nucleic acids. Reactions of
appropriate antibodies can detect changes associated with physiological
modulation of protein structure (such as occurs on oxygenation and de-
oxygenation of hemoglobin or allosteric modification of enzymes),
evolutionary substitutions of amino acids, and the much larger changes
associated with denaturation (reviewed by Crumpton, 1974). Antibodies
have been important reagents for structural analysis of polysaccharides
as well, being sensitive to conformation as well to the presence of
different monosaccharide constituents and glycosidic linkages
(Heidelberger, 1982 Kabat et al., 1981). Although nucleic acid immunochemistry has a shorter history, it has provided valuable reagents for
studies of both purified nucleic acids and complex biological material
such as whole nuclei or chromatin (reviewed in Stollar, 1975;1986).
Antibodies distinguish among various nucleic acid structures by
recognizing either specific bases or conformations.

The ability to define novel nucleic acid structures immunologically is
enhanced by the fact that the most abundant nucleic acids are only
poorly immunogenic, whereas minor forms with different structures are
more effective in inducing antibody formation. It has been very
difficult to induce antibodies to native B-DNA, tRNA or ribosomal RNA
by immunization of normal subjects, but it has been possible to obtain
strong responses to physically or chemically modified DNA. Antbodies
that are induced by immunogenic helical analogues of DNA recognize only

what is different between the inducing immunogen and B-DNA. For example, normal rabbits and mice respond to poly(dG)·poly(dC) (Lafer and Stollar, Lee et al., 1984); they make antibodies that react with the inducing polymer but not with native DNA. A number of other helical nucleic acids are also immunogenic and their antibodies also fail to react with native DNA (table 1).

DENATURED DNA

Local denaturation and modified bases in unusual DNA structures can be recognized immunologically. Anti-denatured DNA antibodies induced by insoluble complexes of denatured DNA and methylated bovine serum albumin recognize short sequences of purine and pyrimidines and do not react with helical base-paired regions (Stollar, 1980; Wakizaka and Okuhara, 1975). Again, modified bases in the DNA provide a much stronger immune stimulus than do the normal bases. For example, denatured T4 bacteriophage DNA induces several-fold higher concentrations of antibody than does calf thymus or E. coli DNA, and the anti-T4 DNA antibodies are directed overwhelmingly toward the glucosylated hydroxymethylcytosine of the phage DNA (Townsend et al., 1965; Gruenewald and Stollar, 1973). Similarly, DNA bases modified by carcinogens (Sage and Leng, 1980), ultraviolet irradiation (Seaman et al., 1972), or photo-oxidation (Seaman et al., 1974), and tRNA bases subjected to biochemical postsynthetic modification (Aharonov et al., 1974) all act as dominating determinants of both antibody production and antibody specificity on immunization with nucleic acid-MBSA complexes. In a second approach to production of anti-denatured DNA antibodies, normal and modified bases, nucleosides or nucleotides can be conjugated covalently to protein carriers; the conjugates induce antibodies that react selectively with the immunizing base (reviewed in Munns and Liszewski, 1980; Strickland and Boyle, 1984). These antibodies are useful reagents for determining whether the bases are accessible in polymeric nucleic acids of unusual conformation.

Both of the above-described types of antibodies are sensitive indicators of denaturation of DNA. Anti-nucleoside antibodies detect exposure of bases and thus the local denaturation that occurs in regions where cis- or trans-diamminedichloroplatinum(II)(cis-DDP) binds to DNA (Sundquist et al., 1986). They also detect exposure of methyl-

254

TABLE 1

IMMUNOGENIC HELICAL NUCLEIC ACIDS

1. Right-handed double-helical polydeoxyribonucleotides
    a. Poly(dG) poly(dC)
    b. Poly(dG-dA) poly(dT-dC)

2. Left-handed double-helical polydeoxyribonucleotides

3. Double-helical polyribonucleotides

4. Double-helical RNA-DNA hybrid

5. Triple-helical polynucleotides
    a. Poly(dG-dA) poly(dT-dm5C)
    b. Poly(dA) 2poly(U)
    c. Poly(A) 2poly(U)
    d. Poly(A) poly(U) poly(I)

6. Modified DNA
    a. AAF-DNA
    b. Ultraviolet irradiated DNA
    c. Photooxidized DNA
    d. Cis-DDP-DNA
    e. Trans-DDP-DNA

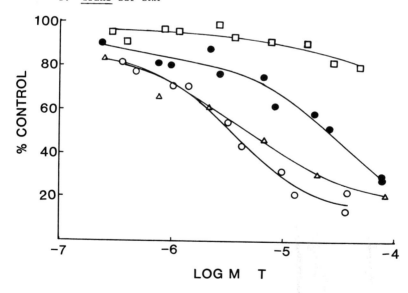

Figure 1.   Competitive immunoassay for measurement of the accessibility
of thymine in oligonucleotides. Antibodies were raised by immunization
of rabbits with a thymidine-protein conjugate (Stollar, 1980). soluble
reactants were tested for binding to the antibodies in competition with
immobilized denatured DNA. The competitors were: thymidine monophosphate (●);poly(dT (▲); d(GCGCGCGCGCGCGCTTTTTTTTTTTTT (○); and
d(CGCGATTCGCG) (◻).

cytosine in regions of DNA altered by photo-oxidation of guanine (Schreck et al., 1973). Anti-thymidine antibodies may be used to test whether thymine is exposed to solvent and accessible to antibody when it occurs in short oligonucleotides. An oligonucleotide in which free oligo(dT) segments extend from a base-paired dG-dC core is reactive with anti-thymidine antibodies (figure 1). An oligonucleotide that forms either a hairpin loop with unpaired T or a complementary pair with a T-T mismatch (Summers et al., 1985) reacts very poorly with these antibodies; if the thymine is looped out of the hairpin or duplex, it is not extended far enough to be accessible to this probe (figure 1).

ANTIBODIES TO Z-DNA

Major immunochemical approaches to unusual DNA structures have involved helical variants. The most prominent subject of such studies has been left-handed Z-DNA. Polyclonal and monoclonal antibodies to Z-DNA have been particularly useful reagents for determination of conditions under which the left-handed helical DNA can be formed in complex materials of biological origin (reviewed in Stollar, 1986).

Antibodies have been induced by several forms of Z-DNA; among the first were those obtained by immunization with poly(dG-dC) stabilized in the Z-DNA conformation by bromination of the polymer in 4M NaCl (Lafer et al., 1981). Others have been induced by poly(dG-dm5C) (Lafer et al., 1983; Zarling et al., 1984a), poly(dG-dbr5C) (Zarling et al., 1984a,b) or DNA modified by chlorodiethylenetriaminoplatinum(II)chloride (Malfoy and Leng, 1981) or N-acetoxy-N-acetyl-2-aminofluorene (AAF) (Hanau et al., 1984). Monoclonal antibodies of varying specificity have been produced; some recognize a wide variety of sequences in the Z-DNA conformation whereas others are specific for one sequence, such as (dG-dC) (Zarling et al., 1984b, Nordheim et al., 1986) or (dG-d5BrC) (Zarling et al., 1984a,b). Comparisons of the reactions of various analogues and consideration of the size of the antibody combining site dimensions indicate that certain of these antibodies recognize a region spanning 2 to 3 base pairs over the surface that replaces the major groove (Lafer et al., 1981). This has been substantiated by chemical protection experiments carried out by A. Nordheim and colleagues (Runkel and Nordheim, 1986).

Although solutions of unphysiologically high ionic strength provided
the first conditions for conversion of linear poly(dG-dC) to the Z-DNA
form (Pohl and Jovin 1972), Z-DNA formation can occur under physiological conditions (Singleton et al., 1982). Driving forces for its formation can be the presence of certain bases, such as me5C or br5C in
place of C; the presence of complex amines or divalent cations; supercoiling; and stabilization by binding proteins (reviewed in Rich et
al., 1984). Antibodies to Z-DNA are useful models of the latter. At
high concentrations, especially if the antibodies are of high affinity,
they can lower the ionic strength or the amount of supercoiling required for Z-DNA formation (Lafer et al., 1986). A sufficently high
concentration can drive formation of the Z-DNA helix of linear
poly(dC-dG) even in physiological ionic strength (Lafer et al., 1985).
This action must be kept in mind when the antibodies are used as
reporters for detection of Z-DNA; for this purpose they should be used
at concentrations that do not themselves drive the formation of the
structure.

Once the specificities of anti-Z-DNA antibodies were established, they
were applied to the search for the structure in nature and to other
questions of biological interest (table 2). Incorporation of DNA
fragments into supercoiled plasmids, combined with an immunochemical
assay, is an excellent test for Z-DNA forming potential (Nordheim et
al., 1982) and an effective method for the isolation of gene segments
that have this potential (Thomae et al., 1983). The immunological
identification of Z-DNA in fixed chromosomes has established that the
appropriate sequences can be converted to Z-DNA within chromatin
structure, that acidifcation during fixation can enhance Z-DNA
formation, and that supercoiling (probably associated with removal of
some chromosomal proteins) is a major driving force for its formation,
even in the absence of acidification (reviewed in Stollar, 1986, 1983).
Experiments with antibodies of restricted specificity indicate that
alternating (dG-dC) sequences are not the major Z-DNA forming sequences
in Drosophila polytene chromosomes; a more likely candidate is the
frequently recurring poly(dT-dG)•poly(dC-dA) sequence (Nordheim et al.,
1986). What is not clear, however, is how much Z-DNA occurs under
truly native conditions; there is little immunofluorescent staining of
isolated polytene chromosomes that are dissected into saline with a low
concentration of protamine and with no exposure to deproteinizing or

257

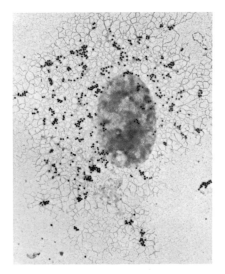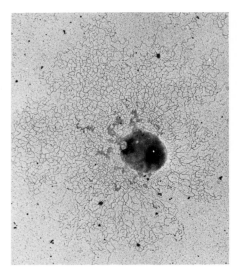

Figure 2. Identification of Z-DNA in the DNA extruded from bacteria. Normal rabbit serum (left panel) or rabbit anti-Z-DNA serum (right panel), diluted 1:250, was incubated with the preparation and bound antibody was detected with gold-labeled protein A. The electron micrograph was provided by W. Earnshaw.

TABLE 2

APPLICATIONS OF ANTI-Z-DNA ANTIBODIES

1. Identification of Z-DNA in:

>   polytene chromosomes
>   metaphase chromosomes
>   protozoal macronuclei
>   Drosophila nuclei
>   mammalian nuclei
>   recombination junctions

2. Mapping of Z-DNA forming sites in supercoiled DNA

3. Measurement of Z-DNA formation driven by torsional strain under physiological conditions in:

>   closed circular dsDNA
>   form V DNA
>   chromatin

4. Measurement of Z-DNA stabilization by protein binding

5. Isolation of Z-DNA-containing plasmids by immunoaffinity chromatography

acidifying fixation (Hill and Stollar, 1983). On the other hand, the antibodies detect Z-DNA in isolated nuclei that are not subjected to acid fixation (Lancillotti et al., 1985; 1987). As chromosomal DNA occurs in constrained loops with many nucleosomes, each of which is associated with supercoiling (Germond et al., 1975), it is likely that Z-DNA is at least a transiently formed structure that would occur when the number of nucleosomes per domain changes. The ready adoption of the Z-DNA conformation under physiological supercoiling conditions is evident from its presence in DNA extruded from bacteria (figure 2).

Antibodies have identified Z-DNA structures in sequences that occur in interesting sites in relation to gene organization and regulation including enhancer sequences and regions upstream from protein coding sequences (reviewed in Rich et al., 1984). These experiments have not identified a regulatory function for Z-DNA in relation to transcription. On the other hand, antibodies have identified Z-DNA in a protein-mediated recombination complex and a recombination protein itself is a Z-DNA-binding protein (Kmiec and Holloman, 1984; 1986). Z-DNA could well be a transient intermediate in the alignment and junction formation in constrained DNA segments.

POLYPURINE SEGMENTS AND TRIPLE-STRANDED DNA

Poly(purine) polymers, alone or with associated poly(pyrimidines) can form multi-strand structures; it has been suggested that such structures could be formed in naturally occurring polypurine sequences within DNA (Lee et al., 1980, 1973). Antibodies to triple-helical poly(purine) poly(pyrimidine) helices have been studied with both polyribonucleotides and polydeoxyribonucleotides (reviewed in Stollar, 1975). Antibodies specific for the triple-helical poly(A).2poly(U) were reported first by Nahon-Merlin and Lacour (reviewed in Lacour et al., 1973). Later, we found that such anti-poly(A).2poly(U) antibodies react equally well with the immunizing polymer and with poly(A).poly(dT).poly(U) (Stollar and Raso, 1974). On the other hand, they do not react at all with a triple-helix in which the polypurine is a polydeoxyribonucleotide, poly(dA).2poly(U). Antibodies to the latter, in turn, fail to react with poly(A).2poly(U) but cross-react completely with poly(dA).2poly(U) and poly(dA).poly(dT).poly(U) (Stollar and Raso,

1974). Poly(dA).2poly(dT), formed at high temperature at high ionic strength, also reacts with this antibody, as does a triple helix formed by poly(TTC)·poly(AAG) (unpublished data). Recently, Lee et al. (1987) have described a monoclonal antibody directed against a triple-helical poly(dG–dA)·poly(dT–dm5C); it also yielded immunofluorescent staining of mammalian interphase nuclei.

IMMUNOLOGICAL DETECTION OF HELIX MODIFICATIONS BY CIS–DDP.

As noted above, anti–nucleoside antibodies can detect regions of local unwinding of the DNA helix adjacent to sites of adduct formation between cis–DDP and guanine (Sundquist et al., 1986). When the cis–DDP–modified native DNA is used as an immunogen, it induces antibodies specific for the modified DNA (Poirier, 1981). Monoclonal antibodies against cis–DDP–DNA do not react with the denatured regions or with unmodified DNA. They do not appear to bind directly to the platinated base, as they react equally well with DNA modified by several analogues of cis–DDP having varied substituents. Furthermore, although they recognize the adduct formed with adjacent guanine residues in cis–DDP–DNA or cis–DDP–poly(dG)·poly(dC), the homopolymer cis–DDP–poly(dG) does not bind to the antibodies. Rather, these antibodies appear to detect a modified DNA conformation that results from the reaction with cis–DDP (W. Sundquist, S. J. Lippard and B. D. Stollar, unpublished data).

RECOGNITION OF B–DNA BY MONOCLONAL AUTOANTIBODIES

Although B–DNA or analogues such as poly(dA)·poly(dT) and poly(dA–dT) are not effective immunogens (Madaio et al., 1984), certain autoanti-bodies in sera of patients with systemic lupus erythematosus (SLE) do react with B–helical native DNA. Monoclonal autoantibodies with this activity have been isolated from hybridomas derived from 'lupus' mice (reviewed in Stollar, 1986; Eilat, 1982), which have a disease similar in several respects to human SLE. Some of these autoantibodies discriminate among B–DNA helices of different base sequences. The murine lupus monoclonal autoantibody H241, for example, reacts with poly(dG–dC) and not poly(dA–dT) (Stollar et al., 1986a). Tests with a series of self-complementary helical oligodeoxyribonucleotides indica-ted that the binding site of this antibody straddles the backbone of one DNA strand. It makes contact with portions of adjacent cytosine and guanine in the major groove and with a small part of the backbone and a 2-amino group of guanine in the minor groove (Stollar et al., 1986b). Other examples of lupus monoclonal autoantibodies prefer an A-T sequence to G-C (Tron et al., 1982; Kubota et al., 1986). An example under current study appears to react preferentially with a (dG-dA)·(dT-dC) sequence (Y. Kanai, G. Zon, S. Lehar and B.D. Stollar, unpublished data). The antibodies, however, do not show as high a selectivity of base sequence as that of restriction endonucleases.

FUTURE PROSPECTS

The library of anti-nucleic acid antibodies is growing, and with it the potential for application of antibodies both as detecting reagents and as models for ways in which proteins recognize nucleic acid structure. The latter question is of growing interest with the identification of an increasing number of nucleotide sequences that serve site-specific functions in gene organization and expression. These functions are mediated by specific interactions with proteins. One mode of such interaction is seen in the binding of helical segments of repressor proteins (Ohlendorf and Matthews, 1983); another is the binding of a restriction nuclease (McClarin et al., 1986). Antibody H241 defines an additional interaction of non-helical protein with a base sequence (Stollar et al., 1986b). Furthermore, antibodies to Z-DNA demonstrate the ability of protein to stabilize an unusual DNA conformation (Lafer et al., 1985;1986), an action that may be characteristic of regulatory DNA-binding proteins.

A question of interest for their projected applications is whether a set of several antibodies could detect different structures coexisting over short intervals within a given DNA molecule. This is being tested with an oligonucleotide that has the potential for containing, within 36 base pairs, two RNA-DNA hybrid segments and a B-DNA or Z-DNA region (figure 3). This oligonucleotide reacts well with anti-B-DNA auto-antibodies and anti-RNA-DNA hybrid antibodies in physiological saline. It also reacts with anti-Z-DNA antibodies in 4M NaCl, but the formation of a covalent junction between the (dG-dC) central core and the flanking hybrid regions inhibits the formation of Z-DNA. These experiments

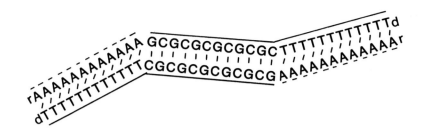

Figure 3. An oligonucleotide that may contain B-DNA or Z-DNA and RNA-DNA hybrid structure.

indicate that it will be possible to use a set of antibodies of varying specificities for detecting unusual DNA conformations in local regions of DNA.

ACKNOWLEDGEMENTS

Current research in the author's laboratory described in this article is supported by grants GM32375 and AM27232 from the National Institutes of Health.

REFERENCES

Aharonov A., Fuchs S., Stollar B. D. and Sela M. (1974) Immunogenicity and antigenic specificity of a glutaraldehyde cross-linked transfer RNA-protein conjugate. Eur. J. Biochem. 42, 73-79.

Crumpton M. J. (1974) Protein antigens: the molecular bases of anti-genicity and immunogenicity. In The Antigens (Edited by Sela M.) pp 1-78. Academic Press, New York.

Eilat D. (1982) Monoclonal autoantibodies: an approach to studying autoimmune disease. Mol. Immunol. 19, 943-955.

Germond J. E., Hirst B., Oudet P., Gross-Bellard M. and Chambon P. (1975) Folding of the DNA double helix in chromatin-like structures from simian virus 40. Proc. Natl. Acad. Sci. U.S.A. 72, 1843.

Gruenewald R. and Stollar B. D. (1973) The role of antigenic determinants in the control of IgM and IgG antibody responses to de-natured DNA. J. Immunol. 111, 106-113.

Hanau L. H., Santella R. M., Grunberger D. and Erlanger B. F. (1984) An immunochemical examination of acetylaminofluorene-modified poly(dG-dC)·poly(dG-dC) in the Z-conformation. J. Biol. Chem. 259, 173-178.

Heidelberger M. (1982) Structure and immunological specificity of poly-saccharides. Prog. Chem. Org. Nat. Prod. 42, 287-296.

Hill R. J. and Stollar B. D. (1983) Dependence of Z-DNA antibody binding to polytene chromosomes on acid fixation and DNA torsional strain. Nature 305, 338-340.

Kabat E. A., Liao J., Burzynska M. H., Wong T. C., Thogersen H. and Lemieux R. U. (1981) Immunochemical studies on blood groups-LXIX. The conformation of the trisaccharide determinant in the combining site of anti-I Ma (Group I). Mol. Immunol. 18, 873-881.

Kmiec E. B. and Holloman W. K. (1984) Synapsis promoted by Ustilago rec1 protein. Cell 36, 593-598.

Kmiec E. B. and Holloman W. K. (1986) Homologous pairing of DNA molecules by Ustilago rec1 protein is promoted by sequences of Z-DNA. Cell 44, 545-554.

Kubota T., Akatsuka T. and Kanai Y. (1986) A monoclonal anti-double-stranded DNA antibody from an autoimmune MRL/Mp-lpr/lpr mouse: specificity and idiotype in serum immunoglobulins. Immunol. Lett. 14, 53-58.

Lacour F., Nahon-Merlin E. and Michelson M. (1973) Immunological recognition of polynucleotide structure. Curr. Top. Microbiol. Immunol. 62, 1-39.

Lafer E. M., Möller A., Nordheim A., Stollar B. D. and Rich A. (1981) Antibodies specific for left-handed Z-DNA. Proc. Natl. Acad. Sci. U.S.A. 78, 3546-3550.

Lafer E. M., Möller A., Valle R. P. C., Nordheim A., Rich A. and Stollar B. D. (1983) Antibody recognition of Z-DNA. Cold Spring Harbor Symp. Quant. Biol. 47, 155-162.

Lafer E. M., Sousa R. and Rich A. (1985) Anti-Z-DNA antibody binding can stabilize Z-DNA in relaxed and linear plasmids under physiological conditions. EMBO J. 4, 3655-3660.

Lafer E. M., Sousa R., Rich A. and Stollar B. D. (1986) The effect of anti-Z-DNA antibodies on the B-DNA-Z-DNA equilibrium. J. Biol. Chem., 261, 6438-6443.

Lafer E. M. and Stollar B. D. (1984) The specificity of antibodies elicited by poly(dG)·poly(dC). J. Biomolec. Struct. Dyn. 2, 487-494.

Lancillotti F., Lopez M. C., Alonso C. and Stollar B. D. (1985) Locations of Z-DNA in polytene chromosomes. J. Cell Biol. 100, 1759-1766.

Lee J. S., Burkholder G. D., Latimer L. J. P., Haug B. L. and Braun R.P. (1987) A monoclonal antibody to triplex DNA binds to eucaryotic chromosomes. Nucleic Acids Res. 15, 1047-1060.

Lee J. S., Evans D. H. and Morgan A. R. (1980) Polypurine DNAs and RNAs form secondary structures which may be tetrastranded. Nucleic Acids Res. 8, 4305-4320.

Lee J. S., Woodsworth M. L. and Latimer L. J. P. (1984) Monoclonal antibodies specific for poly(dG)·poly(dC) and poly(dG-dm5C). Biochemistry 23, 3277-3281.

Madaio M. P., Hodder S., Schwartz R. S. and Stollar B. D. (1984) Responsiveness of autoimmune and normal mice to nucleic acid antigens. J. Immunol. 132, 872-876.

Malfoy B. and Leng M. (1981) Antiserum to Z-DNA. FEBS Lett. 132, 45-48.

McClarin J. A., Frederick C. A., Wang B. C., Greene P., Boyer H. W., Grable J. and Rosenberg J. M. (1986) Structure of the DNA-Eco R1 Endonuclease recognition complex at 3A resolution. Science 234, 1526-1541.

Munns T. W. and Liszewski M. K. (1980) Antibodies specific for modified nucleosides: an immunochemical approach for the isolation and characterization of nucleic acids. Progr. Nucleic Acid res. Mol. Biol. 24, 109–165.

Nordheim A., Lafer E. M., Peck L. J., Wang J. C., Stollar B. D. and Rich A. (1982) Negatively supercoiled plasmids contain left–handed Z–DNA segments as detected by specific antibody binding. Cell 31, 309–318.

Nordheim A., Pardue M. L., Weiner L. M., Lowenhaupt K., Scholten P., Möller A., Rich A. and Stollar B. D. (1986) Analysis of Z–DNA in fixed polytene chromosomes with monoclonal antibodies that show base sequence–dependent selectivity in reactions with supercoiled plasmids and polynucleotides. J. Biol. Chem. 261, 468–476.

Ohlendorf D. H. and Matthews B. W. (1983) Structural studies of protein–nucleic acid interactions. Ann. Rev. Biophys. Bioeng. 12, 259–284.

Pohl F. M. and Jovin T. M. (1972) Salt–induced cooperative conformational change of a synthetic DNA: equilibrium and kinetic studies with poly(dG–dC). J. Mol. Biol. 67, 375–396.

Poirier M. C. (1981) Antibodies to carcinogen–DNA adducts. J. Natl. Canc. Inst. 67, 515–519.

Rich A., Nordheim A. and Wang A. H.–J. (1984) The chemistry and biology of left–handed Z–DNA. Annu. Rev. Biochem., 53, 791–846.

Runkel L. and Nordheim A. (1986) Chemical footprinting of the interaction between left–handed Z–DNA and anti–Z–DNA antibodies by diethylpyrocarbonate carbethoxylation. J. Mol. Biol. 189, 487–501.

Sage E. and Leng M. (1980) Conformation of poly(dG–dC).poly(dG–dC) modified by the carcinogens N–acetoxy–N–acetyl–2–aminofluorene and N–hydroxy–N–2–aminofluorene. Proc. Natl. Acad. Sci. U. S. A. 77, 4597–4601.

Schreck R. R., Warburton D., Miller O. J., Beiser S. M. and Erlanger B. F. (1973) Chromosome structure as revealed by a combined chemical and immunochemical procedure. Proc. Natl. Acad. Sci. U.S.A. 70, 804–807.

Seaman E., Levine L. and Van Vunakis H. (1966) Antibodies to methylene blue photooxidation product in deoxyribonucleic acid. Biochemistry 5, 1216–1223.

Seaman E., Van Vunakis H. and Levine L. (1972) Serologic Estimation of thymine dimers in the deoxyribonucleic acid of bacterial and mammalian cells following irradiation with ultraviolet light and postirradiation repair. J. Biol. Chem. 247, 5709–5715.

Singleton C. K., Klysik J., Stirdivant S. M. and Wells R. D. (1982) Left–handed Z–DNA is induced by supercoiling in physiological ionic conditions. Nature 299, 312–316.

Stollar B. D. (1975) The specificity and applications of antibodies to helical nucleic acids. Crit. Rev. Biochem. 3, 45–69.

Stollar B. D. (1980) The experimental induction of antibodies to nucleic acids. Methods in Enzymol. 70, 70–85.

Stollar B. D. (1986) Antibodies to DNA. CRC Crit. Rev. Biochem. 20, 1–36.

Stollar B. D. and Raso V. (1974) Antibodies recognize specific structures of triple–helical polynucleotides built on poly(A) or poly(dA). Nature 250, 231–234.

Stollar B. D., Huang B. S. and Blumenstein M. (1986a) Recognition of nucleic acid structure by monoclonal autoantibodies reactive with B-DNA and a monoclonal anti-GMP antibody. In Fourth Conversation in Biomolecular Stereodynamics (Edited by Sarma R. H.). pp 69-84. Adenine Press, Albany, NY.

Stollar B. D., Zon G. and Pastor R. W. (1986b) A recognition site in synthetic helical oligonucleotides for monoclonal anti-native DNA autoantibody. Proc. Natl. Acad. Sci. U. S. A. 83, 4469-4473.

Strickland P. T. and Boyle J. M. (1984) Immunoassay of carcinogen-modified DNA. Progr. Nucleic Acid Res. Mol. Biol. 31, 1-58.

Summers M. F., Byrd R. A., Gallo K. A., Samson C. J., Zon G., and Egan W. (1988) Nuclear magnetic resonance and circular dichroism studies of a duplex-single-stranded hairpin loop equilibrium for the oligo-deoxyribonucleotide sequence d(CGCGATTCGCG). Nucleic Acids Res. 13, 6375-6386.

Sundquist W. I., Lippard S. J. and Stollar B. D. (1986) Binding of cis- and trans-Diamminedichloroplatinum(II) to deoxyribonucleic acid exposes nucleosides as measured immunochmically with anti-nucleoside antibodies. Biochemistry, 25, 1520-1524.

Thomae R., Beck S. and Pohl F. M. (1983) Isolation of Z-DNA-containing plasmids. Proc. Natl. Acad. Sci. U.S.A. 80, 5550-5553.

Townsend E., Van Vunakis H. and Levine L. (1965) Immunochemical studies on bacteriophage deoxyribonucleic acid IV. hapten inhibition. Biochemistry 4, 943-948.

Tron F., Jacob L., Lety M. A. and Bach J. F. (1982) Monoclonal anti-double-stranded deoxyribonucleic antibodies II. Indirect immuno-fluorescent studies. Clin. Immunol. Immunopathol. 24, 351-360.

Wakizaka A. and Okuhara E. (1975) Immunological studies on nucleic acids: an investigation of the antigenic determinants of denatured deoxyribonucleic acid (DNA) reactive with rabbit anti-DNA antisera by a radioimmunoassay technique. Immunochemistry 12, 843-845.

Zarling D. A., Arndt-Jovin D. J., McIntosh L. P., Robert-Nicoud M. and Jovin T. M. (1984a) Interactions of anti-poly[d(G-br5C)] IgG with synthetic, viral and cellular Z DNA's. J. Biomolec. Structure Dynamics 1, 1081-1107.

Zarling D. A., Arndt-Jovin D. J., Robert-Nicoud M., McIntosh L. P., Thomae R. and Jovin T. M. (1984b) Immunoglobulin recognition of synthetic and natural left-handed Z-DNA conformation and sequences. J. Mol. Biol. 176, 369-415.

Molecular Mechanical Studies on the B → Z Energetics of DNA Sequences

Shashidhar N. Rao[†], Thomas Jovin[‡], and Peter A. Kollman[*†]

[†]: Department of Pharmaceutical Chemistry, University of California, San Francisco, CA 94143.

[‡]: Abteilung Molekulare Biologie, Max Plank Institut für Biophysikalische Chemie, Postfach 968, D-3400, Göttingen, Federal Republic of Germany.

ABSTRACT

We present molecular mechanical studies on the hexanucleotides d(TATATA)·d(TATATA) and d(CGCGCG)·d(CGCGCG) as well as analogues of the CpG hexanucleotides in which: a thiophosphate replaces the phosphate groups between C and G and a thiophosphate replaces the phosphate groups between G and C, the cytosine is replaced by 5-methyl cytosine and the cytosine is replaced by 5-nitro cytosine and a modified AT hexanucleotide in which a 2-amino group has been added to adenine. The energies are compared for the B and Z forms of the nucleotides and the following results emerge: Consistent with experiment, the B-Z energy gap is greater for the TA hexamer than for the CG hexamer and a 5-methyl group on cytosine potentiates the Z form. The results also suggest that 2-amino substitution on adenines and 5-nitro substitutions on cytosines will make the molecules more "Z philic" than the unsubstituted molecules. On the other hand, although the calculations are ambiguous on whether replacing the phosphate with thiophosphate will potentiate Z formation, they would suggest that a thiophosphate between C and G is more likely to favor the Z form than one between G and C. This result, which seems to be independent of the charge model, is in disagreement with experimental results and suggests that differential solvation effects may be key factors in the Z philic or Z phobic behavior of thiophosphate containing hexanucleotides.

INTRODUCTION

The solution of the X-ray crystal structure of Z-DNA (1-3) has led to many interesting structural, biophysical and immunological studies on its properties (4-7). At this point, its biological relevance is not established, but it is clear that one can grow up antibodies selective to the Z form of DNA (5-7). We have earlier studied B and Z forms of DNA with molecular mechanics methods mainly using the sequence $d(CGCGCG)_2$ as a model (8). A comparative simulation of the dynamics of B and Z forms of DNA and normal modes of these two structural forms have also been published (9,10). The key results of our earlier study were as follows: the B form of DNA was calculated to be more stable than Z, unless counterions were added to the system or the dielectric constant was reduced. This was consistent with the observation of Z-DNA mainly in high salt condition or upon addition of di- and trivalent cations. Substituting a 5-methyl group on cytosine was calculated to potentiate the formation of the Z structure and molecular mechanics simulations on non-alternating sequences such as $dG_6 \cdot dC_6$ indicated that non-alternating pyrimidine-purine sequences might adopt the Z form if they were flanked by alternating CG or 5-methyl-CG containing strands. The latter prediction has been verified (11). One of the puzzles of our earlier simulations was that the Z philic tendency of alternating AT sequences was calculated to be comparable or greater than that of GC sequences. However, there is clear experimental evidence to suggest that this is not the case (12,13). In the studies presented here, we use a different force field and dielectric model than in the earlier study (8) and address some of the above issues as well as new ones.

We first ask whether one can reproduce the following experimental trends with molecular mechanics minimizations: The relative Z philicity of alternating CG versus TA DNA sequences; the relative Z philicity of 5MeC containing CG versus unmethylated CG sequences and the relative Z philicity of thiophosphate substituted DNA. We then address the following mechanistic question: Does 5-methyl cytosine substitutions in $d(CG)_n$ sequences stabilize Z-DNA because of solvation effects (less solvent exposure of the hydrophobic 5-methyl group) or intramolecular energies. The former explanation was suggested by Rich and coworkers (14,15), and the latter was found to be a factor in our earlier molecular mechanics calculations (8). We have carried out calculations on 5-nitro cytosine substituted DNA and they suggest that, based on intramolecular energies, this molecule should be Z philic.

268

The next goal of our study is to see whether relative entropies can explain the Z philicity of CG and TA nucleotide sequences. Briefly stated, the lack of Z philicity of TA compared to CG sequences could come from: (a) intramolecular energies; (b) conformational entropy, that is, a larger entropy loss for the more flexible B form TA sequences to form a Z structure; (c) differential solvation effects due to a minor groove "spine of hydration" stabilizing the B structure of TA sequences compared to the Z structure and (d) configurational entropy, in that TA sequences can sample many other right handed B like structures compared to the more restricted conformiational space in Z-DNA.

Finally, we assess models for thiophosphates substituted DNA. Jovin et al. (16) have shown that substituting thiophosphates in the G-3',5'-C part of CG rich DNA potentiates Z formation whereas a similar substitution at the C-3',5'-C site retards it. What can molecular mechanics calculations say about this issue?

In order to address these questions, we have carried out molecular mechanics simulations on the following deoxyhexanucleotide pentaphosphates: $d(XGXGXG)_2$ with X being cytosine, 5-methyl cytosine and 5-nitro cytosine; $d(TYTYTY)_2$ with Y being adenine and 2-amino adenine; and $d(CGCGCG)_2$ with a thiophosphate replacing the phosphate at the 3' end of each of the cytosines and with a thiophosphate replacing the phosphate at the 3' end of each of the guanines. Except in the case of thiophosphate incorporated hexamers, our results are in qualitative agreement with the experimental results.

METHODS

The molecular mechanical energies were evaluated via equation 1 in ref. 17 using the force field parameters presented by Weiner et al. (18) and the molecular mechanics program AMBER (17,19,20) and the structures were energy refined until a rms gradient of 0.1 kcal/mole-Å was achieved. In all the calculations a distance-dependent dielectric constant, $\varepsilon=4R_{ij}$ was used. This value of dielectric constant was chosen rather than $\varepsilon=R_{ij}$ (used in our previous investigations on hexanucleotide conformations) in view of the distortions which were introduced in the energy minimized structures using the latter value. Typically, in the left-handed Z form structures of the CG hexanucleotides, the amino group of guanine was involved in hydrogen bonding interactions with the neighboring phosphates groups in the $\varepsilon=R_{ij}$ models. Such interactions

269

led to the narrowing of the minor groove and to distortions away from the normal Z form. Since the emphasis of the present investigations is on the relative differences in energies of B and Z forms, we chose to use a dielectric model which would more accurately preserve these structures upon energy minimization and hence the use of $\varepsilon = 4R_{ij}$. In addition to minimizing the total energies with this dielectric constant, we have also evaluated energies at $\varepsilon = R_{ij}$ for these minimized structures. The charges on 5-methyl cytosine, 5-nitro cytosine, 2-amino adenine (schematically illustrated in Fig. 1) and dimethyl thiophosphate were determined using the quantum chemically derived electrostatic potentials (21) with an STO-3G basis set. These have been listed in Appendix 1. The additional species types and force field parameters for the atoms in the modified bases have been listed in Appendix 2.

In the molecular mechanics simulations of thiophosphate containing hexanucleotides, three charge models were used for the thiophosphate group, to estimate the significance of the charges on the thiophosphates on the relative energies. The first model (C1 in Table 4) had charges obtained through quantum chemical calculations as described above (see Appendix 1). In the other two models, the phosphorus atom charges were kept as in the first model, while the oxygen and sulfur charges were varied by equal and opposite amounts. Thus, in the second model (C2 in Table 4) the charges on oxygen and sulfur are -0.3506 and -1.1791, respectively, while in the third model (C3 in Table 4) the corresponding charges are -1.1506 and -0.3791, respectively. In addition to the simulations on the two thiophosphate containing hexamers, we have carried out molecular mechanics studies on two models of $d(GCGCGC)_2$ in which a thiophosphate replaced the phosphate group in the CpG and GpC fragments, respectively.

The energy minimized structures of $d(CGCGCG) \cdot d(CGCGCG)$, $d(TATATA) \cdot d(TATATA)$, $d(C^*GC^*GC^*G) \cdot d(C^*GC^*GC^*G)$, $d(C^{**}GC^{**}GC^{**}G) \cdot d(C^{**}GC^{**}GC^{**}G)$, and $d(TA^*TA^*TA^*) \cdot d(TA^*TA^*TA^*)$ where C^*, C^{**} and A^* represent 5-methyl cytosine, 5-nitro cytosine and 2-NH$_2$ adenine, respectively, have been referred to as CG, TA, C^*G, $C^{**}G$ and TA^*, respectively. The energy minimized structures of the thiophosphate incorporated hexanucleotides have been referred to as TPCG when the thiophosphate is at the 3' end of cytosines and TPGC when this group is at the 3' end of guanine nucleotides. In these cases we report the results of studies on both the $d(CGCGCG)_2$ and $d(GCGCGC)_2$ sequences. Suffixes B and Z indicate the appropriate polymorphic forms.

We have also carried out normal mode analysis and thermochemistry

270

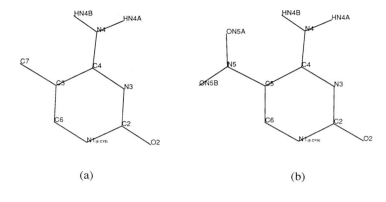

(a)   (b)

(c)

Fig. 1:   Schematic representation of bases in (a) 5-methyl cytosine,
           (b) 5-nitro cytosine and (c) 2-amino adenine residues used in
           the investigations on deoxyhexanucleotides in the present
           study.

calculations on the energy minimized structures of d(TATATA)·d(TATATA) and d(CGCGCG)·d(CGCGCG) in the B and Z polymorphic forms, obtained in the molecular mechanics simulations and have computed the entropy of the four structures using normal mode analysis (20). The energies of the molecular mechanically simulated structures were further refined to very low gradients (10^{-5} kcal/mole-Å) and the resultant structures were input into the normal mode program to determine the normal modes and thermochemical parameters.

The starting structures of each of the hexanucleotides were B–DNA (22) and Z–DNA (4). In the B form, the conformations about C3'–O3' and C5'–O5' are <u>trans</u>, those about C4'–C5', P–O5' and P–O3' are respectively <u>gauche</u>$^+$, <u>gauche</u>$^-$, and <u>gauche</u>$^-$, that about the glycosidic bond is <u>anti</u> and the sugar pucker is C2' endo. In the Z form, the helical repeat has two nucleotides, one corresponding to a purine base and he other to a pyrimidine base. In the purine nucleotide, the conformations about C3'–O3' and O3'–P are <u>gauche</u>$^-$, those about C5'–C4' and C5'–O5' are <u>trans</u> and the conformation about P–O5' is <u>gauche</u>$^+$. The corresponding conformation in the pyrimidine nucleotide are <u>gauche</u>$^-$, <u>gauche</u>$^+$, <u>gauche</u>$^+$, <u>trans</u> and <u>trans</u>, respectively. The conformational combination of sugar pucker and glycosidic torsion in the purine and pyrimidine nucleotides are respectively C3' <u>endo</u> – <u>syn</u> and C2' <u>endo</u> – <u>anti</u>. The nucleotides in the DNA fragment are referred to by the serial number of the base together with the base name with the numbering being continuous into the second strand. Thus, bases in CYT3 and GUA12 are paired in the Watson–Crick configuration.

RESULTS and DISCUSSIONS

The conformational features of the energy refined hexanucleotides are very similar to those in the starting structures of B (22) and Z (4) forms and hence are not discussed. In the light of this situation, it is of interest to compare the relative total energies in these polymorphic forms and understand the preference of either of these two forms in the hexanucleotides investigated in the present study (Table 1).

Our calculations on the CG and TA hexamers indicate that the differences in B and Z form energies are higher in the TA hexamer by 3.6 kcal/mole. This difference can qualitatively be taken to be a measure of the larger preference of the CG hexamers to be in the Z form rather

272

than the TA hexamers. It may be noted that since we are comparing two
chemically non-identical systems, it is not meaningful to directly
compare the total energies and assign any quantitative significance to
the difference in energies noted above. We have calculated the normal
modes and thermochemical parameters for TA and CG in both B and Z forms
and the results are presented in Table 2. As expected, the
conformational entropy of B is higher than that of Z by 17.9 eu (CG) and
15.4 eu (TA). The CG value is qualitatively similar to the difference
calculated earlier by Karplus and coworkers (9,10) and both are
apparently inconsistent with the solution thermodynamic values suggested
by Sturtevant et al. (23). The theoretical values make more intuitive
sense given the structures of B and Z, and it is not clear what
solvent/counterion effects can give rise to the reversal of sign of the
B-Z entropy difference. At first glance, the fact that the absolute
entropy of TA is smaller than CG seems surprising, but normal mode
analyses on the individual nucleosides finds that the entropies of
cytosine plus guanosine is 5.0 eu higher than adenosine plus thymidine
in our model. When one corrects the hexanucleotides for these
differences (6x5.0 = 30.0 eu), one finds, consistent with one's
intuition, that the entropy change upon double helix formation is more
negative for CG than TA.

In any case, one cannot use the B-Z entropy difference as a way to
rationalize the fact that TA is more refractory to go to Z DNA than is
CG.

The salient features of our investigations on TA and CG hexamers
are consistent with an earlier study on the conformational analysis of
guanine and adenine nucleotides (24,25). It was predicted that guanine
prefers C3' endo - syn conformational combination, characteristic of the
Z forms, while the adenine nucleotide prefers a C2' endo - anti
combination of sugar pucker and glycosidic orientation, characteristic
of the B form. However, earlier molecular mechanics simulations on
these two hexamers (8) suggested the Z-philic tendency for TA hexamers
to be either comparable or greater than CG hexamers, in contrast with
the present study and experimental results. The reason for this may be
that, in the earlier minimizations, the TA Z form moved significantly
from the actual Z structure observed in CG polymers.

Our calculations show that introducing a 5-methyl cytosine in place
of cytosine in the CG hexamer reduces the (B-Z) energy difference. An
analysis of the total energies of CG and C^*G in terms of contributions
from van der Waals, electrostatic and dihedral contributions (Table 3)

suggests that the more Z-philic nature of the methyl substituted hexamer arises predominantly from differences in dihedral energy contributions. In the B form, the methyl group causes a larger dihedral distortion, compared to a hydrogen, as in cytosine; whereas in the Z form, the methyl group is accomodated in the major groove with no significant backbone distortions. The electrostatic and van der Waals contributions together are very similar in both the methylated and unmethylated hexamers.

Figures 2a and b show stereo pairs of the superpositon of CG and $C^{**}G$ in the B and Z forms, respectively. It is clear that there are no significant relative distoritons in the structures of the two sequences either in the B or Z forms. In fact, the root mean square deviations for the base and backbone atoms are less than 0.2Å, in the two hexamers. Substitution of cytosine by 5-nitro cytosine leads to a greater reduction in the (B-Z) energy difference than in the case of the methyl substitution. A component analysis as described above (Table 3) suggests that unlike the case of $C^{*}G$, in $C^{**}G$, the dihedral distortion relative to CG is minimal. In other words, the nitro group, unlike the methyl group, can be accomodated in the major grooves of both the B and Z forms without causing any significant backbone conformational changes. Further, the main contribution to the Z-philic nature of the 5-nitro substituted hexamer comes surprisingly from van der Waals contributions. As seen from Table 3, this term has a difference of 10 kcal/mole in $C^{**}G$ as compared to 3.2 kcal/mole in CG. This indicates that the nitro group is more "snugly" accomodated in the very shallow major groove of the Z form than either a hydrogen or a methyl group.

Figures 3a and b show stereo pairs of the superposition of TA and TA^{*} in the B and Z forms, respectively. As in the case of $C^{**}G$, here too, the two sets of structures are very similar with no significant relative distortions. The root mean sqaure deviations of the base and the backbone atoms of the two structures are less than 0.2Å. Our calculations also find that 2-amino substitution on adenine also tends to potentiate the Z form, as is reflected by a lower (B-Z) energy difference by 3.6 kcal/mole than in TA. Since the 2-amino substitution is in the deep minor grooves of B and Z forms, it is not surprising that like in $C^{**}G$ no significant distortion is caused in the oligonucleotide backbone conformations in either form. However, unlike in $C^{**}G$, the chief driving contribution towards the more Z-potentiating nature of TA^{*} compared to TA, seems to be electrostatic energies (Table 3). This is not surprising in view of (a) the dipolar amino group being in the

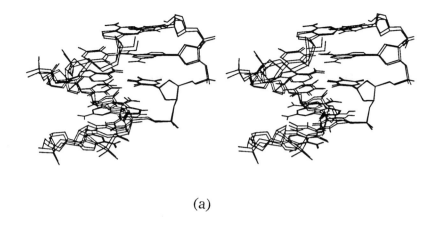

(a)

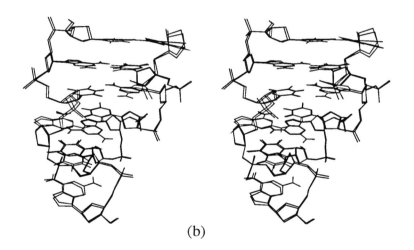

(b)

Fig. 2:   Stereo view of superposition of d(CGCGCG)$_2$ and d(C**GC**G C**G)$_2$ in (a) the B form and (b) the Z form of DNA.

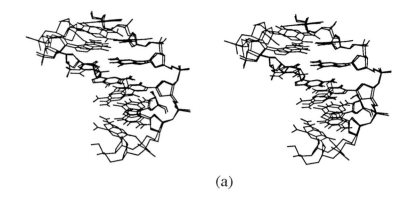

(a)

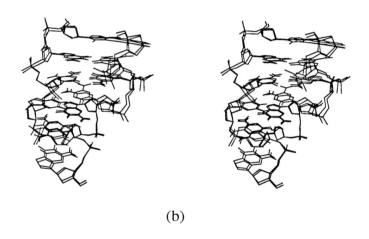

(b)

Fig. 3:   Stereo view of superposition of d(TATATA)$_2$ and d(TA*TA*TA*)$_2$
in (a) the B form and (b) the Z form of DNA.

vicinity of other charged groups and (b) that this group is accommodated very easily in the minor groove, particularly in Z DNA, without causing any stereochemical contacts.

Our results on the CG,TA,C*G and TA* are consistent with the recent experimental studies on these compounds and polymers containing such fragments. Crystal structure analysis has been reported for 2-amino adenine containing the TA sequence and this molecule has been shown to take up a Z form (26). Solution studies (27) on a polymer of TA* also suggests that the molecule is in a Z form. Incorporation of 5-methyl cytosine into CG rich polymers has been shown to cause B → Z transitions at much lower salt concentrations than in the case of CG oligomers and polymers (28-35). Further, the crystal structure of a CG hexanucleotide in which 5-methyl cytosine replaced cytosine, has been shown to be in the Z form of DNA (15). These experimental observations are in agreement with the Z-potentiating nature of 5-methyl (in cytosine) and 2-amino (in adenine) substitutions as obtained in our calculations. Based on this agreement, it is tempting to speculate that CG polymers with 5-nitro cytosine replacing cytosine, will also strongly favor the Z form even at low salt concentrations.

Our studies on the thiophosphate incorporated hexamers suggest that independent of the charge and dielectric model (Tables 4 and 5), the thiophosphate at the 3' end of cytosine favors the Z form over B more than the hexamer with the thiophosphate at the 3' end of guanine. This is in contrast with the experimental studies (16) which suggest that thiophosphate at the 3' of guanine potentiate the Z form over B form more than the thiophosphate at the 3' end of cytosine. This lack of consistency with experimental results could arise due to factors other than the differences in internal energies, such as differential solvation and counterion effects, particularly around the thio group, in B and Z forms, not explicitly considered in the present study.

CONCLUSIONS

We have applied molecular mechanical methods to study various DNA sequences in the B and the Z conformations. The known ability of 5-CH$_3$ substitution on cytosine to potentiate Z formation and the change in sequence from CG → TA to reduce Z formation is reproduced in the calculations. Further, 5-NO$_2$ cytosine substituted poly d(C-G) is predicted to be "Z philic" as 5-CH$_3$ cytosine substituted poly d(C-G).

On the other hand, the calculations are unable to rationalize the difference in Z philicity for DNA modified by thiophosphate linkages in the GpC fragment compared to the thiophosphate linkage in the CpG fragment.

ACKNOWLEDGEMENTS

We are very much thankful to Dr. U. Chandra Singh and Dr. James Caldwell for their help with the normal mode analysis program in AMBER(UCSF). We gratefully acknowledge the support of the National Cancer Institute (grant CA-25644) in the research. The use of the facilities of the UCSF Computer Graphics Laboratory (R. Langridge, director and T. Ferrin, facility manager), supported by NIH RR-1081, is also gratefully acknowledged. The purchase of the FPS-264 array processor was made possible through grants from the NIH (RR-02441) and NSF (DMB-84-13762), and their support for this is much appreciated.

Table 1

Total and relative energies (in Kcal/mole) of B and Z forms of the hexanucleotides investigated in the present study.

Hexamer	E(B)[a]	E(Z)[b]	ΔE(B–Z)[c]
CG[d]	−95.5	−81.5	−14.0
C*G[d]	−131.3	−118.2	−12.9
C**G[d]	−166.9	−158.8	−8.1
TA[d]	−108.7	−91.1	−17.6
TA*[d]	−99.8	−85.8	−14.0

[a] Total energy of the optimized B structure.
[b] Total energy of the optimized Z structure.
[c] Difference in total energies of the optimized B and Z structures.
[d] See Methods for the meaning of these abbreviations.

Table 2

Conformational entropies (eu) of the energy refined hexanucleotides CG and TA in the B and Z polymorphic forms of DNA.

Hexamer	S_B	S_Z
CG	1094.6	1076.7
TA	1072.4	1057.0

Table 3

Component analysis of total energies – van der Waals (VDW), 1-4 van der Waals (VDW14), electrostatic (ELE), 1-4 electrostatic (ELE14) and dihedral (DIH) terms for the hexanucleotides CG, C*G, C**G, TA and TA* in the B and Z polymorphic forms of DNA. Differences in the energy components for these two forms (B-Z) are also tabulated.

Hexamer	Polymorphic form	VDW	VDW14	ELE	ELE14	DIH
CG	B	-206.4	49.5	116.1	-258.1	156.5
	Z	-209.6	43.5	119.4	-251.0	164.2
Difference (B-Z)		3.2	6.0	-3.3	-7.1	-7.7
CG	B	-213.8	53.6	141.1	-318.9	159.8
	Z	-219.5	47.4	149.3	-312.7	164.2
Difference (B-Z)		5.7	6.2	-6.2	-6.2	-4.4
C**G	B	-219.9	54.4	243.2	-457.9	157.0
	Z	-229.9	48.8	248.0	-451.9	164.5
Difference (B-Z)		10.0	5.6	-4.8	-6.0	-7.5
TA	B	-197.0	53.2	104.3	-276.1	156.3
	Z	-202.6	47.9	111.6	-270.0	164.1
Difference (B-Z)		5.6	5.3	-7.3	-6.1	-7.8
TA*	B	-214.7	51.4	114.5	-291.9	155.6
	Z	-219.8	45.8	118.4	-285.1	163.1
Difference (B-Z)		5.1	5.6	-3.9	-6.8	-7.5

Table 4

Total energies for the thiophosphate incorporated hexanucleotide models in the B and Z forms of DNA. C1, C2 and C3 correspond to three charge models for the thiophosphate group (see text for details)

Hexamer	Form	C1	C2	C3	C1	C2	C3
TPGC	B	−92.9	−93.0	−92.8	−521.7	−522.0	−521.2
	Z	−78.1	−75.2	−81.3	−465.3	−455.1	−477.4
TPCG	B	−92.7	−91.3	−94.0	−515.6	−501.2	−520.7
	Z	−78.1	−75.1	−86.0	−463.8	−453.9	−486.3

Table 5

Differences in energies of B and Z forms (E_B-E_Z) in the hexanucleotides CG, TPGC and TPCG in four different dielectric models M1, M2, M3 and M4 (see footnote).

Hexamer	M1	M2	M3	M4
CG	14.0	14.0	46.0	38.4
TPGC	14.8	15.7	56.4	47.4
	17.8	18.4	66.9	57.2
	11.5	12.7	43.7	36.1
TPCG	14.6	13.4	51.8	40.7
	16.2	14.8	47.3	41.6
	8.0	8.2	34.4	26.9

M1: $\varepsilon = 4R_{ij}$, only CpG
M2: $\varepsilon = 4R_{ij}$, average of GpC and CpG
M3: $\varepsilon = R_{ij}$, only CpG
M4: $\varepsilon = R_{ij}$, average of GpC and CpG

REFERENCES

1. Wang, A.H.-J., Quigley, G.J., Kolpak, F.J., Crawford, J.L., van
   Boom, J.H., van der Marel, G.A. and Rich, A. (1979) Nature 282,
   680-686.

2. Drew, H., Takano, T., Tanaka, S., Itakura, K. and Dickerson, R.E.
   (1980) Nature 286, 567-573.

3. Crawford, J.L., Kolpak, F.J., Wang, A.H.-J., Quigley, G.J., van
   Boom, J.H., van der Marel, G. and Rich, A. (1981) Proc. Natl.
   Acad. Sci. USA 77, 4016-4020.

4. Wang, A.H.-J., Quigley, G.J., Kolpak, F.J., van der Marel, G.A.,
   van Boom, J.H. and Rich, A. (1981) Science 211, 171-176.

5. Nordheim, A., Pardue, M.L., Lafer, E.M., Moller, A., Stollar, B.D.
   and Rich A. (1981) Nature 294, 417-422.

6. Nordheim, A., Tesser, P., Azorin, F., Kwon, Y.H., Möller, A. and
   Rich, A. (1982) Proc. Natl. Acad. Sci. USA 79, 7729-7733.

7. Morgenegg, G., Celio, M.R., Malfoy, B., Leng, M. and Kuenzle, C.C.
   (1983) Nature 303, 540-543.

8. Kollman, P.A., Weiner, P.K., Quigley, G. and Wang, A.H.-J. (1982)
   Biopolymers 21, 1945-1969.

9. Tidor, B., Brooks, B. and Karplus, M. (1983) J. Biomolec. Str.
   Dyn. 1, 231-252.

10. Irikura, K.K., Tidor, B., Brooks, B.R. and Karplus, M. (1985)
    Science 229, 571-571.

11. Wang, A.H.-J., Gessner, R.V., van der Marel, G.A., van Boom, J.H.
    and Rich, A. (1985) Proc. Natl. Acad. Sci. USA 82, 3611-3615.

12. Assa-Munt, N. and Kearns, D.R. (1984) Biochemistry 23, 791-796.

13. Behling, R.W. and Kearns, D.R. (1986) Biochemistry 25, 3335-3346.

14. Fujii, S., Wang, A.H.-J., Quigley, G.J., Westerink, G., van der
    Marel, G.A., van Boom, J.H. and Rich, A. (1985) Biopolymers 24,
    243-250.

15. Fujii, S., Wang, A.H.-J., van der Marel, G., van Boom, J.H. and
    Rich, A. (1982) Nucleic Acids Res. 10, 7879-7892.

16. Jovin, T.M., McIntosch, L.P., Arndt-Jovin, D.J., Zarling, D.A.,
    Robert-Nicoud, M., van de Sande, J.H., Jorgenson, K.F. and
    Eckstein, F. (1983) J. Biomolec. Str. Dyn. 1, 21-57.

17. Weiner, P.K. and Kollman, P.A. (1981) J. Comp. Chem. 2, 287-310.

18. Weiner, S.J., Kollman, P.A., Case, D., Singh, U.C., Ghio, C.,
    Alagona, J., Profeta, Jr., S. and Weiner, P.K. (1984) J. Am. Chem.
    Soc., 106, 765-785.

19. Weiner, P.K., Singh, U.C., Kollman, P.A., Caldwell, J.W. and Case, D. AMBER(UCSF) version 2.0 Department of Pharmaceutical Chemistry, University of California, San Francisco, CA 94143 (1985).

20. Singh, U.C., Weiner, P.K., Caldwell, J.W. and Kollman, P.A. AMBER(UCSF) version 3.0 (developed for FPS) Department of Pharmaceutical Chemistry, University of California, San Francisco, CA 94143 (1985).

21. Singh, U.C. and Kollman, P.A. (1984) J. Comp. Chem. **5**, 129-144.

22. Arnott, S., Campbell-Smith, P. and Chandrasekaran, R. (1976) in "CRC Handbook of Biochemistry", vol II, G.D. Fasman, Ed., CRC Cleveland, Ohio, pp. 411-422.

23. Chaires, J.B. and Sturtevant, J.M. (1986) Proc. Natl. Acad. Sci. USA **83**, 5479-5483.

24. Sasisekharan, V. (1973) in V Jerusalem Symp. Quantum Chem. Biochem. (E.D. Bergmann and B. Pullman, eds.) pp. 247-260, Academic Press, New York.

25. Rao, S.N. and Sasisekharan, V. (1981) Ind. J. Biochem. Biophys. **18**, 303-310.

26. Coll, M., Wang, A.H.-J., van der Marel, G.A., van Boom, J.H. and Rich, A. (1986) J. Biomolec. Str. Dyn. **4**, 157-172.

27. Howard, F.B., Chen, C.W., Cohen, J.S. and Miles, H.T. (1984) Biochem. Biophys. Res. Commun. **118**, 848-853.

28. Roy, K.B. and Miles, H.T. (1983) Biochem. Biophys. Res. Commun. **115**, 100-105.

29. Behe, M.J., Felsenfeld, G., Szu, S.C. and Charney, E. (1985) Biopolymers **24**, 289-300.

30. Feuerstein, B.G., Marton, L.J., Keniry, M.A., Wade, D.L. and Shafer, R.H. (1985) Nucleic Acids Res. **13**, 4133-4141.

31. Thomas, T.J., Bloomfield, V.A. and Cannellakis, Z.N. (1985) Biopolymers **24**, 725-729.

32. Kruegar, W.C. and Prairie, M.D. (1985) Biopolymers **24**, 905-910.

33. Behe, M.J. and Felsenfeld, G. (1981) Proc. Natl. Acad. Sci. USA **78**, 1619-1623.

34. Giessner-Prettre, C., Pullman, B., Tran-Dinh, S. Neumann, J.-M., Huynh-Dinh, T. and Igolen, J. (1984) Nucleic Acids Res. **12**, 3271-3281.

35. Behe, M.J. (1986) Biopolymers **25**, 519-523.

APPENDIX 1

Quantum chemically derived electrostatic charges on the modified bases and 2-amino adenine and 5-methyl cytosine, thiophosphate, used in the present investigations (see Fig. 1 for nomenclature of the modified bases

5-Methyl cytosine		5-Nitro cytosine	
N1	-0.6631	N1	-0.6513
C6	0.5387	C6	0.6757
C5	-0.5776	C5	-0.8931
C7	0.0888	N5	1.0042
C4	0.9392	ON5A	-0.4643
N4	-0.8170	ON5B	-0.4643
HN4A	0.3380	C4	1.1458
HN4B	0.3380	N4	-0.9349
N3	-0.9251	HN4A	0.3877
C2	1.0318	HN4B	0.3877
O2	-0.5277	C2	1.0438
		O2	-0.5053

2-Amino adenine		Thiophosphate	
N9	-0.4130	P	1.2588
C8	0.4420		
N7	-0.6270	OA	-0.7506
C5	0.0600		
C6	0.6070	SB	-0.7791
N6	-0.6590		
HN6A	0.3160		
HN6B	0.3160		
N1	-0.7270		
C2	0.8950		
N2	-0.8160		
HN2A	0.3240		
HN2B	0.3240		
N3	-0.6710		
C4	0.3930		

APPENDIX 2

The additional bond length, bond angle and dihedral parameters used in
the molecular mechanics simulations of the hexanucleotides discussed in
text. In 5-methyl cytosine, C3 was assigned to the methyl group (in the
united atom representation), while in 5-nitro cytosine, species types NQ
and OQ were assigned to the nitrogen and oxygens of the nitro group.
The sulfur aom in the thiophosphate moiety was assigned the species type
SP

Bond	K_r	r_{eq}
CM - NQ	350.0	1.486
OQ - NQ	550.0	1.208
P - SP	200.0	1.940

Angle	K_θ	θ_{eq}
C3 - CM - CA	85.0	119.7
C3 - CM - CJ	85.0	120.7
OQ - NQ - CM	85.0	118.0
OQ - NQ - OQ	85.0	124.0
CJ - CM - NQ	85.0	120.7
CA - CM - NQ	85.0	119.7
SP - P - OS	120.0	112.0
SP - P - O2	140.0	115.1

Dihedral	m	$V_{n/2}$	r	n
X - CM - NQ - X	4	20.0	180	2

USUAL AND UNUSUAL DNA STRUCTURES: A SUMMING-UP

Richard E. Dickerson
Molecular Biology Institute
University of California at Los Angeles
Los Angeles, CA 90024

Being asked to give the final talk at a conference such as this is both an honor and a formidable task. After three days of intensive exchange of ideas, the audience is verging on battle fatigue; thinking of the undone tasks waiting at home, and of the possibility of missing the shuttle service to the Mobile or Pensacola airports. The problem under such circumstances is not one of knowing what to say, but how to make it interesting. It would not be very interesting to use the talk merely as a platform for my own latest results, nor to give a four minute potted summary of each of the previous speakers' talks. Instead, I am going to use my allotted time (and no more) to say a little about where I think we stand today on DNA structure and where we should be going.

The symposium has been entitled "Unusual DNA Structures". To me, the most important single question that a molecular biologist can ask about DNA structure today is: <u>How does base sequence influence helix geometry?</u> If you think back to the early years of this century, the gene then was little more than an hereditary marker; a colored pebble in a statistical grid of boxes. One enormous step was taken in modern genetics when it was established that genes were arranged linearly along some kind of one-dimensional storage material; and molecular biology really emerged only when it was realized that that storage material was a DNA double helix. The model of DNA as a straight line in a genetic map has been of enormous practical utility. But DNA is a real molecule, with real structure, and with behavior that follows from that structure. This is what our symposium has been about.

The State of DNA Oligonucleotide Crystal Structure Today

The origins of single-crystal DNA structure analysis, although relatively recent, have been recounted so often that little more than an outline is necessary: how Andrew H.-J. Wang in Alex Rich's labora-

Table I. SINGLE CRYSTAL X-RAY STRUCTURE ANALYSES OF SYNTHETIC OLIGONUCLEOTIDES

Sequence	T^1	Space Group	Cell Dimensions			N^2	Vol per Base Pair	d^3	R^4
			a	b	c				
I. A-Helical Family									
1. ^{i}C-C-G-C	5°C	P4$_3$2$_1$2	41.1	41.1	26.7	4	1409 Å	2.0	16.5%
2. G-G-C-C-G-G-C-C	-8°C	P4$_3$2$_1$2	42.06	42.06	25.17	4	1391	2.25	17.4%
3. G-G-C-C-G-G-C-C	-18°C	P4$_3$2$_1$2	40.51	40.51	24.67	4	1265	2.25	15.8%
4. C-C-C-C-G-G-G-G	?	P4$_3$2$_1$2	42.1	42.1	25.1	4	1390		
5. G-G-C-C-G-C-C-C	RT	P4$_3$2$_1$2	43.28	43.28	24.66	4	1444	1.7	18%
6. G-G-A-T-G-G-G-A-G	RT	P4$_3$	45.29	45.29	24.73	9	1409	3.0	33%
7. G-G-T-A-T-A-C-C	RT	P6$_1$	45.01	45.01	41.55	8	1519	1.8	19.8%
8. G-G-^{br}U-T-^{br}U-T-C-C	RT	P6$_1$	45.05	45.05	41.27	8	1528	1.7	14%
9. G-G-G-G-C-C-C-C	18°C	P6$_1$	45.32	45.32	42.25	8	1566	2.5	14%
10. G-G-G̊-G-C-T̊-C-C	2°C	P6$_1$	45.20	45.20	42.97	8	1584	2.25	13.6%
11. G-G-G-G̊-T̊-C-C-C	2°C	P6$_1$	44.71	44.71	42.40	8	1529	2.1	14.5%
12. G-G-G-T̊-G̊-C-C-C	RT	P6$_1$	45.62	45.62	40.99	8	1529	2.5	15%
13. r(G-C-G)d(T-A-T-A-C-G-C)	0°C	P2$_1$2$_1$2$_1$	24.20	43.46	49.40	10	1299	2.0	16%
II. B-Helical Family									
14. C-G-C-G-A-A-T-T-C-G-C-G	RT	P2$_1$2$_1$2$_1$	24.87	40.39	66.20	12	1385	1.9	17.8%
15. C-G-C-G-A-A-T-T-C-G-C-G	16K	P2$_1$2$_1$2$_1$	23.44	39.31	65.26	12	1253	2.7	15.1%
16. C-G-C-G-A-A-T-T-^{br}C-G-C-G	7°C	P2$_1$2$_1$2$_1$	24.20	40.09	63.95	12	1293	2.3	17.3%
17. C-G-C-G̊-A-A-T-T-T̊-G-C-G	4°C	P2$_1$2$_1$2$_1$	25.53	41.22	65.63	12	1433	2.5	13%
18. C-G-C-G̊-A-A-T-T-Å-G-C-G	5°C	P2$_1$2$_1$2$_1$	25.69	41.96	65.19	12	1464	2.5	17.0%
19. C-G-C-Å-A-A-T-T-C̊-G-C-G	4°C	P2$_1$2$_1$2$_1$	25.37	41.44	65.20	12	1428	2.5	19.0%
20. C-C-A-A-G̊-Å-T-T-G-G	5°C	C2	32.52	26.17	34.30	5	1278	1.3	17.1%
21. Gp^{S}CpGp^{S}CpGp^{S}C	18°C	P2$_1$2$_1$2$_1$	34.90	39.15	20.64	6	1175	2.17	14.5%
III. Z-Helical Family									
22. C-G-C-G	RT	C222$_1$	19.50	31.27	64.67	4	1232	1.5	21%
23. C-G-C-G-C-G	RT	P2$_1$2$_1$2$_1$	17.88	31.55	44.58	6	1048	0.9	14%
24. ^{m}C-G-^{m}C-G-^{m}C-G	-8°C	P2$_1$2$_1$2$_1$	17.76	30.57	45.42	6	1027	1.3	15.6%
25. ^{br}C-G-^{br}C-G-^{br}C-G	18°C	P2$_1$2$_1$2$_1$	18.01	30.88	44.76	6	1037	1.4	13.3%
26. ^{br}C-G-^{br}C-G-^{br}C-G	37°C	P2$_1$2$_1$2$_1$	17.93	30.83	44.73	6	1030	1.4	12.5%
27. ^{m}C-G-T-A-^{m}C-G	-10°C	P2$_1$2$_1$2$_1$	17.91	30.43	44.96	6	1021	1.2	16%
28. ^{m}C-G-A-T-^{m}C-G	?	P2$_1$2$_1$2$_1$	18.3	31.1	44.1	6	1046	1.54	19.3%
29. C-G̊-C-G-T̊-G	15°C	P2$_1$2$_1$2$_1$	17.45	31.63	45.56	6	1048	1.0	19.5%
30. ^{br}Ů-G-C-G-C-G̊	4°C	P2$_1$2$_1$2$_1$	17.94	30.85	49.94	6	1152	2.25	15.6%
31. C-G-C-G	?	P6$_5$	31.25	31.25	44.06	6	1035	1.5	19.3%
32. C-G-C-G-C-G	?	P6$_5$	31.27	31.27	43.56	6	1025	1.6	19%
33. C-G-C-A-T-G-C-G	?	P6$_5$	30.90	30.90	43.14	6	991	2.5	16%
34. C-G-T-A-C-G-T-A-C-G	?	P6$_5$	17.93	17.93	43.41	2	1007	1.5	25.5%
IV. Irregular Structures									
35. pA-T-A-T	?	P2$_1$	21.12	21.29	8.77	2	986	1.0	15.2%

Table I. (continued) Survey as of April 1987

	sig^5	N^6_{data}	N^7_{solv}	Remarks	Date	Research Group	Refs.	Coordinates in the Public Domain?	Original Data in Public Domain?
1.	2σ	1486	86		1981	UCLA(CIT)	1-3	Yes, Brookhaven	Yes, Brookhaven
2.	3σ	920	75		1982	MIT	4,5	No	No
3.	3σ	874	84		1982	MIT	4,5	No	No
4.					1987	Weiz/MIT	6	No	No
5.	?	2455	80		1987	Weizmann	43	No	No
6.	0σ	1032	0	Disordered?	1986	Cambridge	7	No	No
7.	2σ	3251	66		1981	Weiz/Camb	8,9	No	No
8.	2σ	4669	84		1981	Weiz/Camb	8-10	No	No
9.	2σ	1283	106		1985	Cambridge	11	Yes, Brookhaven	Yes, Brookhaven
10.	2σ	1924	52	GT mismatch	1985	Camb/Weiz	12,13	No	No
11.	2σ	2486	104	GT mismatch	1985	Camb/Weiz	14	No	No
12.	?	2000	75	GT mismatch	1987	Weizmann	43	No	No
13.	1.5σ	2521	179	DNA/RNA	1982	MIT	15,5	No	No
14.	2σ	2725	80	Bent Axis	1980	UCLA(CIT)	16-19	Yes, Brookhaven	Yes, Brookhaven
15.	2σ	1051	83	Bent Axis	1982	UCLA(CIT)	20	Yes, Brookhaven	Yes, Brookhaven
16.	2σ		114	Unbent Axis	1982	UCLA	21,22	Yes, Brookhaven	Yes, Brookhaven
17.	2σ	2000	71	GT mismatch	1985	Cambridge	23	No	No
18.	2σ	2028	83	GA mismatch	1986	Cambridge	24	No	No
19.	2σ	2029	82	AC mismatch	1986	Cambridge	25	No	No
20.	2σ	2639	48	GA mismatch $\beta = 118.9°$	1987	UCLA	26	**Yes, Brookhaven**	Yes, Brookhaven
21.	2σ	1327	72	phosphoro-thioate	1987	Cambridge	27	No	No
22.	2σ	1900	84		1980	UCLA(CIT)	28,29	Yes, Brookhaven	Yes, Brookhaven
23.	2σ	15000	62		1979	MIT	30,31	No	No
24.	1.5σ	4208	98		1982	MIT	31	Yes, in paper	No
25.	2σ	2919	61		1986	Strasbourg	32	Yes, Brookhaven	No
26.	2σ	3765	83		1986	Strasbourg	32	Yes, Brookhaven	No
27.	1.5σ	5412	98		1984	MIT	33	Yes, in paper	No
28.	1.5σ	2386	88		1985	MIT	34	No	No
29.	1σ	10964	91		1985	MIT	35	No	No
30.	2σ	1105	64		1986	Cambridge	36	No	No
31.	?	?	85	Disordered	1980	MIT	37	No	No
32.	?	?	85	Disordered	1985	MIT	38	No	No
33.	?	?	?	Disordered	1985	MIT	38	No	No
34.	2σ	506	7	Disordered	1985	Wis/Stras	39,40	No	No
35.	1.5σ	2717	42	$\beta = 97.84°$	1978	Camb/Weiz	41,42	No	No

^{1}T = Temperature of data collection

^{2}N = Number of unique base pairs in structure

^{3}d = Resolution in Å

^{4}R = Residual error or R factor after refinement

^{5}sig = Sigma level of data refined

^{6}N$_{data}$ = Number of reflections refined

^{7}N$_{solv}$ = Number of solvent molecules located

i, br, m = iodo, bromo or methyl at C5 of cytosine

* = participant in mismatch base pair

p^S = phosphorothioate group, -POS-

- = normal phosphate group, -PO$_2$-

RT = room temperature (or unspecified)

Weiz = Weizmann Institute; Wis = U. of Wisconsin

Brookhaven = Brookhaven Protein Data Bank

References for Table I:

1. R.E. Dickerson, H.R. Drew, and B.N. Conner (1981). Biomolecular Stereodynamics (R.H. Sarma, Ed.), Adenine Press, New York, Vol. 1, pp. 1-34.

2. B.N. Conner, T. Takano, S. Tanaka, K. Itakura and R.E. Dickerson (1982). Nature 295, 294-299.

3. B.N. Conner, C. Yoon, J.L. Dickerson and R.E. Dickerson (1984). J. Mol. Biol. 174, 663-695.

4. A.H.-J. Wang, S. Fujii, J.H. van Boom and A. Rich (1982). Proc. Natl. Acad. Sci. U.S. 79, 3968-3972.

5. S. Fujii, A.H.-J. Wang, J. van Boom and A. Rich (1982). Nucl. Acids Res. Symp. Series 11, 109-112.

6. T.E. Haran, Z. Shakked, A.H.-J. Wang and A. Rich (1987). J. Mol. Biol.-- in press.

7. M. McCall, T. Brown, W.N. Hunter and O. Kennard (1986). Nature 322, 661-664.

8. Z. Shakked, D. Rabinovich, W.B.T. Cruse, E. Egert, O. Kennard, G. Sala, S.A. Salisbury and M.A. Viswamitra (1981). Proc. Roy. Soc. Lond. B213, 479-487.

9. Z. Shakked, D. Rabinovich, O. Kennard, W.B.T. Cruse, S.A. Salisbury and M.A. Viswamitra (1983). J. Mol. Biol. 166, 183-201.

10. O. Kennard, W.B.T. Cruse, J. Nachman, T. Prange, Z. Shakked and D. Rabinovich (1986). J. Biomol. Str. Dyn. 3, 623-647.

11. M. McCall, T. Brown and O. Kennard (1985). J. Mol. Biol. 183, 385-396.

12. T. Brown, O. Kennard, G. Kneale and D. Rabinovich (1985). Nature 315, 604-606.

13. W.N. Hunter, G. Kneale, T. Brown, D. Rabinovich and O. Kennard (1986). J. Mol. Biol. 190, 605-618.

14. G. Kneale, T. Brown, O. Kennard and D. Rabinovich (1985). J. Mol. Biol. 186, 805-814.

15. A.H.-J. Wang, S. Fujii, J.H. van Boom, G.A. van der Marel, S.A.A. van Boeckel and A. Rich (1982). Nature 299, 601-604.

16. R.M. Wing, H.R. Drew, T. Takano, C. Broka, S. Tanaka, K. Itakura and R.E. Dickerson (1980). Nature 287, 755-758.

17. H.R. Drew, R.M. Wing, T. Takano, C. Broka, S. Tanaka, I. Itakura and R.E. Dickerson (1981). Proc. Natl. Acad. Sci. U.S. 78, 2179-2183.

References for Table I:

18. R.E. Dickerson and H.R. Drew (1981). J. Mol. Biol. 149, 761-786.

19. H.R. Drew and R.E. Dickerson (1981). J. Mol. Biol. 151, 535-556.

20. H.R. Drew, S. Samson and R.E. Dickerson (1982). Proc. Natl. Acad. Sci. U.S. 79, 4040-4044.

21. A.V. Fratini, M.L. Kopka, H.R. Drew and R.E. Dickerson (1982). J. Biol. Chem. 257, 14686-14707.

22. M.L. Kopka, A.V. Fratini, H.R. Drew and R.E. Dickerson (1983). J. Mol. Biol. 163, 129-146.

23. W.N. Hunter, T. Brown, G. Kneale, N.N. Anand, D. Rabinovich and O. Kennard (1987). J. Biol. Chem.--submitted.

24. T. Brown, W.N. Hunter, G. Kneale and O. Kennard (1986). Proc. Natl. Acad. Sci. U.S. 83, 2402-2406.

25. W.N. Hunter, T. Brown, N.N. Anand and O. Kennard (1986). Nature 320, 552-555.

26. G.G. Privé, U. Heinemann, S. Chandrasegaran, L.-S. Kan, M.L. Kopka and R.E. Dickerson--in preparation.

27. W.B.T. Cruse, S.A. Salisbury, T. Brown, R. Cosstick, F. Eckstein and O. Kennard (1987). J. Mol. Biol. 192, 891-905.

28. H.R. Drew, T. Takano, S. Tanaka, K. Itakura and R.E. Dickerson (1980). Nature 286, 567-573.

29. H.R. Drew and R.E. Dickerson (1981). J. Mol. Biol. 152, 723-736.

30. A.H.-J. Wang, G.J. Quigley, F.J. Kolpak, J.L. Crawford, H.H. van Boom, G. van der Marel and A. Rich (1979). Nature 282, 680-686.

31. S. Fujii, A.H.-J. Wang, G. van der Marel, J.H. van Boom and A. Rich (1982). Nucl. Acids Res. 10, 7879-7892.

32. B. Chevrier, A.C. Dock, B. Hartmann, M. Leng, D. Moras, M.T. Thuong and E. Westhof (1986). J. Mol. Biol. 188, 707-719.

33. A.H.-J. Wang, T. Hakoshima, G. van der Marel, J.H. van Boom and A. Rich (1984). Cell 37, 321-331.

34. A.H.-J. Wang, R.V. Gessner, G.A. van der Marel, J.H. van Boom and A. Rich (1985). Proc. Natl. Acad. Sci. U.S. 82, 3611-3615.

35. P.S. Ho, C.A. Frederick, G.J. Quigley, G.A. van der Marel, J.H. van Boom, A.H.-J. Wang and A. Rich (1985). EMBO J. 4, 3617-3623.

36. T. Brown, G. Kneale, W.N. Hunter and O. Kennard (1986). Nucl. Acids Res. 14, 1801-1809.

37. J.L. Crawford, F.J. Kolpak, A.H.-J. Wang, G.J. Quigley, J.H. van Boom, G. van der Marel and A. Rich (1980). Proc. Natl. Acad. Sci. U.S. 77, 4016-4020.

38. S. Fujii, A.H.-J. Wang, G.J. Quigley, H. Westerink, G. van der Marel, J.H. van Boom and A. Rich (1985). Biopolymers 24, 243-250.

39. R.G. Brennan and M. Sundaralingam (1985). J. Mol. Biol. 181, 561-563.

40. R.G. Brennan, E. Westhof and M. Sundaralingam (1986). J. Biomol. Str. Dynam. 3, 649-665.

41. M.A. Viswamitra, O. Kennard, P.G. Jones, G.M. Sheldrick, S. Salisbury, L. Falvello, and Z. Shakked (1978). Nature 273, 687-688.

42. M.A. Viswamitra, Z. Shakked, P.G. Jones, G.M. Sheldrick, S.A. Salisbury and O. Kennard (1982). Biopolymers 21, 513-533.

43. D. Rabinovich, T. Haran, M. Eisenstein and Z. Shakked (1987). J. Mol. Biol.--submitted.

tory at MIT and Horace Drew in my own laboratory, then at Caltech, were working in parallel in 1977-8 on crystal structures of short C-G oligomers because of the apparent phase transition in poly(dC-dG)·poly(dC-dG) that could be induced by salt or ethanol (1), how Wang solved C-G-C-G-C-G in mid-1979 and Drew followed with C-G-C-G a few months later, and how both of these structures proved to be neither of the expected A or B helices, but an entirely new left-handed Z structure.

To show how rapidly the field has grown in eight years, Table I lists all single-crystal x-ray structure analyses of oligomers that were known to the author as of April 1987. Such tables go out of date rapidly, but still give an idea of the state of the field, and are a convenient starting point for searching the subsequent literature. Crystal data--space group and cell dimensions--are included not merely for completeness, but because they convey valuable information to the noncrystallographic reader. They show how many structures of a given helix family--A, B or Z--are truly independent of one another, and how many are variations of a common helix. A-helical structures numbered 1 - 5, for example, all have the same space group and roughly the same cell dimensions. In crystallographic terms they are isomorphous. They are variants of a single A-helix type. Similarly, structures 7 - 12 constitute an isomorphous set, as do B-helical structures 14 - 19. Although the various structure analyses reported as Nos. 14 - 19 contain much detailed information about temperature, base edge substitution and mispairings, from the standpoint of learning about the B helix they are virtually all the same structure. From 1980, when the first crystal structure of a B-DNA double helix appeared (No. 14), to 1987, when structures 20 and 21 appeared, everything that we knew about single crystals of B-DNA was derived from a single sequence: C-G-C-G-A-A-T-T-C-G-C-G. As will be seen, many of the conclusions arrived at with confidence from this single sequence have had to be modified in the light of the two new B helices.

In the same way, Z-helical structures 23-30 simply ring the changes on one particular Z helix. These are important changes; they establish whether A·T base pairs can be substituted for G·C, and whether the strict alternation of purine and pyrimidine can be broken (Yes, in both cases, with a price paid in stability of Z relative to the B form). But the Z helix, at this juncture, does not appear to exhibit the sequence-based polymorphism that is found to a limited extent in A and in full-blown form in the B helix.

The reason for solving structures of small synthetic DNA oligomers,

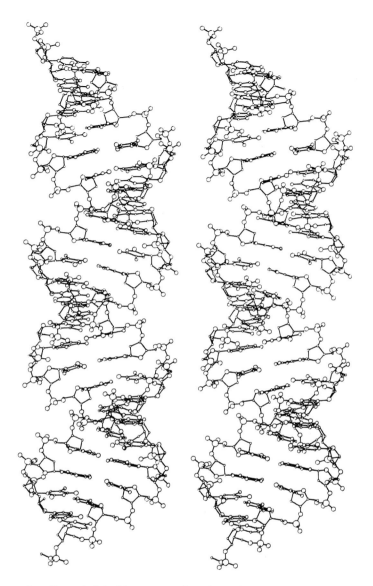

Figure 1. Extended-helix stereo drawing of A-DNA, generated from
structure 8 of Table I, by deleting the top and bottom base pairs to minimize end effects, and then repeating the resulting octamer up and down the
helix axis with the proper rotation and displacement. Note how the double
helix appears as a double ribbon, wrapped around an invisible central core.
Because of the displacement of the base pairs off the helix axis (Fig. 5),
the major groove is very deep, whereas the minor groove is shallow. Sugar
conformations cluster around C3'-<u>endo</u>, with appreciable scatter (Fig. 4).
From References 2 and 3.

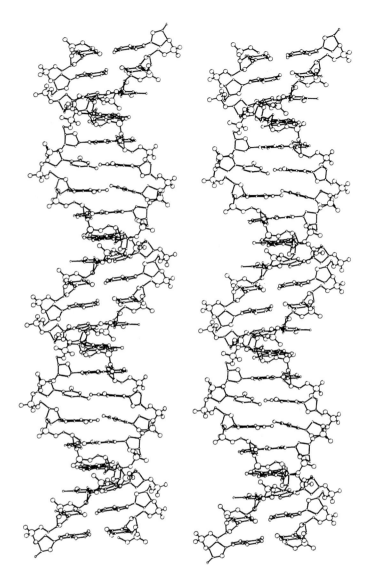

Figure 2. Extended-helix drawing of B-DNA, generated by helical repetition of the central ten base pairs of structure 16 of Table I. Since
base pairs now sit on the helix axis, major and minor grooves are of equal
depth, although the major groove is considerably wider. There is a mild
preference of sugar puckering for the "classical" C2'-endo, but the distribution is much broader in B-DNA than for A-DNA (Fig. 4), an observation
that may have biological significance in terms of expressing the base
sequence as local helix structure variations. From References 2 and 3.

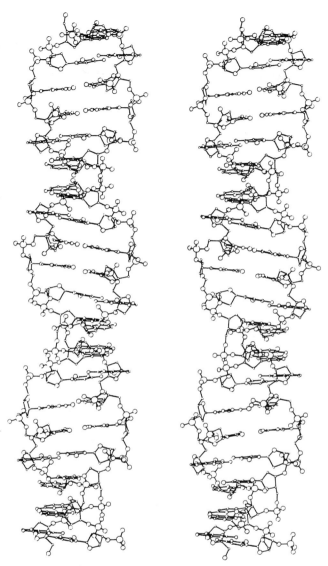

Figure 3. Extended-helix drawing of Z-DNA, generated by helical repetition of the central four base pairs of structure 23 of Table I. Now
the base pairs are displaced to the other side of the helix axis from A,
so the axis passes by the minor groove edge of each base pair. This makes
the minor groove extremely deep, and the major groove little more than a
surface winding around the helix cylinder. The backbone chains have a zig-
zag course because pyrimidines (C) have the customary <u>anti</u> glycosyl bonds
whereas purines (G) have an unusual <u>syn</u> geometry. Sugar puckering is
C2'-<u>endo</u> at pyrimidines and O4'-<u>endo</u> (not C3'-<u>endo</u>) at purines (Fig. 4).
From References 2 and 3.

of course, is to understand how natural, long-chain double-helical DNA behaves. Figures 1 - 3 are infinite-chain stereo drawings of A, B or Z-DNA, generated by helical extension of structures 8, 16 and 23 of Table I. They provide a correct impression of the appearance of extended helices of these three types, but preserve the sequence-induced local variations in helix structure that are important in drug and control protein recognition.

What are the structural features that define the three families of double helix, A, B, and Z, in addition to the obvious fact that A and B are right-handed whereas Z is left-handed? The ideal A and B helices have a single base pair as their helical repeat unit, whereas the zig-zag backbone pathway produced by alternating syn and anti glycosyl bonds (between base and sugar) at purines and pyrimidines in Z-DNA makes the helical repeat unit two successive base pairs: one purine·pyrimidine, followed by a pyrimidine·purine. It once was maintained that A-DNA sugars had exclusively C3'-endo conformations while B-DNA had C2'-endo, and it frequently is stated in the literature that in Z-DNA the sugars are C3'-endo at purines (mainly G) and C2'-endo at pyrimidines (mainly C). Neither of these statements is accurate. Figure 4a shows that sugar conformations in A-helical oligonucleotides do cluster around the C3'-endo conformation, although with a significant distribution about the ideal $\delta = 82°$. But B-helical oligomers are distributed quite differently with respect to their "classical" C2'-endo value of $\delta = 144°$ (Figure 4b). A much broader distribution of conformations is seen, peaking at C2'-endo but spreading out to C1'-exo, O4'-endo and even into the C3'-endo region. Sugar conformations in the B helix are more flexible than in the A helix, a structural feature that may have biological significance in that the B helix finds it easier to "express" its base sequence via local helix deformations than does A (or the even more rigid Z).

Figure 4c shows that the common attribution of purine sugars in Z-DNA to the C3'-endo conformation is inaccurate. The distribution centers symmetrically about the O4'-endo conformation, with only a single example of a guanine sugar that is truly C3'-endo. Hence a more proper description of Z-DNA is that it has C2'-endo sugars at pyrimidines and O4'-endo sugars at purines.

Figure 4 shows that sugar conformations, per se, are a poor defining character for helix families. But base pair displacement from the helix axis, Figure 5, is an excellent defining property. All known A-helical oligonucleotides have strong positive base pair dis-

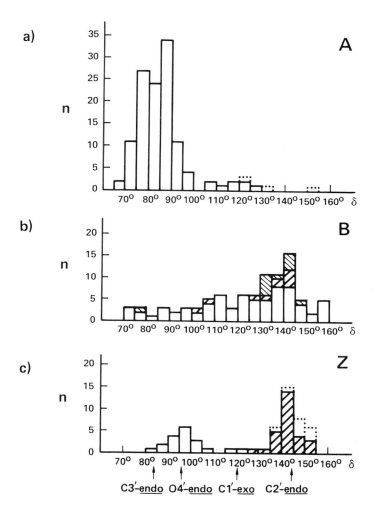

Figure 4. Frequency histogram of observed occurrence of torsion
angle delta (C5'-C4'-C3'-O3'), in published x-ray structures of A, B and
Z oligomers. (a) A helices, from structures 1-3, 7-11 and 13 of Table I.
Dotted bars at right represent atypical 3'-terminal sugars, which often
adopt a C2'-endo-like geometry in the absence of a continuation of the
chain. (b) B̄ helices, from structures 14-16 (open bars), 20 (dark crosshatching, and 21 (light crosshatching). Note the broader distribution of
conformation than for the A helices, and the less narrow preference for
the "classical" conformation, C2'-endo in this case. (c) Z helices, from
structures 22, 23 and 25-27. Open bars are purines (generally G) and
crosshatched bars are pyrimidines (generally C). Dotted bars at right
are atypical 3'-terminal purines, which frequently adopt a C2'-endo-like
geometry in the absence of a continuation of the chain. Note that the
purine conformations center around O4'-endo rather than C3'-endo.

placements, around +4 Å, meaning that the helix axis runs by the major groove edge of each base pair. B-helices have zero or a small negative displacement; and intermediate displacements between +1 Å and +3 Å are never encountered in any type of double helix. Something holds the base pairs of B-DNA on-axis; and when this feature is removed from the helix, it slumps all the way from the B to the A displacement region, without intermediate structures. We have proposed (5-7) that the B helix is stabilized by a spine of hydration running down the minor groove in A·T-rich regions, or two strings of water molecules lining the walls of the minor groove in general-sequence DNA, and that destruction of this ordered minor groove water is a necessary first step for a B-to-A helix conversion. This accounts for the fact that the B helix is the high humidity form, most stable under in vivo conditions. In the left-handed Z helix, base pairs are displaced off-axis in the opposite direction than in A, with little overlap between the B and Z regions.

Hence there are three "players" in the DNA structure game: A, B and Z-DNA, each with defining characteristics that differentiate it sharply from members of the other two families. The B form is the stable form under in vivo conditions; the A is a low-humidity form of uncertain biological relevance, and Z is a possibility in sequences that approximate a regular alternation of purines (mainly G) and pyrimidines (mainly C). Bulky side groups at guanine C8 and cytosine C5 positions favor the Z state, as do negative supercoiling and low water activity conditions. A proven in vivo role for Z-DNA has yet to be established.

Base Sequence and Helix Geometry

At the beginning of this presentation it was asserted that the most important single question that a molecular biologist can ask about DNA structure today was: How does base sequence influence helix geometry? Now we shall try to provide at least a partial answer to this question. Base sequence influences helix geometry in three principal ways:

1. Helix Families: A, B, and Z and transitions between them.

Certain sequences are known to favor certain types of helix. The most familiar example is the just-discussed requirement of approximate alternation of purines and pyrimidines for adoption of the left-handed Z structure. Introduction of A·T base pairs, and deviation from strict

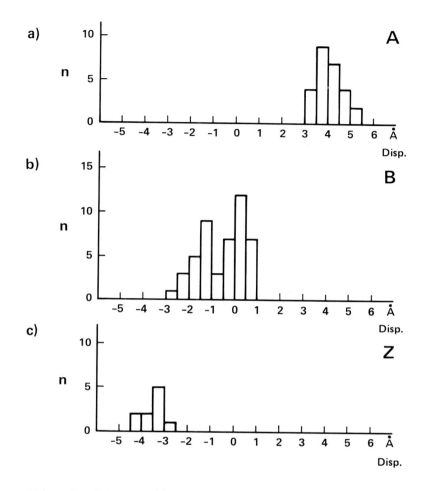

Figure 5. Frequency histogram of base pair displacements from helix axis, as seen in published x-ray structures of the three types of helical oligomers. Displacement is defined as the perpendicular distance to the helix axis from the long axis of the base pair as drawn from purine C8 to pyrimidine C6. Displacement is positive if the axis passes by the major groove side of the base pair, and negative if it passes by the minor groove side. (a) A helices, from structures 1, 2, 8 and 13 of Table I. (b) B helices, from structures 14-16, 20 and 21 of Table I. (c) Z helices, from structures 22 and 23 of Table I. Note that there is no overlap whatsoever in the distributions of displacements for the two right-handed helices, A and B, and virtually none between right-handed B and left-handed Z. Displacement is a better defining character for the three helix families than is sugar conformation.

Y-R alternation, both exact a price in free energy or in stability of
the Z form relative to B or A. Another example of sequence-derived
preference for helix type is the stabilization of B-DNA relative to
A-DNA by runs of consecutive A·T base pairs, a consequence of the
ordered spine of hydration running down the minor groove (5,6,8).
Possibly related to this is the observation that poly(dA)·poly(dT)
has a helical structure which, although still within the B family, is
distinctly different from the structure of poly(dA-dT)·poly(dA-dT) and
of general mixed-sequence B-DNA.

2. Unusual Secondary Structures.

David Lilley, earlier in this volume, points out that embedding
potentially cruciform sequences within a region rich in A·T base pairs
facilitates formation of the cruciforms, probably because of the lower
melting temperature of A·T regions. This lower T_m itself is a consequence of A·T base pairs having only two hydrogen bonds rather than
three as in G·C pairs. This is an example of how base sequence (or
at least base composition) can promote unusual secondary structures.
Bob Wells has also pointed out in this volume that, while palindromic or inverted repeats favor the bulging-out of hairpin arms of
cruciform structures, parallel repeats facilitate slips and loopouts,
in which a given copy of a repeated sequence on one strand becomes
paired with a different complementary copy on the other strand, farther
down the helix.

3. Local Helix Deformation and Bending.

Many proteins recognize and bind to specific DNA sequences in a
double helix--repressors, restriction enzymes, promoters, etc.--and
there is evidence to suggest that in some cases this recognition
involves not merely a passive reading of hydrogen bond patterns down
the floors of the grooves, but an actual sensing of small perturbations
in helix geometry that are produced by base sequence. DNase I, for
example, cleaves C-G-C-G-A-A-T-T-C-G-C-G best at just those steps
where the x-ray analysis reveals the greatest local helix twist angle
from one base pair to the next (9,10). DNase I also is thought to be
sensitive to minor groove width (11,12), an effect that itself is
sequence-dependent (13).
Two principal models have been proposed for sequence-specific

300

local deformations of the double helix: Calladine purine-purine clash (14,15) and the Trifonov wedge (16,17 and this volume). Calladine proposed that local perturbations in helix twist angle and base roll angle may arise because they relieve steric clash between purines in adjacent base pairs on opposite helix strands (Figure 6). For Y-R steps (Y = pyrimidine; R = purine) the clash occurs in the minor groove, and for R-Y steps it is found in the major groove. Such a clash exists because purine double rings are large enough to extend past the center line of each base pair, and their edges are brought into proximity by the significant large positive propeller twist of each base pair of a right-handed double helix. This propeller twist, in turn, is observed in all right-handed helices because it improves the base overlap and stacking of bases in each individual strand of helix (Figure 7).

Calladine's rules did an excellent job of explaining the observed twist and roll angle variations in about half the structures listed in Table I--all those that were known in 1982. But it has failed with several of the structures reported since that time. In part this is because of the greater variety of sequences examined. The cross-chain purine-purine clash model would not be expected to work well for structure No. 9, G-G-G-G-C-C-C-C, since the simple model only considers what happens at Y-R and R-Y steps, and has nothing to say about Y or R homopolymer regions. The purine clash model is not so much wrong, as it is incomplete. It fails to take into account other ways in which stress can be placed on an idealized helix, in a manner that produces local structure variations.

The Trifinov wedge model in its original form (16,17) was greeted with scepticism by DNA structuralists, because it appeared to require that two stacked base pairs be lifted apart at one of their ends. This is defined as _tilt_ (10,13), and is energetically disfavored because it requires greater separation of stacked base pairs. It is inherently easier to roll two stacked planks relative to one another about their long axes, than it is to pull the planks apart at one end. The current version of the Trifonov model (20, and this volume) proposes that an 8.7°wedge-like opening-up of base pair stacking occurs at consecutive A·T base pairs having A's on the same chain, but that the roll component predominates over tilt by a ratio of about 3.5 to 1. Diekmann and coworkers (21) have provided an elegant physical explanation for the roll and tilt components of AA wedges, as diagrammed in Figure 8. It is assumed that in two parallel, stacked adenines, the bulky N6

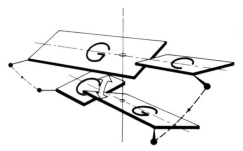

PYRIMIDINE - PURINE

Figure 6. Calladine purine-purine cross-chain steric clash (double-headed arrow) produced in the minor groove at Y-R steps. Each G·C base pair has positive propeller twist about its long axis. Because purines extend past the helix axis, they are brought into too-close contact by the propeller twist. The opposite sequence, R-Y (i.e.--G-C), would lead to a similar purine-purine clash in the major groove. Calladine proposed that local helix twist and roll angles adjust themselves in order to minimize this clash, thus producing sequence-dependent local perturbations in helix geometry (14, 15).

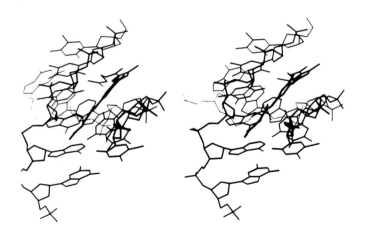

Figure 7. Oblique stereo view of the six central base pairs in B-helical C-G-C-G-A-A-T-T-^{br}C-G-C-G, in its complex with the antitumor drug netropsin (18, 19). The crescent-shaped drug molecule curves down the narrow minor groove, here seen in profile and opening to the upper right. Note especially in this unconventional helix view, how the six base pairs of each independent strand are stacked efficiently atop one another, almost as though the other strand were not there. When two such base stacks are wrapped around one another to form a right-handed helix, each individual base pair then is observed to possess positive propeller twist.

302

amine groups (marked by * in Figure 8) repel one another in a way that
opens up the roll angle between base pairs toward the top of the Figure,
and causes a lesser opening up of a tilt angle to the right, or adenine,
side of the A·T pairs. No comparable wedge opening is seen in G·C pairs,
both because the N6 carbonyl oxygens of guanines are smaller and have
less steric repulsion, and because the N2 amines on the minor groove
edge of the base pairs act as a block to wedge rotation. I·C base pairs,
which have the smaller carbonyls on the major groove edge, but no
blocking amines on the minor groove edge, show wedge bending intermediate
between that of A·T and G·C.

Hence the Calladine cross-chain purine-purine clash, and the
Trifonov-Diekmann A-A wedge, are complementary models of two different
steric repulsions between purines, that in different circumstances
tend to produce local, sequence-specific deformations in the DNA
double helix. Other factors undoubtedly exist, of which we are unaware
to date because we have not had examples of their particular sequences
to study in single-crystal x-ray analyses.

Another structural feature that could be regarded as an alternative
response to Calladine-like purine-purine clash is the B_I vs. B_{II}
dichotomy of phosphate conformations (Figure 9). Of the 22 phosphates
along the two chains of the dodecamer C-G-C-G-A-A-T-T-^{br}C-G-C-G
(structure No. 16), only the penultimate phosphate on each strand was
in the B_{II} conformation, with the C3'-O3'-P "elbow" of the main chain
pointing inward toward the helix axis. The other 20 phosphates re-
mained in the more common B_I state, with the "elbow" lying in the
plane of the helix cylinder, pointing toward the minor groove. In
contrast, in the B-DNA hexamer, structure No. 21, four of the ten
phosphates are B_{II}, as are four of the nine space group independent
phosphates in the decamer, structure No. 20. As mentioned in the
caption to Figure 9, a B_{II} phosphate tends to force a more C2'-<u>endo</u>-like
sugar pucker. The overwhelming predominance of B_I in the dodecamer
may account for its broad range of sugar pucker or torsion angle δ
(open bars in Figure 4b), whereas the more frequent occurrence of B_{II}
phosphates in the hexamer and decamer may explain why their sugars cluster more toward the classical B-DNA C2'-<u>endo</u> conformation (crosshatched additions to Figure 4b).

The foregoing, while it accounts for observed sugar puckering
behavior, begs the question as to <u>why</u> a B_{II} conformation should be
adopted over a B_I. The decamer structure (7) suggests that a B_{II}
phosphate may be an alternative strategy for coping with Calladine-
like cross-chain purine clash. The second step in the

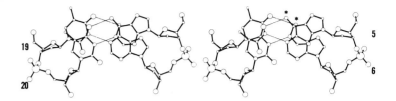

19

20

5

6

Figure 8. Base pairs A5·T20 and A6·T19, from the B-DNA dodecamer C-G-C-G-A-A-T-T-C-G-C-G (10). Asterisks mark the two N6 amine groups which, in the Trifonov-Diekmann wedge model, lie in too-close contact. Mutual repulsion tends to open the roll angle between base pairs toward the top of the stereo drawing, and open the tilt angle toward the right side. This effect, involving two successive bases on the <u>same</u> helix strand, is independent of propeller twist.

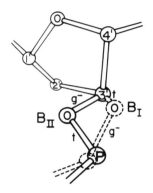

Figure 9. Alternative conformations of the DNA backbone chain at phosphates. In the major, B_I conformation (dotted), the C4'-C3'-O3'-P torsion angle is <u>trans</u> and the following C3'-O3'-P-C5' angle is <u>gauche</u> In the less commin, $\overline{B_{II}}$ conformation (solid lines), the first bond is g while the second is <u>t</u>. Rotation of the C3'-O3' bond from its B_I to its B_{II} position also rotates bond C3'-C2' back behind the plane of the paper, favoring the "classical" C2'-<u>endo</u> sugar pucker for B-DNA. From Reference 13.

C-C-A-A-G-A-T-T-G-G helix is C-A, and the purine clash model would predict too-close contact between A and G (paired with C) in the minor groove. This should decrease the local twist angle and produce a positive local roll angle. Exactly the opposite is observed from the crystal structure: the roll angle is -7° (opening toward the major groove) and the helix twist angle is a mammoth 51°! But one effect of adoption of the B_{II} backbone conformation is to stretch out the sugar-phosphate chain to its maximum extent, allowing the bases to either side of the phosphate in question to swing farther away from one another. B_{II} conformations in the decamer are uniformly associated with higher than average helix twist angles. The step in question has B_{II} phosphates on both strands, to both sides of the pair of base pairs, and this causes a maximal increase in twist angle. In this particular instance, purine-purine clash is avoided, not by the traditional Calladine strategies, but by allowing the backbone to change from B_I to B_{II}, and the base planes to slip sideway away from one another.

In sum, Calladine clash, Trifonov wedges, and B_I/B_{II} phosphates, all seem to be independent, complementary strategies for dealing with steric stress in the helix by permitting local helix deformations. All of these are sequence-related, and the deformations that they cause may be important in the recognition of base sequence by control proteins such as repressors, restriction enzymes, and promotors. This is a structure-based language that complements the codon-based language of the genetic message. The codon language is used for coding; the structure-based language apparently is used for control. We need to find out more about how the base sequence of double-helical DNA affects that helix structure, and the extent to which this is a factor in operator and binding site recognition.

GENERAL REFERENCES

1. H.R. Drew, R.E. Dickerson and K. Itakura (1978). J. Mol. Biol. 125, 535-543.

2. R.E. Dickerson (1983). Scientific American (December), pp. 94-111.

3. R.E. Dickerson, M.L. Kopka and P. Pjura (1985). In "Biological Macromolecules and Assemblies: Vol. 2--Nucleic acids and Interactive Proteins", (F. Jurnak and A. McPherson, Eds.), Wylie, New York, pp. 38-126.

4. R.E. Dickerson, H.R. Drew, B.N. Conner, R.M. Wing, A.V. Fratini and M.L. Kopka (1982). Science 216, 475-485.

5. H.R. Drew and R.E. Dickerson (1981). J. Mol. Biol. 151, 535-556.

6. M.L. Kopka, A.V. Fratini, H.R. Drew and R.E. Dickerson (1983). J. Mol. Biol. 163, 129-146.

7. G.G. Privé, U. Heinemann, S. Chandrasegaran, L.-S. Kan, M.L. Kopka and R.E. Dickerson (1987)--in preparation.

8. A.G.W. Leslie, S. Arnott, R. Chandrasekaran and R.L. Ratliff (1980). J. Mol. Biol. 143, 49-72.

9. G.P. Lomonossoff, P.J.G. Butler and A. Klug (1981). J. Mol. Biol. 149, 745-760.

10. R.E. Dickerson and H.R. Drew (1981). J. Mol. Biol. 149, 761-786.

11. H.R. Drew (1984). J. Mol. Biol. 176, 535-557.

12. D. Suck and C. Oefner (1987). Nature 321, 620-625.

13. A.V. Fratini, M.L. Kopka, H.R. Drew and R.E. Dickerson (1982). J. Biol. Chem. 257, 14686-14707.

14. C.R. Calladine (1982). J. Mol. Biol. 161, 343-352.

15. R.E. Dickerson (1983). J. Mol. Biol. 166, 419-441.

16. E.N. Trifonov and J.L. Sussman (1980). Proc. Natl. Acad. Sci. U.S. 77, 3816-3820.

17. E.N. Trifonov (1980). Nucl. Acids Res. 8, 4041-4053.

18. M.L. Kopka, C. Yoon, D. Goodsell, P. Pjura and R.E. Dickerson (1985). Proc. Natl. Acad. Sci. U.S. 82, 1376-1380.

19. M.L. Kopka, C. Yoon, D. Goodsell, P. Pjura and R.E. Dickerson (1985). J. Mol. Biol. 183, 553-563.

20. L. Ulanovsky, M. Bodner, E.N. Trifonov and M. Choder (1986). Proc. Natl. Acad. Sci. U.S. 83, 862-866.

21. S. Diekmann, E. von Kitzing, L. McLaughlin, J. Ott and F. Eckstein (1987). Proc. Natl. Acad. Sci. U.S.--submitted.

dG•dT mismatch, 121

diepoxybutane, 13

diethylpyrocarbonate, 8, 24

differential decay of birefringence, 228

dimethylsulfate, 9

distance geometry, 117

DNase, 7

DNA oligonucleotide crystal structure, 287

double helix, A, B, and Z, 296

<u>Eco</u>RI methylase, 7

electron microscopy, 13, 181, 182

electrophoretic anomaly, 182

extrahelical bases, 131

form V DNA, 76

gel retardation, 227

gene 3 product, 7

heat shock consensus sequences, 46

heat shock genes, 45, 50

helical repeat, 13

helicoidal variables, 193

herpes simplex virus type 1, 4

<u>Hha</u>I methylase, 7

Holliday junction, 1, 57

homopurine•homopyrimidine tracts, 137

Hoogsteen base pairs, 36

hydrogen bonding interactions, 241

hydrophobic effect, 238, 241

hydroxylamine, 9

OsO$_4$, 8

permanganate, 24
phage T4 endonuclease VII, 8
polyelectrolyte effect, 241
polypurine•polypyrimidine structures, 3, 4, 26, 34, 35, 36, 40, 41, 259
polytene chromosomes, 49
psoralen, 12, 103, 107, 110
pulsed field gel electrophoresis, 85
purine-clash model, 229

roll, 177, 180, 184, 208
rotational isomeric state modeling, 217

S1 nuclease, 6
single-crystal x-ray structure analyses, 292
single strand nuclease sensitivity, 23, 26, 30, 32, 41
SIR (Successive Infinitesimal Rotations) model, 190
site-specific binding of RNA polymerase, 245
slipped structures, 2
sodium bisulfite, 9
stable looped complex, 239, 240
S-type cruciforms, 60
supercoil relaxation, 10
supercoiled DNA, 55, 84, 85, 91, 92, 103
systemic lupus erythematosus (SLE), 260

telestability, 2
tilt, 180, 184, 208
tilt angle, 177
topoisomerase I, 49
toroidal form, 91
torsional stress, 73, 74, 75, 76, 78, 79, 109